9th Mathematical Physics Meeting: Summer School and Conference on Modern Mathematical Physics 2017

Belgrade, Serbia
18 – 23 September 2017

Editors:

B. Dragovich
I. Salom
M. Vojinovic

ISBN: 978-1-5108-6540-2

Printed from e-media with permission by:

Curran Associates, Inc.
57 Morehouse Lane
Red Hook, NY 12571

Some format issues inherent in the e-media version may also appear in this print version.

Printed by Curran Associates, Inc. (2018)

Belgrade Institute of Physics
Pregrevica 118
11080 Zemun – Belgrade
Serbia

Phone: +381 (0)11 37 13 000
Fax: +381 (0)11 31 62 190

info@ipb.ac.rs

Additional copies of this publication are available from:

Curran Associates, Inc.
57 Morehouse Lane
Red Hook, NY 12571 USA
Phone: 845-758-0400
Fax: 845-758-2633
Email: curran@proceedings.com
Web: www.proceedings.com

Proceedings

of the

9th MATHEMATICAL PHYSICS MEETING: School and Conference on Modern Mathematical Physics

September 18 – 23, 2017, Belgrade, Serbia

Editors

B. Dragovich, I. Salom and M. Vojinović

Institute of Physics
Belgrade, 2018
SERBIA

9th MATHEMATICAL PHYSICS MEETING: School and Conference on Modern Mathematical Physics

Belgrade, September 18 - 23, 2017

Organizers

- Institute of Physics, Belgrade, Serbia
- Mathematical Institute, Belgrade, Serbia
- Faculty of Mathematics, Belgrade, Serbia

International Advisory Committee

Ignatios Antoniadis (Paris, France)
Irina Arefeva (Moscow, Russia)
Alexander Belavin (Moscow, Russia)
Loriano Bonora (Trieste, Italy)
Maja Buric (Belgrade, Serbia)
Nemanja Kaloper (Davis, USA)
Viatcheslav Mukhanov (Munich, Germany)
Emil Nissimov (Sofia, Bulgaria)
Sergei Odintsov (Barcelona, Spain)
Dmitri Polyakov (Chengdu, China; Moscow, Russia)
Branislav Sazdovic (Belgrade, Serbia)
Djordje Sijacki (Belgrade, Serbia)
Alexei Starobinsky (Moscow, Russia)
Dejan Stojkovic (Buffalo, USA)
Mihai Visinescu (Bucharest, Romania)

International Organizing Committee

Vladimir Dobrev (Sofia, Bulgaria)
Branko Dragovich (Belgrade, Serbia)
Goran Djordjevic (Nis, Serbia)
Zoran Rakic (Belgrade, Serbia)
Igor Volovich (Moscow, Russia)

Local Organizing Committee

Branko Dragovich (Co-Chairman, Institute of Physics, Belgrade)
Igor Salom (Co-Chairman, Institute of Physics, Belgrade)
Sanja Ćirković (Institute of Physics, Belgrade)
Ljubica Davidović (Institute of Physics, Belgrade)
Marko Vojinović (Institute of Physics, Belgrade)

Sponsors

- **Ministry of Education, Science and Technological Development, Republic of Serbia**
- **The Abdus Salam International Centre for Theoretical Physics**

PREFACE

This volume contains some reviews and original research contributions, which are related to the **9th Mathematical Physics Meeting: School and Conference on Modern Mathematical Physics**, organized by the Institute of Physics, Belgrade (Serbia), September 18–23, 2017. The programme of this meeting was mainly oriented towards some recent developments in gravity and cosmology, string and quantum field theory, and some relevant mathematical methods. We hope that articles presented here will be valuable literature not only for the participants of this meeting but also for many other PhD students and researchers in modern mathematical and theoretical physics. We are grateful to all authors for writing their contributions for these proceedings.

The previous eight meetings in this series of schools and conferences on modern mathematical physics were also held in Serbia: Sokobanja, 13–25 August 2001; Kopaonik, 1–12 September 2002; Zlatibor, 20–31 August 2004; Belgrade, 3–14 September 2006; Belgrade, 6–17 July 2008; Belgrade, 14–23 September 2010; Belgrade, 9–19 September 2012; and Belgrade, 24–31 August 2014. The corresponding proceedings of all these meetings were published by the Institute of Physics, Belgrade, and are available in the printed form as well as online at the websites.

This ninth meeting took place at the Mathematical Institute, Serbian Academy of Sciences and Arts, located in Belgrade downtown. We hope that all attending this meeting will recall it as a useful and pleasant event, and will wish to participate again in the future.

We wish to thank all lecturers and other speakers for their interesting and valuable talks. We also thank all participants for their active participation. Financial support of our sponsors, *Ministry of Education, Science and Technological Development of the Republic of Serbia, Belgrade; and the Abdus Salam International Centre for Theoretical Physics* was very significant for realization of this activity.

April 2018

Editors

B. Dragovich

I. Salom

M. Vojinović

CONTENTS

Review and Research Works

Scale hierarchies and string cosmology

Ignatios Antoniadis*

LPTHE, UMR CNRS 7589

Sorbonne Universités, UPMC Paris 6, 75005 Paris, France

and

Albert Einstein Center, Institute for Theoretical Physics

University of Bern, Sidlerstrasse 5, CH-3012 Bern, Switzerland

Abstract

I describe the phenomenology of a model of supersymmetry breaking in the presence of a tiny (tuneable) positive cosmological constant. It utilises a single chiral multiplet with a gauged shift symmetry, that can be identified with the string dilaton (or an appropriate compactification modulus). The model is coupled to the MSSM, leading to calculable soft supersymmetry breaking masses and a distinct low energy phenomenology that allows to differentiate it from other models of supersymmetry breaking and mediation mechanisms. We also study the question if this model can lead to inflation by identifying the dilaton with the inflaton. We find that this is possible if the Kähler potential is modified by a term that has the form of NS5-brane instantons, leading to an appropriate inflationary plateau around the maximum of the scalar potential, depending on two extra parameters. We then generalise this model to a general class where the inflation is driven by supersymmetry breaking with the superpartner of the goldstino (sgoldstino) playing the role of the inflaton. Imposing an R-symmetry allows to satisfy easily the slow-roll conditions, avoiding the so-called η-problem, and leads to two different classes of small field inflation models; they are characterised by an inflationary plateau around the maximum of the scalar potential, where R-symmetry is either restored or spontaneously broken, with the inflaton rolling down to a minimum describing the present phase of our Universe. The models agree with cosmological observations and predict a tensor-to-scalar ratio of primordial perturbations $10^{-9} \lesssim r \lesssim 10^{-4}$ and an inflation scale 10^{10} GeV $\lesssim H_* \lesssim 10^{12}$ GeV.

1. Introduction

If String Theory is a fundamental theory of Nature and not just a tool for studying systems with strongly coupled dynamics, it should be able to describe at the same

* e-mail address: ignatios.antoniadis@upmc.fr

time particle physics and cosmology, which are phenomena that involve very different scales from the microscopic four-dimensional (4d) quantum gravity length of 10^{-33} cm to large macroscopic distances of the size of the observable Universe $\sim 10^{28}$ cm spanned a region of about 60 orders of magnitude. In particular, besides the 4d Planck mass, there are three very different scales with very different physics corresponding to the electroweak, dark energy and inflation. These scales might be related via the scale of the underlying fundamental theory, such as string theory, or they might be independent in the sense that their origin could be based on different and independent dynamics. An example of the former constraint and more predictive possibility is provided by TeV strings with a fundamental scale at low energies due for instance to large extra dimensions transverse to a four-dimensional braneworld forming our Universe [1]. In this case, the 4d Planck mass is emergent from the fundamental string scale and inflation should also happen around the same scale [2].

Here, we will adopt the second more conservative approach, assuming that all three scales have an independent dynamical origin. Moreover, we will assume the presence of low energy supersymmetry that allows for an elegant solution of the mass hierarchy problem, a unification of fundamental forces as indicated by low energy data and a natural dark matter candidate due to an unbroken R-parity. The assumption of independent scales implies that supersymmetry breaking should be realized in a metastable de Sitter vacuum with an infinitesimally small (tunable) cosmological constant independent of the supersymmetry breaking scale that should be in the TeV region. In a recent work [3], we studied a simple $N = 1$ supergravity model having this property and motivated by string theory. Besides the gravity multiplet, the minimal field content consists of a chiral multiplet with a shift symmetry promoted to a gauged R-symmetry using a vector multiplet. In the string theory context, the chiral multiplet can be identified with the string dilaton (or an appropriate compactification modulus) and the shift symmetry associated to the gauge invariance of a two-index antisymmetric tensor that can be dualized to a (pseudo)scalar. The shift symmetry fixes the form of the superpotential and the gauging allows for the presence of a Fayet-Iliopoulos (FI) term, leading to a supergravity action with two independent parameters that can be tuned so that the scalar potential possesses a metastable de Sitter minimum with a tiny vacuum energy (essentially the relative strength between the F- and D-term contributions). A third parameter fixes the Vacuum Expectation Value (VEV) of the string dilaton at the desired (phenomenologically) weak coupling regime. An important consistency constraint of our model is anomaly cancellation which has been studied in [5] and implies the existence of additional charged fields under the gauged R-symmetry.

In a more recent work [6], we analyzed a small variation of this model which is manifestly anomaly free without additional charged fields and allows to couple in a straight forward way a visible sector containing the minimal supersymmetric extension of the Standard Model (MSSM) and studied the mediation of supersymmetry breaking and its phenomenological consequences. It turns out that an additional ‘hidden sector’ field z is needed to be added for the matter soft scalar masses to be non-tachyonic; although this field participates in the supersymmetry breaking and is similar to the so-called Polonyi field, it does not modify the main properties of the metastable de Sitter (dS) vacuum. All soft scalar masses, as well as trilinear A-terms, are generated at the tree level and are universal under the assumption that matter kinetic terms are independent of the 'Polonyi' field, since matter fields are neutral under the shift symmetry and supersymmetry breaking is driven by a combination of the $U(1)$ D-term and the dilaton and z-field F-term. Alternatively, a way to avoid the tachyonic scalar masses without adding the extra field z is to modify the matter kinetic terms by a dilaton dependent factor.

A main difference of the second analysis from the first work is that we use a field representation in which the gauged shift symmetry corresponds to an ordinary $U(1)$ and not an R-symmetry. The two representations differ by a Kähler transformation that leaves the classical supergravity action invariant. However, at the quantum level, there is a Green-Schwarz term generated that amounts an extra dilaton dependent contribution to the gauge kinetic terms needed to cancel the anomalies of the R-symmetry. This creates an apparent puzzle with the gaugino masses that vanish in the first representation but not in the latter. The resolution to the puzzle is based to the so called anomaly mediation contributions [7, 8] that explain precisely the above apparent discrepancy. It turns out that gaugino masses are generated at the quantum level and are thus suppressed compared to the scalar masses (and A-terms).

This model has the necessary ingredients to be obtained as a remnant of moduli stabilisation within the framework of internal magnetic fluxes in type I string theory, turned on along the compact directions for several abelian factors of the gauge group. All geometric moduli can in principle be fixed in a supersymmetric way, while the shift symmetry is associated to the 4d axion and its gauging is a consequence of anomaly cancellation [9, 10].

We then make an attempt to connect the scale of inflation with the electroweak and supersymmetry breaking scales within the same effective field theory, that at the same time allows the existence of an infinitesimally small (tuneable) positive cosmological constant describing the present dark energy of the universe. We thus address the question whether the same scalar potential can provide inflation with the dilaton playing also the role of the inflaton at an earlier stage of the universe evolution [11]. We show that this is possible if one modifies the Kähler potential by a correction that plays no role around the minimum, but creates an appropriate plateau around the maximum. In general, the Kähler potential receives perturbative and non-perturbative corrections that vanish in the weak coupling limit. After analysing all such corrections, we find that only those that have the form of (Neveu-Schwarz) NS5-brane instantons can lead to an inflationary period compatible with cosmological observations. The scale of inflation turns out then to be of the order of low energy supersymmetry breaking, in the TeV region. On the other hand, the predicted tensor-to-scalar ratio is too small to be observed.

Inflationary models [12] in supergravity[1] suffer in general from several problems, such as fine-tuning to satisfy the slow-roll conditions, large field initial conditions that break the validity of the effective field theory, and stabilisation of the

[1]For reviews on supersymmetric models of inflation, see for example [13].

(pseudo) scalar companion of the inflaton arising from the fact that bosonic components of superfields are always even. The simplest argument to see the fine tuning of the potential is that a canonically normalised kinetic term of a complex scalar field X corresponds to a quadratic Kähler potential $K = X\bar{X}$ that brings one unit contribution to the slow-roll parameter $\eta = V''/V$, arising from the e^K proportionality factor in the expression of the scalar potential V. This problem can be avoided in models with no-scale structure where cancellations arise naturally due to non-canonical kinetic terms leading to potentials with flat directions (at the classical level). However, such models require often trans-Planckian initial conditions that invalidate the effective supergravity description during inflation. A concrete example where all these problems appear is the Starobinsky model of inflation [14], despite its phenomenological success.

All three problems above are solved when the inflaton is identified with the scalar component of the goldstino superfield[2], in the presence of a gauged R-symmetry [16]. Indeed, the superpotential is in that case linear and the big contribution to η described above cancels exactly. Since inflation arises in a plateau around the maximum of the scalar potential (hill-top) no large field initial conditions are needed, while the pseudo-scalar companion of the inflaton is absorbed into the R-gauge field that becomes massive, leading the inflaton as a single scalar field present in the spectrum. This model provides therefore a minimal realisation of natural small-field inflation in supergravity, compatible with present observations, as we show below. Moreover, it allows the presence of a realistic minimum describing our present Universe with an infinitesimal positive vacuum energy arising due to a cancellation between an F- and D-term contributions to the scalar potential, without affecting the properties of the inflationary plateau, along the lines of Ref. [3, 11, 4].

On general grounds, there are two classes of such models depending on whether the maximum corresponds to a point of unbroken (case 1) or broken (case 2) R-symmetry. The latter corresponds actually to a generalisation of the model we discussed above [11], inspired by string theory [3]. It has the same field content but in a different field basis with a chiral multiplet $S \propto \ln X$ playing the role of the string dilaton. Thus, S has a shift symmetry which is actually an R-symmetry gauged by a vector multiplet and the superpotential is a single exponential. The scalar potential has a minimum with a tuneable vacuum energy and a maximum that can produce inflation when appropriate corrections are included in the Kähler potential. In these coordinates R-symmetry is restored at infinity, corresponding to the weak coupling limit. Small field inflation is again guaranteed consistently with the validity of the effective field theory.

2. Conventions

Throughout this paper we use the conventions of [17]. A supergravity theory is specified (up to Chern-Simons terms) by a Kähler potential $\mathcal{K}$, a superpotential W, and the gauge kinetic functions $f_{AB}(z)$. The chiral multiplets z^α, χ^α are enumerated

[2] See [15] for earlier work relating supersymmetry and inflation.

by the index α and the indices A, B indicate the different gauge groups. Classically, a supergravity theory is invariant under Kähler tranformations, viz.

$$\begin{aligned} \mathcal{K}(z,\bar{z}) &\longrightarrow \mathcal{K}(z,\bar{z}) + J(z) + \bar{J}(\bar{z}), \\ W(z) &\longrightarrow e^{-\kappa^2 J(z)} W(z), \end{aligned} \tag{1}$$

where κ is the inverse of the reduced Planck mass, $m_p = \kappa^{-1} = 2.4 \times 10^{15}$ TeV. The gauge transformations of chiral multiplet scalars are given by holomorphic Killing vectors, i.e. $\delta z^\alpha = \theta^A k_A^\alpha(z)$, where θ^A is the gauge parameter of the gauge group A. The Kähler potential and superpotential need not be invariant under this gauge transformation, but can change by a Kähler transformation

$$\delta\mathcal{K} = \theta^A \left[r_A(z) + \bar{r}_A(\bar{z}) \right], \tag{2}$$

provided that the gauge transformation of the superpotential satisfies $\delta W = -\theta^A \kappa^2 r_A(z) W$. One then has from $\delta W = W_\alpha \delta z^\alpha$

$$W_\alpha k_A^\alpha = -\kappa^2 r_A W, \tag{3}$$

where $W_\alpha = \partial_\alpha W$ and α labels the chiral multiplets. The supergravity theory can then be described by a gauge invariant function

$$\mathcal{G} = \kappa^2 \mathcal{K} + \log(\kappa^6 W \bar{W}). \tag{4}$$

The scalar potential is given by

$$\begin{aligned} V &= V_F + V_D \\ V_F &= e^{\kappa^2 \mathcal{K}} \left(-3\kappa^2 W \bar{W} + \nabla_\alpha W g^{\alpha\bar{\beta}} \bar{\nabla}_{\bar{\beta}} \bar{W} \right) \\ V_D &= \frac{1}{2} (\mathrm{Re} f)^{-1\,AB} \mathcal{P}_A \mathcal{P}_B, \end{aligned} \tag{5}$$

where W appears with its Kähler covariant derivative

$$\nabla_\alpha W = \partial_\alpha W(z) + \kappa^2 (\partial_\alpha \mathcal{K}) W(z). \tag{6}$$

The moment maps $\mathcal{P}_A$ are given by

$$\mathcal{P}_A = i(k_A^\alpha \partial_\alpha \mathcal{K} - r_A). \tag{7}$$

In this paper we will be concerned with theories having a gauged R-symmetry, for which $r_A(z)$ is given by an imaginary constant $r_A(z) = i\kappa^{-2}\xi$. In this case, $\kappa^{-2}\xi$ is a Fayet-Iliopoulos [18] constant parameter.

3. The model

The starting point is a chiral multiplet S and a vector multiplet associated with a shift symmetry of the scalar component s of the chiral multiplet S

$$\delta s = -ic\theta\,, \tag{8}$$

and a string-inspired Kähler potential of the form $-p\log(s+\bar{s})$. The most general superpotential is either a constant $W = \kappa^{-3}a$ or an exponential superpotential $W = \kappa^{-3}ae^{bs}$ (where a and b are constants). A constant superpotential is (obviously) invariant under the shift symmetry, while an exponential superpotential transforms as $W \to We^{-ibc\theta}$, as in eq. (3). In this case the shift symmetry becomes a gauged R-symmetry and the scalar potential contains a Fayet-Iliopoulos term. Note however that by performing a Kähler transformation (1) with $J = \kappa^{-2}bs$, the model can be recast into a constant superpotential at the cost of introducing a linear term in the Kähler potential $\delta K = b(s+\bar{s})$. Even though in this representation, the shift symmetry is not an R-symmetry, we will still refer to it as $U(1)_R$. The most general gauge kinetic function has a constant term and a term linear in s, $f(s) = \delta + \beta s$.

To summarise,[3]

$$\begin{aligned} \mathcal{K}(s,\bar{s}) &= -p\log(s+\bar{s}) + b(s+\bar{s}), \\ W(s) &= a, \\ f(s) &= \delta + \beta s\,, \end{aligned} \tag{9}$$

where we have set the mass units $\kappa = 1$. The constants a and b together with the constant c in eq. (8) can be tuned to allow for an infinitesimally small cosmological constant and a TeV gravitino mass. For $b > 0$, there always exists a supersymmetric AdS (anti-de Sitter) vacuum at $\langle s+\bar{s}\rangle = b/p$, while for $b = 0$ (and $p < 3$) there is an AdS vacuum with broken supersymmetry. We therefore focus on $b < 0$. In the context of string theory, S can be identified with a compactification modulus or the universal dilaton and (for negative b) the exponential superpotential may be generated by non-perturbative effects.

The scalar potential is given by:

$$\begin{aligned} V &= V_F + V_D \\ V_F &= a^2 e^{\frac{b}{l}} l^{p-2}\left\{\frac{1}{p}(pl-b)^2 - 3l^2\right\} \qquad l = 1/(s+\bar{s}) \\ V_D &= c^2\frac{l}{\beta + 2\delta l}(pl-b)^2 \end{aligned} \tag{10}$$

In the case where S is the string dilaton, V_D can be identified as the contribution of a magnetized D-brane, while V_F for $b = 0$ and $p = 2$ coincides with the

[3]In superfields the shift symmetry (8) is given by $\delta S = -ic\Lambda$, where Λ is the superfield generalization of the gauge parameter. The gauge invariant Kähler potential is then given by $\mathcal{K}(S,\bar{S}) = -p\kappa^{-2}\log(S+\bar{S}+cV_R) + \kappa^{-2}b(S+\bar{S}+cV_R)$, where V_R is the gauge superfield of the shift symmetry.

tree-level dilaton potential obtained by considering string theory away its critical dimension [19]. For $p \geq 3$ the scalar potential V is positive and monotonically decreasing, while for $p < 3$, its F-term part V_F is unbounded from below when $s + \bar{s} \to 0$. On the other hand, the D-term part of the scalar potential V_D is positive and diverges when $s + \bar{s} \to 0$ and for various values for the parameters an (infinitesimally small) positive (local) minimum of the potential can be found.

If we restrict ourselves to integer p, tunability of the vacuum energy restricts $p = 2$ or $p = 1$ when $f(s) = s$, or $p = 1$ when the gauge kinetic function is constant. For $p = 2$ and $f(s) = s$, the minimization of V yields:

$$b/l = -\rho_0 \approx -0.183268 \quad , \quad p = 2 \tag{11}$$

$$\frac{a^2}{bc^2} = A_2(-\rho_0) + B_2(-\rho_0)\frac{\Lambda}{b^3c^2} \approx -50.6602 + O(\Lambda), \tag{12}$$

where Λ is the value of V at the minimum (i.e. the cosmological constant), $-\rho_0$ is the negative root of the polynomial $-x^5 + 7x^4 - 10x^3 - 22x^2 + 40x + 8$ compatible with (12) for $\Lambda = 0$ and $A_2(\alpha)$, $B_2(\alpha)$ are given by

$$A_2(\alpha) = 2e^{-\alpha}\frac{-4 + 4\alpha - \alpha^2}{\alpha^3 - 4\alpha^2 - 2\alpha} \quad ; \quad B_2(\alpha) = 2\frac{\alpha^2 e^{-\alpha}}{\alpha^2 - 4\alpha - 2} \tag{13}$$

It follows that by carefully tuning a and c, Λ can be made positive and arbitrarily small independently of the supersymmetry breaking scale. A plot of the scalar potential for certain values of the parameters is shown in figure 1.

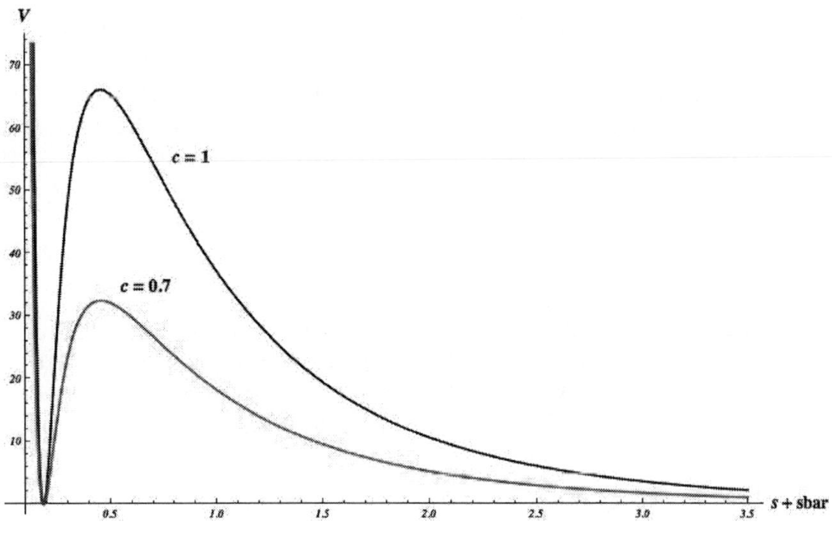

Figure 1: A plot of the scalar potential for $p = 2$, $b = -1$, $\delta = 0$, $\beta = 1$ and a given by equation (12) for $c = 1$ (black curve) and $c = 0.7$ (red curve).

At the minimum of the scalar potential, for nonzero a and $b < 0$, supersymmetry is broken by expectation values of both an F and D-term. Indeed the F-term and D-term contributions to the scalar potential are

$$
\begin{aligned}
V_F\big|_{s+\bar{s}=\frac{-\rho_0}{b}} &= \frac{1}{2}a^2b^2e^{-}\rho_0\left(1+\frac{2}{\rho_0}\right)^2 > 0, \\
V_D\big|_{s+\bar{s}=\frac{-\rho_0}{b}} &= -\frac{b^3c^2}{\rho_0}\left(1+\frac{2}{\rho_0}\right)^2 > 0\,.
\end{aligned} \tag{14}
$$

The gravitino mass term is given by

$$
(m_{3/2})^2 = e^{\mathcal{G}} = \frac{a^2b^2}{\rho_0^2}e^{-\rho_0}\,. \tag{15}
$$

Due to the Stueckelberg coupling, the imaginary part of s (the axion) gets eaten by the gauge field, which acquires a mass. On the other hand, the Goldstino, which is a linear combination of the fermion of the chiral multiplet χ and the gaugino λ gets eaten by the gravitino. As a result, the physical spectrum of the theory consists (besides the graviton) of a massive scalar, namely the dilaton, a Majorana fermion, a massive gauge field and a massive gravitino. All the masses are of the same order of magnitude as the gravitino mass, proportional to the same constant a (or c related by eq. (12) where b is fixed by eq. (11)), which is a free parameter of the model. Thus, they vanish in the same way in the supersymmetric limit $a \to 0$.

The local dS minimum is metastable since it can tunnel to the supersymmetric ground state at infinity in the s-field space (zero coupling). It turns out however that it is extremely long lived for realistic perturbative values of the gauge coupling $l \simeq 0.02$ and TeV gravitino mass and, thus, practically stable; its decay rate is [5]:

$$
\Gamma \sim e^{-B} \quad \text{with} \quad B \approx 10^{300}\,. \tag{16}
$$

4. Coupling a visible sector

The guideline to construct a realistic model keeping the properties of the toy model described above is to assume that matter fields are invariant under the shift symmetry (8) and do not participate in the supersymmetry breaking. In the simplest case of a canonical Kähler potential, MSSM-like fields ϕ can then be added as:

$$
\begin{aligned}
\mathcal{K} &= -\kappa^{-2}\log(s+\bar{s}) + \kappa^{-2}b(s+\bar{s}) + \sum \varphi\bar{\varphi}, \\
W &= \kappa^{-3}a + W_{MSSM},
\end{aligned} \tag{17}
$$

where $W_{MSSM}(\phi)$ is the usual MSSM superpotential. The squared soft scalar masses of such a model can be shown to be positive and close to the square of the gravitino mass (TeV^2). On the other hand, for a gauge kinetic function with a linear term in s, $\beta \neq 0$ in eq. (9), the Lagrangian is not invariant under the shift symmetry

$$\delta\mathcal{L} = -\theta\frac{\beta c}{8}\,\epsilon^{\mu\nu\rho\sigma}F_{\mu\nu}F_{\rho\sigma}. \tag{18}$$

and its variation should be canceled. As explained in Ref. [5], in the 'frame' with an exponential superpotential the R-charges of the fermions in the model can give an anomalous contribution to the Lagrangian. In this case the 'Green-Schwarz' term $\mathrm{Im}sF\tilde{F}$ can cancel quantum anomalies. However as shown in [5], with the minimal MSSM spectrum, the presence of this term requires the existence of additional fields in the theory charged under the shift symmetry.

Instead, to avoid the discussion of anomalies, we focus on models with a constant gauge kinetic function. In this case the only (integer) possibility[4] is $p = 1$. The scalar potential is given by (10) with $\beta = 0$, $\delta = p = 1$. The minimization yields to equations similar to (11), (12) and (13) with a different value of ρ_0 and functions A_1 and B_1 given by:

$$\begin{aligned}
b\langle s+\bar{s}\rangle &= -\rho_0 \approx -0.233153 \\
\frac{bc^2}{a^2} &= A_1(-\rho_0) + B_1(-\rho_0)\frac{\Lambda}{a^2 b} \approx -0.359291 + O(\Lambda) \\
A_1(\alpha) &= 2e^{\alpha}\alpha\frac{3-(\alpha-1)^2}{(\alpha-1)^2} \quad , \quad B_1(\alpha) = \frac{2\alpha^2}{(\alpha-1)^2},
\end{aligned} \tag{19}$$

where $-\rho_0$ is the negative root of $-3 + (\rho - 1)^2(2 - \rho^2/2) = 0$ close to -0.23, compatible with the second constraint for $\Lambda = 0$. However, this model suffers from tachyonic soft masses when it is coupled to the MSSM, as in (17). To circumvent this problem, one can add an extra hidden sector field which contributes to (F-term) supersymmetry breaking. Alternatively, the problem of tachyonic soft masses can also be solved if one allows for a non-canonical Kähler potential in the visible sector, which gives an additional contribution to the masses through the D-term.

Let us discuss first the addition of an extra hidden sector field z (similar to the so-called Polonyi field [20]). The Kähler potential, superpotential and gauge kinetic function are given by

$$\begin{aligned}
\mathcal{K} &= -\kappa^{-2}\log(s+\bar{s}) + \kappa^{-2}b(s+\bar{s}) + z\bar{z} + \sum \varphi\bar{\varphi}, \\
W &= \kappa^{-3}a(1+\gamma\kappa z) + W_{\mathrm{MSSM}}(\varphi), \\
f(s) &= 1 \quad , \quad f_A = 1/g_A^2,
\end{aligned} \tag{20}$$

where A labels the Standard Model gauge group factors and γ is an additional constant parameter. The existence of a tuneable dS vacuum with supersymmetry

[4]If $f(s)$ is constant, the leading contribution to V_D when $s + \bar{s} \to 0$ is proportional to $1/(s+\bar{s})^2$, while the leading contribution to V_F is proportional to $1/(s+\bar{s})^p$. It follows that $p < 2$; if $p > 2$, the potential is unbounded from below, while if $p = 2$, the potential is either positive and monotonically decreasing or unbounded from below when $s + \bar{s} \to 0$ depending on the values of the parameters.

breaking and non-tachyonic scalar masses implies that γ must be in a narrow region:

$$0.5 \lesssim \gamma \lesssim 1.7\,. \tag{21}$$

In the above range of γ the main properties of the toy model described in the previous section remain, while Rez and its F-auxiliary component acquire non vanishing VEVs. All MSSM soft scalar masses are then equal to a universal value m_0 of the order of the gravitino mass, while the B_0 Higgs mixing parameter is also of the same order:

$$\begin{aligned} m_0^2 &= m_{3/2}^2\left[(\sigma_s+1)+\frac{(\gamma+t+\gamma t)^2}{(1+\gamma t)^2}\right], \\ A_0 &= m_{3/2}\left[(\sigma_s+3)+t\frac{(\gamma+t+\gamma t^2)}{1+\gamma t}\right], \\ B_0 &= m_{3/2}\left[(\sigma_s+2)+t\frac{(\gamma+t+\gamma t^2)}{(1+\gamma t)}\right], \end{aligned} \tag{22}$$

where $\sigma_s = -3+(\rho+1)^2$ with $\rho = -b(s+\bar{s})$ and $t \equiv \langle \mathrm{Re}\, z\rangle$ determined by the minimization conditions as functions of γ. Also, A_0 is the soft trilinear scalar coupling in the standard notation, satisfying the relation [21]

$$A_0 = B_0 + m_{3/2}\,. \tag{23}$$

On the other hand, the gaugino masses appear to vanish at tree-level since the gauge kinetic functions are constants (see (10.)). However, as mentioned in Section 3., this model is classically equivalent to the theory[5]

$$\begin{aligned} \mathcal{K} &= -\kappa^{-2}\log(s+\bar{s}) + z\bar{z} + \sum_{\varphi}\varphi\bar{\varphi}, \\ W &= \left(\kappa^{-3}a(1+z) + W_{\mathrm{MSSM}}(\varphi)\right)e^{bs}, \end{aligned} \tag{24}$$

obtained by applying a Kähler transformation (1) with $J = -\kappa^{-2}bs$. All classical results remain the same, such as the expressions for the scalar potential and the soft scalar masses (22), but now the shift symmetry (8) of s became a gauged R-symmetry since the superpotential transforms as $W \longrightarrow We^{-ibc\theta}$. Therefore, all fermions (including the gauginos and the gravitino) transform[6] as well under this $U(1)_R$, leading to cubic $U(1)_R^3$ and mixed $U(1)\times G_{\mathrm{MSSM}}$ anomalies. These anomalies are cancelled by a Green-Schwarz (GS) counter term that arises from a quantum correction to the gauge kinetic functions:

$$f_A(s) = 1/g_A^2 + \beta_A s \quad \text{with} \quad \beta_A = \frac{b}{8\pi^2}\left(T_{R_A} - T_{G_A}\right), \tag{25}$$

[5]This statement is only true for supergravity theories with a non-vanishing superpotential where everything can be defined in terms of a gauge invariant function $G = \kappa^2\mathcal{K} + \log(\kappa^6 W\bar{W})$ [22].

[6]The chiral fermions, the gauginos and the gravitino carry a charge $bc/2$, $-bc/2$ and $-bc/2$ respectively.

where T_G is the Dynkin index of the adjoint representation, normalized to N for $SU(N)$, and T_R is the Dynkin index associated with the representation R of dimension d_R, equal to $1/2$ for the $SU(N)$ fundamental. An implicit sum over all matter representations is understood. It follows that gaugino masses are non-vanishing in this representation, creating a puzzle on the quantum equivalence of the two classically equivalent representations. The answer to this puzzle is based on the fact that gaugino masses are present in both representations and are generated at one-loop level by an effect called Anomaly Mediation [7, 8]. Indeed, it has been argued that gaugino masses receive a one-loop contribution due to the super-Weyl-Kähler and sigma-model anomalies, given by [8]:

$$M_{1/2} = -\frac{g^2}{16\pi^2}\left[(3T_G - T_R)m_{3/2} + (T_G - T_R)\mathcal{K}_\alpha F^\alpha + 2\frac{T_R}{d_R}(\log\det\mathcal{K}|_R{}'')_{,\alpha}F^\alpha\right]. \tag{26}$$

The expectation value of the auxiliary field F^α, evaluated in the Einstein frame is given by

$$F^\alpha = -e^{\kappa^2\mathcal{K}/2}g^{\alpha\bar{\beta}}\bar{\nabla}_{\bar{\beta}}\bar{W}. \tag{27}$$

Clearly, for the Kähler potential (10.) or (24) the last term in eq. (26) vanishes. However, the second term survives due to the presence of Planck scale VEVs for the hidden sector fields s and z. Since the Kähler potential between the two representations differs by a linear term $b(s + \bar{s})$, the contribution of the second term in eq. (26) differs by a factor

$$\delta m_A = \frac{g_A^2}{16\pi^2}(T_G - T_R)be^{\kappa^2\mathcal{K}/2}g^{\alpha\bar{\beta}}\bar{\nabla}_{\bar{\beta}}\bar{W}, \tag{28}$$

which exactly coincides with the 'direct' contribution to the gaugino masses due to the field dependent gauge kinetic function (25) (taking into account a rescaling proportional to g_A^2 due to the non-canonical kinetic terms).

We conclude that even though the models (10.) and (24) differ by a (classical) Kähler transformation, they generate the same gaugino masses at one-loop. While the one-loop gaugino masses for the model (10.) are generated entirely by eq. (26), the gaugino masses for the model (24) after a Kähler transformation have a contribution from eq. (26) as well as from a field dependent gauge kinetic term whose presence is necessary to cancel the mixed $U(1)_R \times G$ anomalies due to the fact that the extra $U(1)$ has become an R-symmetry giving an R-charge to all fermions in the theory. Using (26), one finds:

$$M_{1/2} = -\frac{g^2}{16\pi^2}m_{3/2}\left[(3T_G - T_R) - (T_G - T_R)\left((\rho + 1)^2 + t\frac{\gamma + t + \gamma t^2}{1 + \gamma t}\right)\right]. \tag{29}$$

For $U(1)_Y$ we have $T_G = 0$ and $T_R = 11$, for $SU(2)$ we have $T_G = 2$ and $T_R = 7$, and for $SU(3)$ we have $T_G = 3$ and $T_R = 6$, such that for the different gaugino masses this gives (in a self-explanatory notation):

$$M_1 = 11\frac{g_Y^2}{16\pi^2}m_{3/2}\left[1 - (\rho + 1)^2 - \frac{t(\gamma + t + \gamma t)}{1 + \gamma t}\right],$$

$$M_2 = \frac{g_2^2}{16\pi^2} m_{3/2} \left[1 - 5(\rho + 1)^2 - 5\frac{t(\gamma + t + \gamma t^2)}{1 + \gamma t} \right],$$
$$M_3 = -3\frac{g_3^2}{16\pi^2} m_{3/2} \left[1 + (\rho + 1)^2 + \frac{t(\gamma + t + \gamma t^2)}{1 + \gamma t} \right]. \quad (30)$$

5. Phenomenology

The results for the soft terms calculated in the previous section, evaluated for different values of the parameter γ are summarised in Table 1. For every γ, the corresponding t and ρ are calculated by imposing a vanishing cosmological constant at the minimum of the potential. The scalar soft masses and trilinear terms are then evaluated by eqs. (22) and the gaugino masses by eqs. (30). Note that the relation (23) is valid for all γ. We therefore do not list the parameter B_0.

Table 1: The soft terms (in terms of $m_{3/2}$) for various values of γ. If a solution to the RGE exists, the value of $\tan\beta$ is shown in the last columns for $\mu > 0$ and $\mu < 0$.

γ	t	ρ	m_0	A_0	M_1	M_2	M_3	$\tan\beta$ ($\mu > 0$)	$\tan\beta$ ($\mu < 0$)
0.6	0.446	0.175	0.475	1.791	0.017	0.026	0.027		
1	0.409	0.134	0.719	1.719	0.015	0.025	0.026		
1.1	0.386	0.120	0.772	1.701	0.015	0.024	0.026	46	29
1.4	0.390	0.068	0.905	1.646	0.014	0.023	0.026	40	23
1.7	0.414	0.002	0.998	1.588	0.013	0.022	0.025	36	19

In most phenomenological studies, B_0 is substituted for $\tan\beta$, the ratio between the two Higgs VEVs, as an input parameter for the renormalization group equations (RGE) that determine the low energy spectrum of the theory. Since B_0 is not a free parameter in our theory, but is fixed by eq. (23), this corresponds to a definite value of $\tan\beta$. For more details see [23] (and references therein). The corresponding $\tan\beta$ for a few particular choices for γ are listed in the last two columns of table 1 for $\mu > 0$ and $\mu < 0$ respectively. No solutions were found for $\gamma \lesssim 1.1$, for both signs of μ. The lighest supersymmetric particle (LSP) is given by the lightest neutralino and since $M_1 < M_2$ (see table 1) the lightest neutralino is mostly Bino-like, in contrast with a typical mAMSB (minimal anomaly mediation supersymmetry breaking) scenario, where the lightest neutralino is mostly Wino-like [24].

To get a lower bound on the stop mass, the sparticle spectrum is plotted in Figure 2 as a function of the gravitino mass for $\gamma = 1.1$ and $\mu > 0$ (for $\mu < 0$ the bound is higher). The experimental limit on the gluino mass forces $m_{3/2} \gtrsim 15$ TeV. In this limit the stop mass can be as low as 2 TeV. To conclude, the lower end mass spectrum consists of (very) light charginos (with a lightest chargino between 250 and 800 GeV) and neutralinos, with a mostly Bino-like neutralino as LSP (80 – 230 GeV), which would distinguish this model from the mAMSB where the

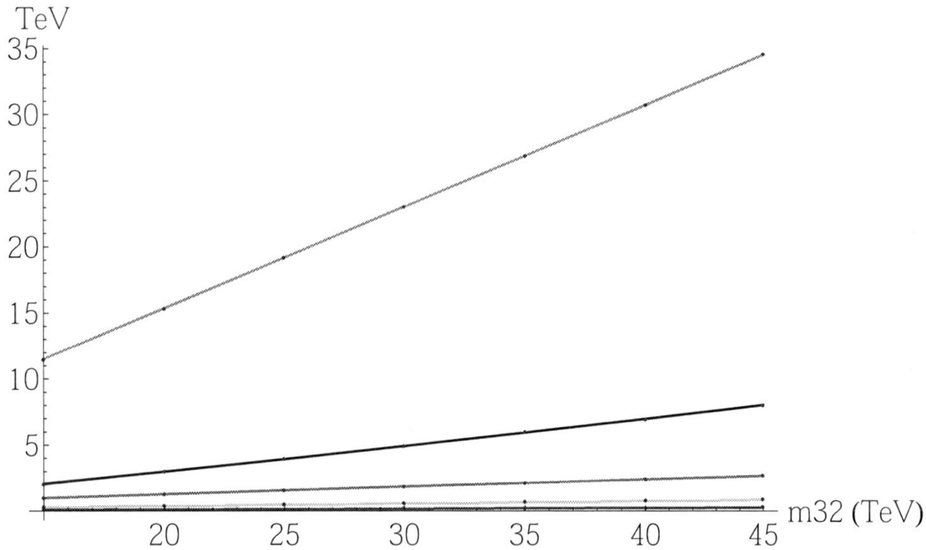

Figure 2: The masses of the sbottom (yellow), stop (black), gluino (red), lightest chargino (green) and lightest neutralino (blue) as a function of $m_{3/2}$ for $\gamma = 1.1$ and for $\mu > 0$. No solutions to the RGE were found when $m_{3/2} \gtrsim 45$ TeV. The lower bound corresponds to a gluino mass of 1 TeV.

LSP is mostly Wino-like. These upper limits on the LSP and the lightest chargino imply that this model could in principle be excluded in the next LHC run. In order for the gluino to escape experimental bounds, the lower limit on the gravitino mass is about 15 TeV. The gluino mass is then between 1-3 TeV. This however forces the squark masses to be very high (10 – 35 TeV), with the exception of the stop mass which can be relatively light (2 – 15 TeV).

6. Non-canonical Kähler potential for the visible sector

As mentioned already in Section 4, an alternative way to avoid tachyonic soft scalar masses for the MSSM fields in the model (17), instead of adding the extra Palonyi-type field z in the hidden sector, is by introducing non-canonical kinetic terms for the MSSM fields, such as:

$$\begin{aligned} \mathcal{K} &= -\kappa^{-2}\log(s+\bar{s}) + \kappa^{-2}b(s+\bar{s}) + (s+\bar{s})^{-\nu}\sum\varphi\bar{\varphi}, \\ W &= \kappa^{-3}a + W_{MSSM}, \\ f(s) &= 1, \qquad f_A(s) = 1/g_A^2\,, \end{aligned} \tag{31}$$

where ν is an additional parameter of the theory, with $\nu = 1$ corresponding to the leading term in the Taylor expansion of $-\log(s+\bar{s}-\varphi\bar{\varphi})$. Since the visible sector fields appear only in the combination $\varphi\bar{\varphi}$, their VEVs vanish provided that the scalar soft masses squared are positive. Moreover, for vanishing visible sector VEVs, the scalar potential and is minimization remains the same as in eqs. (ref-bsalpha). Therefore, the non-canonical Kähler potential does not change the fact that the F-term contribution to the soft scalar masses squared is negative. On the other hand, the visible fields enter in the D-term scalar potential through the derivative of the Kähler potential with respect to s. Even though this has no effect on the ground state of the potential, the φ-dependence of the D-term scalar potential does result in an extra contribution to the scalar masses squared which become positive

$$\nu > -\frac{e^{\alpha}(\sigma_s+1)\alpha}{A(\alpha)(1-\alpha)} \approx 2.6\,. \tag{32}$$

The soft MSSM scalar masses and trilinear couplings in this model are:

$$\begin{aligned} m_0^2 &= \kappa^2 a^2 \left(\frac{b}{\alpha}\right)\left(e^{\alpha}(\sigma_s+1) + \nu\frac{A(\alpha)}{\alpha}(1-\alpha)\right) \\ A_0 &= m_{3/2}(s+\bar{s})^{\nu/2}\,(\sigma_s+3) \\ B_0 &= m_{3/2}(s+\bar{s})^{\nu/2}\,(\sigma_s+2) \end{aligned} \tag{33}$$

where σ_s is defined as in (22), eq. (20) has been used to relate the constants a and c, and corrections due to a small cosmological constant have been neglected. A field redefinition due to a non-canonical kinetic term $g_{\varphi\bar{\varphi}} = (s+\bar{s})^{-\nu}$ is also taken into account. The main phenomenological properties of this model are not expected to be different from the one we analyzed in section 5. with the parameter ν replacing γ. Gaugino masses are still generated at one-loop level while mSUGRA applies to the soft scalar sector. We therefore do not repeat the phenomenological analysis for this model.

7. Identifying the dilaton with the inflaton

In the following, we study the possibility to identify the dilaton with the inflaton. We will show first that the above model does not allow slow roll inflation.

Indeed, the kinetic terms in the model (9-10) for the scalar $\phi \equiv s + \bar{s} = 1/l$ are given by

$$\mathcal{L}_s/e = -g_{s\bar{s}}\partial_\mu s \partial^\mu \bar{s} = -\frac{p\kappa^{-2}}{4}\frac{1}{\phi^2}\partial_\mu\phi\partial^\mu\phi. \tag{34}$$

The canonically normalised field χ therefore satisfies $\chi = \kappa^{-1}\sqrt{\frac{p}{2}}\log\phi$, where we re-introduce the gravitational coupling κ.

The slow roll parameters are given by

$$\begin{aligned} \epsilon &= \frac{1}{2\kappa^2}\left(\frac{dV/d\chi}{V}\right)^2 = \frac{1}{2\kappa^2}\left[\frac{1}{V}\frac{dV}{d\phi}\left(\frac{d\chi}{d\phi}\right)^{-1}\right]^2, \\ \eta &= \frac{1}{\kappa^2}\frac{V''(\chi)}{V} = \frac{1}{\kappa^2}\frac{1}{V}\left[\frac{d^2V}{d\phi^2}\left(\frac{d\chi}{d\phi}\right)^{-2} - \frac{dV}{d\phi}\frac{d^2\chi}{d\phi^2}\left(\frac{d\chi}{d\phi}\right)^{-3}\right], \end{aligned} \tag{35}$$

It can be shown that, when the conditions (11) and (12) are satisfied, the slow roll parameters and the potential depend only on $\rho = -b\phi$; indeed

$$\frac{\kappa^4 V(\rho)}{b^3 c^2} = \frac{e^{-\rho}\left(A_2(\alpha)\rho\left(\rho^2 + 4\rho - 2\right) - 2e^{\rho}(\rho + 2)^2\right)}{2\rho^3}, \tag{36}$$

where $A_2(\alpha) \approx -50.66$ as in eq. (12). In Fig. 3, a plot is shown of $\frac{\kappa^4 V(\rho)}{|b|^3 c^2}$ as a function of ρ. The minimum of the potential is at $\rho_{\min} \approx 0.1832$ (see eq. (11)), while the potential has a local maximum at $\rho_{\max} \approx 0.4551$. A plot of the slow roll parameter η (also in Fig. 3) shows that $|\eta| \ll 1$ is not satisfied. This result holds for any parameters a, b, c satisfying eqs. (11) and (12). A similar analysis to the

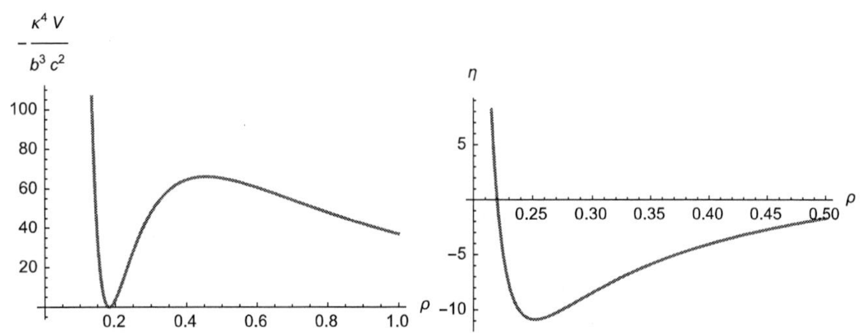

Figure 3: A plot of $-\frac{\kappa^4 V(\rho)}{b^3 c^2}$ as a function of $\rho = -b\phi$ (left), and a plot of the slow roll parameter η as a function of ρ (right). The slow roll condition $|\eta| \ll 1$ is not satisfied for any value of the parameters a, b, c.

one above can be performed for $p = 1$, showing that the slow roll condition $\eta \ll 1$ can not be satisfied.

8. Extensions of the model that satisfy the slow roll conditions

In the previous section we showed that the slow roll conditions can not be satisfied in the minimal versions of the model. In this section we modify the above model by modifying the Kähler potential. While the superpotential is uniquely fixed (up to a Kähler transformation), the Kähler potential admits corrections that can always be put in the form

$$\mathcal{K} = -p\kappa^{-2}\log\left(s + \bar{s} + \frac{\xi}{b}F(s + \bar{s})\right) + \kappa^{-2}b(s + \bar{s}), \tag{37}$$

while the superpotential, the gauge kinetic function and moment map are given by

$$W = \kappa^{-3}a,$$

$$\begin{aligned} f(s) &= \delta + \beta s, \\ \mathcal{P} &= \kappa^{-2} c \left(b - p \frac{1 + \frac{\xi}{b} F_s}{s + \bar{s} + \frac{\xi}{b} F} \right), \end{aligned} \tag{38}$$

where $\mathcal{P}$ is the $U(1)$ moment map (7) and $F_s = \partial_s F(s + \bar{s})$. The scalar potential is given by ($\phi = s + \bar{s}$)

$$\begin{aligned} V &= V_F + V_D, \\ V_F &= \kappa^{-4} \frac{|a|^2 e^{b\phi}}{(\phi + \frac{\xi}{b} F)^p} \left[-3 - \frac{1}{p} \frac{\left(b\,(b\phi + \xi F) - p(b + \xi F_\phi) \right)^2}{\xi F_{\phi\phi}(b\phi + \xi F) - (b + \xi F_\phi)^2} \right], \\ V_D &= \kappa^{-4} \frac{b^2 c^2}{2\delta + \beta\phi} \left[1 - p \frac{1 + \frac{\xi}{b} F_\phi}{b\phi + \xi F} \right]^2 . \end{aligned} \tag{39}$$

As was discussed above, we take $\delta = 1, \beta = 0$ for $p = 1$ and $\delta = 0, \beta = 1$ for $p = 2$.

Identifying $\mathrm{Re}(s)$ with the inverse string coupling, the function F may contain perturbative contributions that can be expressed as power series of $1/(s + \bar{s})$, as well as non-perturbative corrections which are exponentially suppressed in the weak coupling limit. The later can be either of the form $e^{-\lambda(s+\bar{s})}$ for $\lambda > 0$ in the case of D-brane instantons, or of the form $e^{-\lambda(s+\bar{s})^2}$ in the case of (Neveu-Schwarz) NS5-brane instantons (since the closed string coupling is the square of the open string coupling). We have considered a generic contribution of these three different types of corrections and we found that only the last type of contributions can lead to an inflationary plateau providing sufficient inflation. The other corrections can be present but do not modify the main properties of the model (as long as weak coupling description holds). In the following section, we analyse in detailed a function F describing a generic NS5-brane instanton correction to the Kähler potential.

9. Slow-roll Inflation

9.1. p=2 case

We now consider the case with

$$F(\phi) = \exp(\alpha b^2 \phi^2), \tag{40}$$

where $b < 0$ and $\alpha < 0$. $F(\phi)$ vanishes asymptotically at large ϕ. In this case, we obtain

$$V_D = \frac{\kappa^{-4} b^3 c^2}{b\phi} \left[\frac{b\phi - 2 + \xi e^{\alpha b^2 \phi^2}(1 - 4\alpha b\phi)}{b\phi + \xi e^{\alpha b^2 \phi^2}} \right]^2 , \tag{41}$$

and

$$V_F = -\frac{\kappa^{-4}|a|^2 b^2 e^{b\phi}}{2\left(\xi e^{\alpha b^2\phi^2} + b\phi\right)^2}\left[\frac{\left(b\phi + \xi e^{\alpha b^2\phi^2}(1-4\alpha b\phi) - 2\right)^2}{2\alpha\xi e^{\alpha b^2\phi^2}\left(2\alpha b^3\phi^3 + \xi e^{\alpha b^2\phi^2} - b\phi\right) - 1} + 6\right]. \quad (42)$$

There are four parameters in this model namely α, ξ, b and c. The first two parameters α and ξ control the shape of the potential. There are some regions in the parameter space of α and ξ that the potential satisfies the slow-roll conditions i.e. $\epsilon \ll 1$ and $|\eta| \ll 1$. In order to obtain the potential with flat plateau shape which is suitable for inflation and in agreement with Planck '15 data, we choose

$$\alpha \simeq -4.84 \quad \text{and} \quad \xi \simeq 0.025 \quad (43)$$

Note that in the case of $\xi = 0$ and $b < 0$, we can find the Minkowski minimum by solving the equations $V(\phi_{min}) = 0$ and $dV(\phi_{min})/d\phi = 0$, where $\phi_{min} = s_{min} + \bar{s}_{min}$ is the value of ϕ at the minimum of the potential. In the case of $\xi \neq 0$, we can not solve the equations analytically and the relations (11), (12) are not valid. We can always assume that they are modified into

$$b\phi_{min} = -\rho(\xi,\alpha) \quad \text{and} \quad \frac{a^2}{bc^2} = -50.66 \times \lambda(\xi,\alpha,\Lambda)^2, \quad (44)$$

where λ takes positive values and satisfies $|\lambda - 1| \ll 1$. For any given value of parameters ξ, α and the cosmological constant Λ, one can numerically fix the value of ρ and λ. By fine-tuning the cosmological constant Λ to be very close to zero, we can numerically solve the equations $V = 0$ and $dV/d\phi = 0$ for the value of ρ and λ in (44) as:

$$\rho \approx 0.18, \quad (45)$$
$$\lambda \approx 1.017 \quad (46)$$

From eq. (44), we can see that the third parameter, b, controls the vacuum expectation value ϕ_{min}. This can be shown in Fig. 4 where we compare the scalar potential for different values of b. Motivated by string theory, we have the identification $\phi \sim 1/g_s$. We can choose the value of the parameter b such that ϕ_{min} is of the order of 10 to make sure that we are in the perturbative regime in g_s. The last parameter, c, controls the overall scale of the potential but does not change its minimum and its shape. In the following, we will fix b and c by using the cosmological data.

In order to compare the predictions of our models with Planck '15 data, we choose the following boundary conditions:

$$\phi_{int} = 27.32 \qquad \phi_{end} = 22.68 \quad (47)$$

The initial conditions are chosen very near the maximum on the (left) side, so that the field rolls down towards the electroweak minimum. Any initial condition

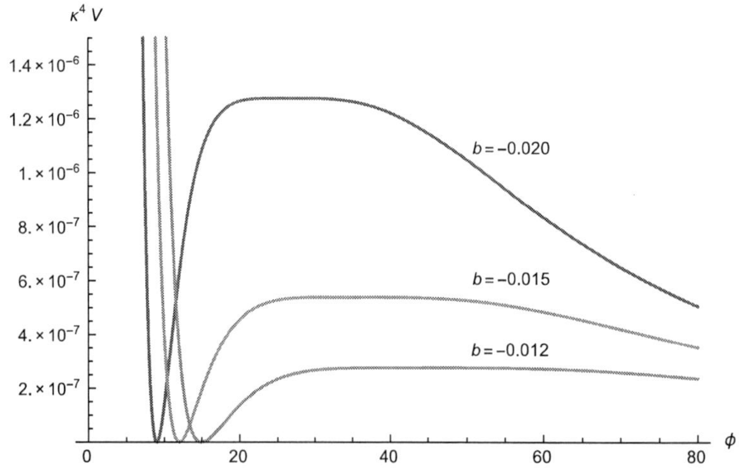

Figure 4: A plot of the scalar potential for $p = 2$, with $b = -0.020$, $b = -0.015$ and $b = -0.012$. Note that we choose the parameters α and ξ as in eq. (43) with $c = 0.06$.

on the right of the maximum may produce also inflation, but the field will roll towards the SUSY vacuum at infinity. The results are therefore very sensitive to the initial conditions (47) of the inflaton field.

The slow roll parameters are given as in equation (35). The total number of e-folds N can be determined by

$$N = \kappa^2 \int_{\chi_{end}}^{\chi_{int}} \frac{V}{\partial_\chi V} d\chi = \kappa^2 \int_{\phi_{end}}^{\phi_{int}} \frac{V}{\partial_\phi V} \left(\frac{d\chi}{d\phi}\right)^2 d\phi. \tag{48}$$

Note that we choose $|\eta(\chi_{end})| = 1$. We can compare the theoretical predictions of our model to the experimental results via the power spectrum of scalar perturbations of the CMB, namely the amplitude A_s and tilt n_s, and the relative strength of tensor perturbations, i.e. the tensor-to-scalar ratio r. In terms of slow roll parameters, these are given by

$$A_s = \frac{\kappa^4 V_*}{24\pi^2 \epsilon_*}, \tag{49}$$

$$n_s = 1 + 2\eta_* - 6\epsilon_*, \tag{50}$$

$$r = 16\epsilon_*, \tag{51}$$

where all parameters are evaluated at the field value χ_{int}.

In order to satisfy Planck '15 data, we choose the parameters $b = -0.0182$, $c = 0.61 \times 10^{-13}$. The value of the slow-roll parameters at the beginning of inflation are

$$\epsilon(\phi_{int}) \simeq 1.86 \times 10^{-24} \quad \text{and} \quad \eta(\phi_{int}) \simeq -1.74 \times 10^{-2}. \tag{52}$$

The total number of e-folds N, the scalar power spectrum amplitude A_s, the spectral index of curvature perturbation n_s and the tensor-to-scalar ratio r are calculated and summarised in Table 2, in agreement with Planck '15 data [25]. Fig. 5 shows that our predictions for n_s and r are within 1σ C.L. of Planck '15 contours with the total number of e-folds $N \approx 1075$. Note that N is the total number of e-folds from ϕ_{int} to ϕ_{end}. However the number of e-folds associated with the CMB observation corresponds to a period between the time of horizon crossing and the end of inflation, which is much smaller than 1075. According to general formula in [25], the number of e-folds between the horizon crossing and the end of inflation is roughly estimated to be around 50-60.

Table 2: The theoretical predictions for $p = 2$, with $b = -0.0182$, $c = 0.61 \times 10^{-13}$, and α, ξ given in eq. (43).

n_s	r	A_s
0.965	2.969×10^{-23}	2.259×10^{-9}

We would like to remark that the parameter c also controls the gravitino mass at the minimum of the potential around $O(10)$ TeV. Indeed, the gravitino mass is written as

$$m_{3/2} = \kappa^2 e^{\kappa^2 \mathcal{K}/2} W = \frac{1}{\kappa}\left(\frac{abe^{b\phi/2}}{b\phi + \xi F(\phi)}\right). \tag{53}$$

For $b = -0.0182$, we get $\phi_{\min} \approx 9.91134$ and the gravitino mass at the minimum of the potential

$$\langle m_{3/2} \rangle \approx 14.98 \quad \text{TeV}. \tag{54}$$

The Hubble parameter during inflation (evaluated at $\phi_* = \phi_{int}$) is

$$H_* = \kappa\sqrt{V_*/3} = 1.38 \quad \text{TeV}. \tag{55}$$

This shows that our predicted scale for inflation is of the order of TeV. The mass of gravitino during the inflation $m^*_{3/2} = 4.15$ TeV is higher than the inflation scale, and the gauge boson mass is $M^*_{A_\mu} = 3.12$ TeV.[7] In fact, the gauge boson acquires a mass due to a Stueckelberg mechanism by eating the imaginary component of s, where its mass at the minimum of the potential is given by

$$\langle M_{A_\mu} \rangle = 15.48 \text{TeV}. \tag{56}$$

As a result, the model essentially contains only one scalar field Re(s), which is the inflaton. This is in contrast with other supersymmetric models of inflation, which usually contain at least two real scalars [26].[8]

[7]The gauge boson mass is given by $m_{A_\mu} = \sqrt{2g_{s\bar{s}}c^2/\text{Re}(s)}$.

[8]This is because a chiral multiplet contains a complex scalar.

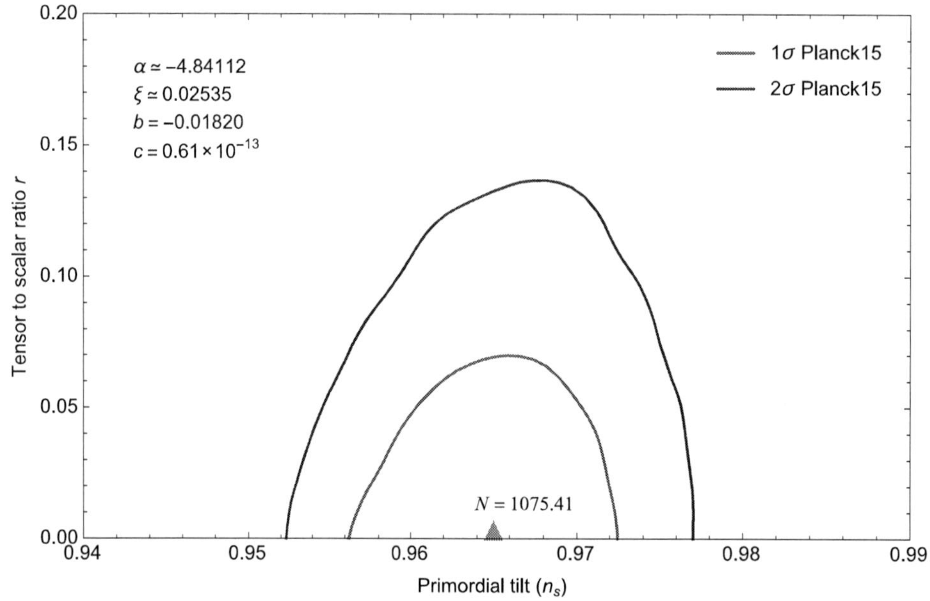

Figure 5: We plot the theoretical predictions for the case $p = 2$, shown in Table 2, in the n_s - r plane together with the Planck '15 results for TT, TE, EE, + lowP and assuming ΛCDM + r.

9.2. p=1 case

In this case, we obtain

$$V_D = \frac{\kappa^{-4} b^2 c^2}{2} \left[\frac{b\phi - 1 + \xi e^{\alpha b^2 \phi^2}(1 - 2\alpha b\phi)}{b\phi + \xi e^{\alpha b^2 \phi^2}} \right]^2 , \tag{57}$$

and

$$V_F = -\frac{\kappa^{-4} |a|^2 b e^{b\phi}}{\xi e^{\alpha b^2 \phi^2} + b\phi} \left[\frac{\left(b\phi + \xi e^{\alpha b^2 \phi^2}(1 - 2\alpha b\phi) - 1\right)^2}{2\alpha\xi e^{\alpha b^2 \phi^2}\left(2\alpha b^3 \phi^3 + \xi e^{\alpha b^2 \phi^2} - b\phi\right) - 1} + 3 \right] . \tag{58}$$

The potential has similar properties with the $p = 2$ case although it may give different phenomenological results at low energy. Similar to the previous case, the relations (20) are not valid when $\xi \neq 0$ and we assume that they are modified into

$$b\phi_{min} = -\rho(\xi, \alpha) \quad \text{and} \quad \frac{bc^2}{a^2} \simeq -0.359 \times \lambda(\xi, \alpha, \Lambda)^{-2} . \tag{59}$$

By choosing $\alpha = -0.781$ and $\xi = 0.3023$ and tuning the cosmological constant Λ to be very close to zero, we can numerically fix $\rho \approx 0.56$ and $\lambda \approx 1.29$ for this

Table 3: The theoretical predictions for $p = 1$ with $b = -0.0234$, $c = 1 \times 10^{-13}$, $\alpha = -0.781$ and $\xi = 0.3023$.

n_s	r	A_s
0.959	4.143×10^{-22}	2.205×10^{-9}

case. The gravitino mass for $p = 1$ case can be written as

$$m_{3/2} = \kappa^2 e^{\kappa^2 \mathcal{K}/2} W = \frac{1}{\kappa}\left(\frac{a\sqrt{b}e^{b\phi/2}}{\sqrt{b\phi + \xi F(\phi)}}\right). \tag{60}$$

By choosing the parameters $b = -0.0234$, $c = 1 \times 10^{-13}$, the gravitino mass at the minimum of the potential is

$$\langle m_{3/2}\rangle = 18.36 \quad \text{TeV}. \tag{61}$$

with $\phi_{\rm min} \approx 21.53$, and

$$\langle M_{A_\mu}\rangle = 36.18 \quad \text{TeV}. \tag{62}$$

By choosing appropriate boundary conditions, we find

$$\phi_{int} = 64.53 \quad \text{and} \quad \phi_{end} = 50.99 \tag{63}$$

As summarised in Table 3, the predictions for the $p = 1$ case are similar to those of $p = 2$, in agreement with Planck '15 data with the total number of e-folds $N \approx 888$. In this case, the Hubble parameter during inflation is

$$H_* = \kappa\sqrt{V_*/3} = 5.09 \quad \text{TeV}. \tag{64}$$

Note that for the $p = 1$ case, the mass of the gauge boson is $M^*_{A_\mu} = 6.78$ TeV, and the mass of the gravitino during inflation is $m^*_{3/2} = 4.72$ TeV.

9.3. SUGRA spectrum

The above model can be coupled to MSSM, as described in section 4.:

$$\begin{aligned} \mathcal{K} &= \mathcal{K}(s+\bar{s}) + \sum \varphi\bar{\varphi}, \\ W &= W_h(s) + W_{\rm MSSM}\,. \end{aligned} \tag{65}$$

The soft supersymmetry breaking terms can then be calculated as follows

$$\begin{aligned} m_0^2 &= e^{\kappa^2\mathcal{K}}\left(-2\kappa^4 W_h(s)\bar{W}_h(s) + \kappa^2 g^{s\bar{s}}\left|\nabla_s W_h\right|^2\right), \\ A_0 &= \kappa^2 e^{\kappa^2\mathcal{K}/2} g^{s\bar{s}} K_s\left(\bar{W}_{\bar{s}} + \kappa^2 K_s \bar{W}\right), \\ B_0 &= \kappa^2 e^{\kappa^2\mathcal{K}/2}\left(g^{s\bar{s}} K_s\left(\bar{W}_{\bar{s}} + \kappa^2 K_s\bar{W}\right) - \bar{W}\right). \end{aligned} \tag{66}$$

For $p = 2$ the Lagrangian contains a Green-Schwarz term eq. (18), and the theory is not gauge invariant (without the inclusion of extra fields that are charged under the $U(1)$). We therefore focus on $p = 1$. The soft terms can be written in terms of the gravitino mass (see eq. (53))

$$\begin{aligned} m_0^2 &= m_{3/2}^2 \left[-2 + C\right], \\ A_0 &= m_{3/2}\, C, \\ B_0 &= A_0 - m_{3/2}, \end{aligned} \tag{67}$$

where

$$C = -\left.\frac{\left(-\xi e^{\alpha b^2\phi^2} + b\phi\left(4\alpha\xi e^{\alpha b^2\phi^2} - 1\right) + 2\right)^2}{4\alpha\xi^2 e^{2\alpha b^2\phi^2} - 4\alpha b\xi\phi e^{\alpha b^2\phi^2} + 8\alpha^2 b^3\xi\phi^3 e^{\alpha b^2\phi^2} - 2}\right|_{\phi=\phi_{\min}}. \tag{68}$$

Using the parameters presented in section 9.2., we find $m_{3/2} = 18.36$ TeV and $C = 1.53$. For $\xi = 0$ the model reduces to the one analysed in section 4., where one has $C = 1.52$ and $m_{3/2} = 17.27$ TeV (with $\phi_{\min} = 9.96$). Moreover, the scalar soft mass is tachyonic. This can be solved either by introducing an extra Polonyi-like field, or by allowing a non-canonical Kähler potential for the MSSM-like fields φ. The resulting low energy spectrum is expected to be similar to the one described in sections 4. and 5.. We do not perform this analysis, but only summarise the results.

Since the tree-level contribution to the gaugino masses vanishes, their mass is generated at one-loop by the so-called 'Anomaly Mediation' contribution (26). As a result, the spectrum consists of very light neutralinos ($O(10^2)$ GeV), of which the lightest (a mostly Bino-like neutralino) is the LSP dark matter candidate, slightly heavier charginos and a gluino in the $1-3$ TeV range. The squarks are of the order of the gravitino mass (~ 10 TeV), with the exception of the stop squark which can be as light as 2 TeV.

10. Symmetric versus non-symmetric point

Here, we generalise the above model of inflation and we are interested in supergravity theories containing a single chiral multiplet transforming under a gauged R-symmetry with a corresponding abelian vector multiplet [16]. We assume that the chiral multiplet X (with scalar component X) transforms as:

$$X \longrightarrow X e^{-iq\omega}. \tag{69}$$

where q is its charge, and ω is the gauge parameter.

The Kähler potential is therefore a function of $X\bar{X}$, while the superpotential is constrained to be of the form X^b:

$$\begin{aligned} \mathcal{K} &= \mathcal{K}(X\bar{X}), \\ W &= \kappa^{-3} f X^b, \end{aligned} \tag{70}$$

where X is a dimensionless field and $\kappa^{-1} = m_p = 2.4 \times 10^{15}$ TeV is the (reduced) Planck mass. For $b \neq 0$, the gauge symmetry eq. 69 becomes a gauged R-symmetry. The gauge kinetic function can have a constant contribution as well as a contribution proportional to $\ln X$

$$f(X) = \gamma + \beta \ln X. \tag{71}$$

The latter contribution proportional to β is not gauge invariant and can be used as a Green-Schwarz counter term to cancel possible anomalies. One can show however that the constant β is fixed to be very small by anomaly cancellation conditions and does not change our results [16]. We will therefore omit this term in our analysis below.

We are interested in the general properties of supergravity theories of inflation that are of the above form. Before performing our analysis, a distinction should be made concerning the initial point where slow-roll inflation starts. The inflaton field (which will turn out to be ρ, where $X = \rho e^{i\theta}$) can either have its initial value close to the symmetric point where $X = 0$, or at a generic point $X \neq 0$. The minimum of the potential, however, is always at a nonzero point $X \neq 0$. This is because at $X = 0$ the negative contribution to the scalar potential vanishes and no cancellation between F-term and D-term is possible. The supersymmetry breaking scale is therefore related to the cosmological constant as $\kappa^{-2} m_{3/2}^2 \approx \Lambda$. One could in principle assume that the value of the potential at its minimum is of the order of the supersymmetry breaking scale. However, in this case additional corrections are needed to bring down the minimum of the potential to the present value of the cosmological constant, and we therefore do not discuss this possibility.

In the first case, inflation starts near $X = 0$, and the inflaton field will roll towards a minimum of the potential at $X \neq 0$. On the other hand, in the second case inflation will start at a generic point $X \neq 0$. In order to make easier contact with the model discussed in the previous sections, it is convenient to work with another chiral superfield S, which is invariant under a shift symmetry

$$S \longrightarrow S - ic\alpha \tag{72}$$

by performing a field redefinition

$$X = e^S. \tag{73}$$

In this case the most general Kähler potential and superpotential are of the form

$$\begin{aligned} \mathcal{K} &= \mathcal{K}(S + \bar{S}), \\ W &= \kappa^{-3} a e^{bS}. \end{aligned} \tag{74}$$

Note that this field redefinition is not valid at the symmetric point $X = 0$ for the first case.

11. Case 1: Inflation near the symmetric point

11.1. Slow roll parameters

In this section we derive the conditions that lead to slow-roll inflation scenarios, where the start of inflation is near a local maximum of the potential at $X = 0$. Since the superpotential has charge 2 under R-symmetry, one has $\langle W \rangle = 0$ as long as R-symmetry is preserved. Therefore, $\langle W \rangle$ can be regarded as the order parameter of R-symmetry breaking. On the other hand, the minimum of the potential requires $\langle W \rangle \neq 0$ and broken R-symmetry. It is therefore attractive to assume that at earlier times R-symmetry was a good symmetry, switching off dangerous corrections to the potential. As similar approach was followed in [27], where a discrete R-symmetry is assumed. Instead, we assume a gauged R-symmetry which is spontaneously broken at the minimum of the potential.

While the superpotential is uniquely fixed in eq. (70), the Kähler potential is only fixed to be of the form $\mathcal{K}(X\bar{X})$. We expand the Kähler potential as follows

$$\begin{aligned} \mathcal{K}(X,\bar{X}) &= \kappa^{-2}X\bar{X} + \kappa^{-2}A(X\bar{X})^2, \\ W(X) &= \kappa^{-3}fX^b, \\ f(X) &= 1, \end{aligned} \tag{75}$$

where A and f are constants. The gauge kinetic function is taken to be constant since it was shown that the coefficient β in front of the logarithmic term in eq. (71) is fixed to be very small by anomaly cancellation conditions [16]. As far as the scalar potential is concerned, the coefficient γ can be absorbed in other parameters of the theory. We therefore take $\gamma = 1$.

The scalar potential is given by

$$\mathcal{V} = \mathcal{V}_F + \mathcal{V}_D, \tag{76}$$

where

$$\mathcal{V}_F = \kappa^{-4} f^2 (X\bar{X})^{b-1} e^{X\bar{X}(1+AX\bar{X})} \left[-3X\bar{X} + \frac{\left(b + X\bar{X}(1+2AX\bar{X})\right)^2}{1+4AX\bar{X}} \right], \tag{77}$$

and

$$\mathcal{V}_D = \kappa^{-4}\frac{q^2}{2}\left[b + X\bar{X}(1+2AX\bar{X})\right]^2. \tag{78}$$

The superpotential is not gauge invariant under the $U(1)$ gauge symmetry. Instead it transforms as

$$W \to We^{-iqbw}. \tag{79}$$

Therefore, the $U(1)$ is a gauged R-symmetry which we will further denote as $U(1)_R$. From $W_X k_R^X = -r_R\kappa^2 W$, where $k_R^X = -iqX$ is the Killing vector for the field

X under the R-symmetry, $r_R = i\kappa^{-2}\xi_R$ with $\kappa^{-2}\xi_R$ the Fayet-Iliopoulos contribution to the scalar potential, and W_X is short-hand for $\partial W/\partial X$, we find

$$r_R \quad = i\kappa^{-2}qb. \tag{80}$$

A consequence of the gauged R-symmetry is that the superpotential coupling b enters the D-term contribution of the scalar potential as a constant Fayet-Iliopoulos contribution.[9]

Note that the scalar potential is only a function of the modulus of X and that the potential contains a Fayet-Iliopoulos contribution for $b \neq 0$. Moreover, its phase will be 'eaten' by the $U(1)$ gauge boson upon a field redefinition of the gauge potential similarly to the standard Higgs mechanism. After performing a change of field variables

$$X = \rho e^{i\theta}, \quad \bar{X} = \rho e^{-i\theta}, \quad (\rho \geq 0) \tag{81}$$

the scalar potential is a function of ρ,

$$\kappa^4 \mathcal{V} = f^2 \rho^{2(b-1)} e^{\rho^2 + A\rho^4} \left(-3\rho^2 + \frac{\left(b + \rho^2 + 2A\rho^4\right)^2}{1 + 4A\rho^2} \right) + \frac{q^2}{2} \left(b + \rho^2 + 2A\rho^4\right)^2. \tag{82}$$

Since we assume that inflation starts near $\rho = 0$, we require that the potential eq. (82) has a local maximum at this point. It turns out that the potential only allows for a local maximum at $\rho = 0$ when $b = 1$. For $b < 1$ the potential diverges when ρ goes to zero. For $1 < b < 1.5$ the first derivative of the potential diverges, while for $b = 1.5$, one has $V'(0) = \frac{9}{4}f^2 + \frac{3}{2}q^2 > 0$, and for $b > 1.5$, on has $V''(0) > 0$. We thus take $b = 1$ and the scalar potential reduces to

$$\kappa^4 \mathcal{V} = f^2 e^{\rho^2 + A\rho^4} \left(-3\rho^2 + \frac{\left(1 + \rho^2 + 2A\rho^4\right)^2}{1 + 4A\rho^2} \right) + \frac{q^2}{2} \left(1 + \rho^2 + 2A\rho^4\right)^2. \tag{83}$$

Note that in this case the the superpotential is linear $W = fX$, describing the sgoldstino (up to an additional low-energy constraint) [30]. Indeed, modulo a D-term contribution, the inflaton in this model is the superpartner of the goldstino. In fact, for $q = 0$ the inflaton reduces to the partner of the goldstino as in Minimal Inflation models [31]. The important difference however is that this is a microscopic realisation of the identification of the inflaton with the sgoldstino, and that the so-called η-problem is avoided (see discussion below).

The kinetic terms for the scalars can be written as[10]

$$\mathcal{L}_{\text{kin}} = -g_{X\bar{X}} \hat{\partial}_\mu X \hat{\partial}^\mu X = -g_{X\bar{X}} \left[\partial_\mu \rho \partial^\mu \rho + \rho^2 \left(\partial_\mu \theta + q A_\mu \right) \left(\partial^\mu \theta + q A^\mu \right) \right]. \tag{84}$$

[9]For other studies of inflation involving Fayet-Iliopoulos terms see for example [28], or [29] for more recent work. Moreover, our motivations have some overlap with [27], where inflation is also assumed to start near an R-symmetric point at $X = 0$. However, this work uses a discrete R-symmetry which does not lead to Fayet-Iliopoulos terms.

[10]The covariant derivative is defined as $\hat{\partial}_\mu X = \partial_\mu X - A_\mu k_R^X$, where $k_R^X = -iqX$ is the Killing vector for the U(1) transformation eq. 69.

It was already anticipated above that the phase θ plays the role of the longitudinal component of the gauge field A_μ, which acquires a mass by a Brout-Englert-Higgs mechanism.

We now interpret the field ρ as the inflaton. It is important to emphasise that, in contrast with usual supersymmetric theories of inflation where one necessarily has two scalar degrees of freedom resulting in multifield inflation [26], our class of models contains only one scalar field ρ as the inflaton. In order to calculate the slow-roll parameters, one needs to work with the canonically normalised field χ satisfying

$$\frac{d\chi}{d\rho} = \sqrt{2g_{X\bar{X}}}. \tag{85}$$

The slow-roll parameters are given in terms of the canonical field χ by

$$\epsilon = \frac{1}{2\kappa^2}\left(\frac{dV/d\chi}{V}\right)^2, \quad \eta = \frac{1}{\kappa^2}\frac{d^2V/d\chi^2}{V}. \tag{86}$$

Since we assume inflation to start near $\rho = 0$, we expand

$$\begin{aligned} \epsilon &= 4\left(\tfrac{-4A+x^2}{2+x^2}\right)^2\rho^2 + O(\rho^4), \\ \eta &= 2\left(\tfrac{-4A+x^2}{2+x^2}\right) + O(\rho^2), \end{aligned} \tag{87}$$

where we defined $q = fx$. Notice that for $\rho \ll 1$ the ϵ parameter is very small, while the η parameter can be made small by carefully tuning the parameter A. Any higher order corrections to the Kähler potential do not contribute to the leading contributions in the expansion near $\rho = 0$ for η and ϵ. Such corrections can therefore be used to alter the potential near its minimum, at some point $X \neq 0$ without influencing the slow-roll parameters.

A comment on the η-problem in Supergravity

A few words are now in order concerning the η-problem [32]. The η problem in $\mathcal{N} = 1$ supergravity is often stated as follows (see for example [33]): If, for instance, a theory with a single chiral multiplet with scalar component φ is taken, then the Kähler potential can be expanded around a reference location $\varphi = 0$ as $\mathcal{K} = \mathcal{K}(0) + \mathcal{K}_{\varphi\bar{\varphi}}(0)\varphi\bar{\varphi} + \ldots$. The Lagrangian becomes

$$\mathcal{L} = -\partial_\mu\phi\partial^\mu\bar{\phi} - \mathcal{V}(0)\left(1 + \kappa^2\phi\bar{\phi} + \cdots\right), \tag{88}$$

where ϕ is the canonically normalised field $\phi\bar{\phi} = \mathcal{K}_{\varphi\bar{\varphi}}(0)\varphi\bar{\varphi}$, and the ellipses stand for extra terms in the expansion coming from $\mathcal{K}$ and W. Following this argument, the mass m_ϕ turns out to be proportional to the Hubble scale

$$m_\phi^2 = \kappa^2\mathcal{V}(0) + \ldots = 3H^2 + \ldots, \tag{89}$$

and therefore

$$\eta = \frac{m_\phi^2}{3H^2} = 1 + \ldots. \tag{90}$$

Or otherwise stated, this leading contribution of order 1 to the η-parameter has its origin from the fact that the F-term contribution to the scalar potential contains an exponential factor $e^{\mathcal{K}}$: $\mathcal{V} = e^{X\bar{X}+\ldots}[\ldots]$ resulting in its second derivative $\mathcal{V}_{X\bar{X}} = V[1+\ldots]$.

However, in our model the factor '1' drops out for the particular choice $b = 1$ in the superpotential[11], resulting in an inflaton mass m_ρ^2 which is determined by the next term $A(X\bar{X})^2$ in the expansion of the Kähler potential,

$$\begin{aligned} m_\chi^2 &= \left(-4A + x^2\right)\kappa^{-2}f^2 + O(\rho^2), \\ H^2 &= \frac{\kappa^{-2}f^2}{6}(2 + x^2) + O(\rho^2). \end{aligned} \tag{91}$$

As a result, there are two ways to evade the η-problem:

- First, one can obtain a small η by having a small $q \ll f$, while A should be of order $O(10^{-1})$. In this case, the rôle of the gauge symmetry is merely to constrain the form of the Kähler potential and the superpotential, and to provide a Higgs mechanism that eliminates the extra scalar (phase) degree of freedom.
- Alternatively there could be a cancellation between q^2 and $4Af^2$.

Since A is the second term in the expansion of the Kähler potential eq. (75), it is natural to be of order $O(10^{-1})$ and therefore providing a solution to the η-problem.

Note that the mass of the inflaton given in eqs. (91) is only valid during inflation at small ρ. The mass of the inflaton at its VEV will be affected by additional corrections that are needed to obtain in particular a vanishing value for the scalar potential at its minimum [16].

The upper bound on the tensor-to-scalar ratio

Before moving on to the next section, let us focus on the approximation at $\rho \ll 1$ where the perturbative expansion of the slow-roll parameters in eqs. (87) is valid, and assume that the horizon exit occurs at the field value ρ_* very close to the maximum $\rho = 0$. In this approximation, eqs. (87) become

$$\epsilon(\rho) \approx \epsilon^{\text{pert}}(\rho) = |\eta_*|^2\rho^2, \quad \eta(\rho) \approx \eta_*, \tag{92}$$

where the asterisk refers to the value of parameters evaluated at the horizon exit.

[11]Note that in hybrid inflation models the η-problem is also evaded by a somewhat similar way, but these models generally include several scalar fields (and superfields) besides the inflaton (see e.g. [34]).

To discuss the upper bound on the tensor-to-scalar ratio, it is convenient to divide the region $[\rho = 0, \rho_{\rm end}]$ into two regions: one is $[0, \rho_{\rm p}]$, where the approximation 92 is valid, and the other is the rest $[\rho_{\rm p}, \rho_{\rm end}]$. Here $\rho_{\rm end}$ means the inflation end. Note that $\rho_{\rm p} < \rho_{\rm end}$ because the approximation 92 breaks down before the end of inflation where $\epsilon(\rho_{\rm end}) = 1$ or $|\eta(\rho_{\rm end})| = 1$. In terms of this division, the number of e-folds from the horizon exit to the end of inflation can be approximated by

$$N_{\rm CMB} \simeq N^{\rm pert}(\rho_*, \rho_{\rm p}) + \kappa \int_{\chi_{\rm p}}^{\chi_{\rm end}} \frac{d\chi}{\sqrt{2\epsilon(\chi)}}, \tag{93}$$

where we introduced

$$N^{\rm pert}(\rho_1, \rho_2) = \kappa \int_{\chi_1}^{\chi_2} \frac{d\chi}{\sqrt{2\epsilon^{\rm pert}(\chi)}} = \frac{1}{|\eta_*|} \ln\left(\frac{\rho_2}{\rho_1}\right). \tag{94}$$

Here χ is the canonically normalised field defined by eq. (85). Let us next focus on the region $[\rho_{\rm p}, \rho_{\rm end}]$. It is natural to expect the following inequality

$$\kappa \int_{\chi_{\rm p}}^{\chi_{\rm end}} \frac{d\chi}{\sqrt{2\epsilon(\chi)}} \lesssim \kappa \int_{\chi_{\rm p}}^{\chi_{\rm end}} \frac{d\chi}{\sqrt{2\epsilon^{\rm pert}(\chi)}}. \tag{95}$$

This is based on the following observation. The right hand side describes a hypothetical situation, as if the slow-roll condition were valid throughout the inflation until its end. But since in the actual inflation the slow-roll condition breaks down in the region $[\rho_{\rm p}, \rho_{\rm end}]$, the actual number of e-folds in this region will be smaller than that in the hypothetical situation. Adding $N^{\rm pert}(\rho_*, \rho_{\rm p})$ to the both hand sides of 95 and using 93, we find

$$N_{\rm CMB} \lesssim \frac{1}{|\eta_*|} \ln\left(\frac{\rho_{\rm end}}{\rho_*}\right). \tag{96}$$

Using 92 and the definition of the tensor-to-scalar ratio $r = 16\epsilon_*$, we obtain the upper bound:

$$r \lesssim 16\left(|\eta_*|\rho_{\rm end} e^{-|\eta_*|N_{\rm CMB}}\right)^2. \tag{97}$$

To satisfy CMB data, let us choose $\eta = -0.02$ and $N_{\rm CMB} \approx 50$. Assuming $\rho_{\rm end} \lesssim 1/2$, we obtain the upper bound $r \lesssim 10^{-4}$. Note that this is a little bit lower than the Lyth bound [35] for small field inflation, $r \lesssim 10^{-3}$. From the upper bound on r, we can also find the upper bound on the Hubble parameter as follows. In general, the power spectrum amplitude A_s is related to the Hubble parameter at horizon exit H_* by

$$A_s = \frac{2\kappa^2 H_*^2}{\pi^2 r}. \tag{98}$$

Combining this with the upper bound $r \lesssim 10^{-4}$ and the value $A_s = 2.2 \times 10^{-9}$ by CMB data, we find the upper bound on the Hubble parameter $H_* \lesssim 10^9$ TeV.

In Ref. [16], we will also find the lower bound $r \gtrsim 10^{-9}$ (equivalently $H_* \gtrsim 10^7$ TeV), based on an model-independent argument. This bound can be lowered at the cost of naturalness between parameters in the potential.

References

[1] I. Antoniadis, N. Arkani-Hamed, S. Dimopoulos and G. R. Dvali, "New dimensions at a millimeter to a Fermi and superstrings at a TeV," Phys. Lett. B **436** (1998) 257 [arXiv:hep-ph/9804398].

[2] I. Antoniadis and S. P. Patil, Eur. Phys. J. **C75** (2015) 182 [arXiv:1410.8845 [hep-th]].

[3] I. Antoniadis and R. Knoops, Nucl. Phys. B **886** (2014) 43 [arXiv:1403.1534 [hep-th]].

[4] F. Catino, G. Villadoro and F. Zwirner, JHEP **1201** (2012) 002 [arXiv:1110.2174 [hep-th]], G. Villadoro and F. Zwirner, Phys. Rev. Lett. **95** (2005) 231602 [hep-th/0508167].

[5] I. Antoniadis, D. M. Ghilencea and R. Knoops, JHEP **1502** (2015) 166 [arXiv:1412.4807 [hep-th]].

[6] I. Antoniadis and R. Knoops, Nucl. Phys. B **902** (2016) 69 [arXiv:1507.06924 [hep-ph]].

[7] L. Randall and R. Sundrum, Nucl. Phys. B **557** (1999) 79 [hep-th/9810155], G. F. Giudice, M. A. Luty, H. Murayama and R. Rattazzi, JHEP **9812** (1998) 027 [hep-ph/9810442].

[8] J. A. Bagger, T. Moroi and E. Poppitz, JHEP **0004** (2000) 009 [hep-th/9911029].

[9] I. Antoniadis and T. Maillard, "Moduli stabilization from magnetic fluxes in type I string theory," Nucl. Phys. B **716** (2005) 3 [hep-th/0412008]; I. Antoniadis, A. Kumar and T. Maillard, "Magnetic fluxes and moduli stabilization," Nucl. Phys. B **767** (2007) 139 [hep-th/0610246].

[10] I. Antoniadis, J.-P. Derendinger and T. Maillard, "Nonlinear N=2 Supersymmetry, Effective Actions and Moduli Stabilization," Nucl. Phys. B **808** (2009) 53 [arXiv:0804.1738 [hep-th]].

[11] I. Antoniadis, A. Chatrabhuti, H. Isono and R. Knoops, "Inflation from Supergravity with Gauged R-symmetry in de Sitter Vacuum," Eur. Phys. J. C **76** (2016) no.12, 680 [arXiv:1608.02121 [hep-ph]].

[12] A. H. Guth, "The Inflationary Universe: A Possible Solution to the Horizon and Flatness Problems," Phys. Rev. D **23** (1981) 347;
A. D. Linde, "A New Inflationary Universe Scenario: A Possible Solution of the Horizon, Flatness, Homogeneity, Isotropy and Primordial Monopole Problems," Phys. Lett. **108B** (1982) 389;
A. Albrecht and P. J. Steinhardt, "Cosmology for Grand Unified Theories with Radiatively Induced Symmetry Breaking," Phys. Rev. Lett. **48** (1982) 1220.

[13] D. H. Lyth and A. Riotto, "Particle physics models of inflation and the cosmological density perturbation," Phys. Rept. **314** (1999) 1 [hep-ph/9807278];
A. D. Linde, "Particle physics and inflationary cosmology," Contemp. Concepts Phys. **5** (1990) 1 [hep-th/0503203].

[14] A. A. Starobinsky,"A New Type of Isotropic Cosmological Models Without Singularity," Phys. Lett. **91B** (1980) 99.

[15] L. Randall and S. D. Thomas, "Solving the cosmological moduli problem with weak scale inflation," Nucl. Phys. B **449** (1995) 229 [hep-ph/9407248];
A. Riotto, "Inflation and the nature of supersymmetry breaking," Nucl. Phys. B **515** (1998) 413 [hep-ph/9707330]:

K. I. Izawa, "Supersymmetry - breaking models of inflation," Prog. Theor. Phys. **99** (1998) 157 [hep-ph/9708315];

W. Buchmuller, L. Covi and D. Delepine, "Inflation and supersymmetry breaking," Phys. Lett. B **491** (2000) 183 [hep-ph/0006168].

[16] I. Antoniadis, A. Chatrabhuti, H. Isono and R. Knoops, "Inflation from Supersymmetry Breaking," Eur. Phys. J. C **77** (2017) no.11, 724 [arXiv:1706.04133 [hep-th]].

[17] D. Z. Freedman and A. Van Proeyen, Cambridge, UK: Cambridge Univ. Pr. (2012) 607 p.

[18] P. Fayet and J. Iliopoulos, Phys. Lett. B **51** (1974) 461;

P. Fayet, Phys. Lett. B **69** (1977) 489.

[19] I. Antoniadis, J. -P. Derendinger and T. Maillard, Nucl. Phys. B **808** (2009) 53 [arXiv:0804.1738 [hep-th]].

[20] J. Polonyi, Hungary Central Inst Res - KFKI-77-93 (77,REC.JUL 78) 5p.

[21] H. P. Nilles, Phys. Rept. **110** (1984) 1,

[22] S. Ferrara, L. Girardello, T. Kugo and A. Van Proeyen, Nucl. Phys. B **223** (1983) 191,

[23] J. R. Ellis, K. A. Olive, Y. Santoso and V. C. Spanos, Phys. Lett. B **573** (2003) 162 [hep-ph/0305212],

[24] T. Gherghetta, G. F. Giudice and J. D. Wells, Nucl. Phys. B **559** (1999) 27 [hep-ph/9904378].

[25] P. A. R. Ade *et al.* [Planck Collaboration], "Planck 2015 results. XX. Constraints on inflation," arXiv:1502.02114 [astro-ph.CO].

[26] D. Baumann and D. Green, "Signatures of Supersymmetry from the Early Universe," Phys. Rev. D **85** (2012) 103520 [arXiv:1109.0292 [hep-th]].

[27] K. Schmitz and T. T. Yanagida, "Dynamical supersymmetry breaking and late-time R symmetry breaking as the origin of cosmic inflation," Phys. Rev. D **94** (2016) no.7, 074021 [arXiv:1604.04911 [hep-ph]].

[28] P. Binetruy and G. R. Dvali, "D term inflation," Phys. Lett. B **388** (1996) 241 [hep-ph/9606342].

[29] C. Wieck and M. W. Winkler, "Inflation with Fayet-Iliopoulos Terms," Phys. Rev. D **90** (2014) no.10, 103507 [arXiv:1408.2826 [hep-th]];

V. Domcke and K. Schmitz,"Unified model of D-term inflation," Phys. Rev. D **95** (2017) no.7, 075020 [arXiv:1702.02173 [hep-ph]].

[30] D. V. Volkov and V. P. Akulov, Is the neutrino a Goldstone particle?; Phys. Lett. B **46** (1973) 109;

M. Roček, Linearizing the Volkov-Akulov model, Phys. Rev. Lett. **41** (1978) 451;

U. Lindström, M. Roček, Constrained local superfields, Phys. Rev. D **19** (1979) 2300;

R. Casalbuoni, S. De Curtis, D. Dominici, F. Feruglio and R. Gatto, Nonlinear realization of supersymmetry algebra from supersymmetric constraint, Phys. Lett. B **220** (1989) 569;

Z. Komargodski and N. Seiberg, From linear SUSY to constrained superfields, JHEP **0909** (2009) 066; [arXiv:0907.2441 [hep-th]];

S. M. Kuzenko and S. J. Tyler, On the Goldstino actions and their symmetries, JHEP **1105**, (2011) 055: [arXiv:1102.3043 [hep-th]].

[31] L. Alvarez-Gaume, C. Gomez and R. Jimenez, "Minimal Inflation," Phys. Lett. B **690** (2010) 68 [arXiv:1001.0010 [hep-th]];

L. Alvarez-Gaume, C. Gomez and R. Jimenez, "A Minimal Inflation Scenario," JCAP **1103** (2011) 027 [arXiv:1101.4948 [hep-th]];

S. Ferrara and D. Roest, "General sGoldstino Inflation," JCAP **1610** (2016) no.10, 038 [arXiv:1608.03709 [hep-th]].

[32] E. J. Copeland, A. R. Liddle, D. H. Lyth, E. D. Stewart and D. Wands, "False vacuum inflation with Einstein gravity," Phys. Rev. D **49** (1994) 6410 [astro-ph/9401011].

[33] D. Baumann and L. McAllister, "Inflation and String Theory," arXiv:1404.2601 [hep-th]; M. Cicoli and F. Quevedo, "String moduli inflation: An overview," Class. Quant. Grav. **28** (2011) 204001 [arXiv:1108.2659 [hep-th]].

[34] G. R. Dvali, Q. Shafi and R. K. Schaefer, "Large scale structure and supersymmetric inflation without fine tuning," Phys. Rev. Lett. **73** (1994) 1886 [hep-ph/9406319].

[35] L. Boubekeur and D. H. Lyth, "Hilltop inflation," JCAP **0507** (2005) 010 [hep-ph/0502047].

Aspects of braneworld cosmology and holography*

Neven Bilić[†]
Division of Theoretical Physics, Rudjer Bošković Institute
P.O. Box 180, 10001 Zagreb, Croatia

ABSTRACT

In a holographic braneworld universe a cosmological fluid occupies a 3+1 dimensional brane located at the boundary of the asymptotic AdS_5 bulk. The AdS/CFT correspondence and the second Randall-Sundrum model are combined to establish a relationship between the RSII braneworld cosmology and the boundary metric induced by the time dependent bulk geometry. Some physically interesting scenarios are discussed in the framework of the Friedmann Robertson Walker cosmology involving the RSII and holographic braneworlds.

Contents

---

* Talk presented at IX Mathematical Physics Meeting: School and Conference on Modern Mathematical Physics, Belgrade, 18-23 September 2017

† e-mail address: bilic@irb.hr

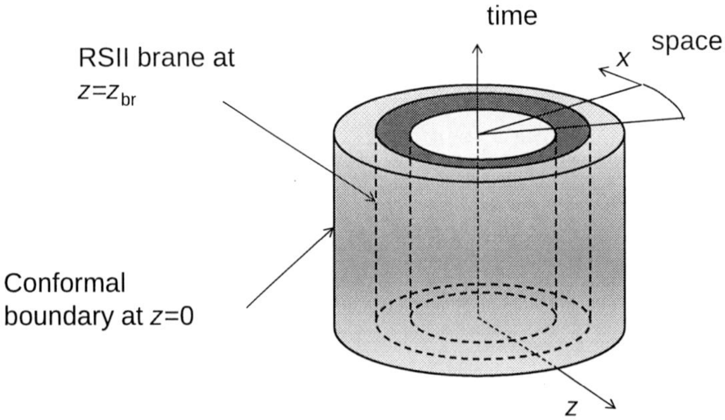

Figure 1: Illustration of the AdS_5 bulk with two branes: RSII brane located at $z = z_{br}$ and the holographic brane at $z = 0$.

1. Introduction

Branewarld cosmology is based on the scenario in which matter is confined on a brane moving in the higher dimensional bulk with only gravity allowed to propagate in the bulk [1, 2, 3, 4]. The brane can be placed, e.g., at the boundary of a 5-dim asymptotically Anti de Sitter space (AdS_5). Anti de Sitter space is dual to a conformal field theory at its boundary through the so called AdS/CFT correspondence [5]. This correspondence reflects an obvious symmetry relationship: On the one hand, AdS_5 is a maximally symmetric solution to Einsteins equations with negative cosmological constant with the symmetry group $AdS_5 \equiv SO(4,2)$. On the other hand, the 3+1 boundary conformal field theory is invariant under conformal transformations: Poincaré + dilatations + special conformal transformation. These transformations constitute the conformal group $\equiv SO(4,2)$.

We will consider two types of braneworlds (Fig. 1): 1) Holographic braneworld in with a 3-brane located at the boundary of the asymptotic AdS_5. The cosmology is governed by matter on the brane in addition to the boundary CFT. 2) Randall-Sundrum braneworld with a 3-brane located at a finite distance from the boundary of AdS_5. We will demonstrate that there exists a map between these two substantially different scenarios. Most of the material presented here is based on [6] and earlier works [7, 8, 9, 10].

We use the metric signature $(+,- - --)$ and curvature convention

$R^a{}_{bcd} = \partial_c \Gamma^a_{db} - \partial_d \Gamma^a_{cb} + \Gamma^e_{db}\Gamma^a_{ce} - \Gamma^e_{cb}\Gamma^a_{de}$ and $R_{ab} = R^s{}_{asb}$, so that Einstein's equations are $R_{ab} - \frac{1}{2}RG_{ab} = +8\pi G T_{ab}$.

2. Randall-Sundrum model

2.1. Basics

The Randall-Sundrum (RS) model [1, 2]. is a simple physically relevant model related to AdS/CFT. The model was originally proposed as a solution to the hierarchy problem in particle physics and as a possible mechanism for localizing gravity on the 3+1 dimensional universe embedded in a 4+1 spacetime without compactification of the extra dimension. It was soon realized that the RS model is deeply rooted in a wider framework of AdS/CFT correspondence [11, 12, 13, 14, 15, 16, 17].

The Randall-Sundrum model is a 4+1-dimensional universe with AdS_5 geometry containing two 3-branes with opposite brane tensions separated in the 5th dimension. The total action is a sum

$$S = S_{\text{bulk}} + S_{\text{GH}} + S_{\text{br1}} + S_{\text{br2}}, \tag{1}$$

where

$$S_{\text{bulk}} = \frac{1}{8\pi G_5} \int d^5x \sqrt{G} \left[-\frac{R^{(5)}}{2} - \Lambda_5 \right], \tag{2}$$

is the bulk action, Λ_5 being the bulk cosmological constant related to the AdS curvature radius as $\Lambda_5 = -6/\ell^2$. The remaining terms are the Gibbons-Hawking boundary term

$$S_{\text{GH}}[h] = \frac{1}{8\pi G_5} \int_\Sigma d^4x \sqrt{-h} K[h]. \tag{3}$$

and two brane actions of the form

$$S_{\text{br}}[h] = \int_\Sigma d^4x \sqrt{-h}(-\sigma + \mathcal{L}^{\text{matt}}[h]). \tag{4}$$

Here we denote by G the determinant of the bulk metric $G_{\mu\nu}$, by h the determinant of the metric $h_{\mu\nu}$ induced on the hypersurface Σ, and by σ the brane tension. Matter on the brane is described by the Lagrangian $\mathcal{L}^{\text{matt}}$.

In the following we will make use of various coordinate systems:

1. Fefferman-Graham coordinates

$$ds^2_{(5)} = G_{ab}dx^a dx^b = \frac{\ell^2}{z^2}\left(g_{\mu\nu}dx^\mu dx^\nu - dz^2\right), \tag{5}$$

2. Gaussian normal coordinates

$$ds^2_{(5)} = e^{-2y/\ell} g_{\mu\nu}dx^\mu dx^\nu - dy^2, \tag{6}$$

3. Schwarzschild coordinates

$$ds^2_{\rm ASch} = f(r)dt^2 - \frac{dr^2}{f(r)} - r^2 d\Omega_\kappa^2, \tag{7}$$

where

$$f(r) = \frac{r^2}{\ell^2} + \kappa - \mu\frac{\ell^2}{r^2}, \tag{8}$$

and

$$d\Omega_\kappa^2 = d\chi^2 + \frac{\sin^2(\sqrt{\kappa}\chi)}{\kappa}(d\vartheta^2 + \sin^2\vartheta d\varphi^2) \tag{9}$$

is the spatial line element for a closed ($\kappa = 1$), open hyperbolic ($\kappa = -1$), or open flat ($\kappa = 0$) space. The dimensionless parameter μ is related to the black-hole mass via [18, 19]

$$\mu = \frac{8G_5 M_{\rm bh}}{3\pi\ell^2}. \tag{10}$$

These coordinate representations are related via simple coordinate transformations

$$z = e^{y/\ell}, \quad \frac{r^2}{\ell^2} = \frac{\ell^2}{z^2} - \frac{\kappa}{2} + \frac{\kappa^2 + 4\mu}{16}\frac{z^2}{\ell^2}. \tag{11}$$

2.2. Second Randall-Sundrum model (RSII)

The RSII model [2] was proposed as an alternative to compactification of extra dimensions. A compactification of extra dimensions is necessary to localize gravity on the 3+1 dimensional universe. If extra dimensions were large that would yield unobserved modification of Newton's gravitational law. Experimental bound on the volume of n extra dimensions is [20]

$$V^{1/n} \leq 0.1\text{mm}. \tag{12}$$

RSII brane-world does not rely on compactification to localize gravity at the brane, but on the curvature of the bulk ("warped compactification). The negative cosmological constant Λ_5 acts to squeeze the gravitational field closer to the brane. One can see this in Gaussian normal coordinates (6) with an exponentially attenuating warp factor $e^{-2\ell y}$.

In RSII observers reside on the positive tension brane at $y = 0$ and the negative tension brane is pushed off to infinity in the fifth dimension. In the original RSII model one assumes the Z_2 symmetry $z \leftrightarrow z_{\rm br}^2/z$, so the region $0 < z \leq z_{\rm br}$ is identified with $z_{\rm br} \leq z < \infty$, with the observer brane at the fixed point $z = z_{\rm br}$. Hence, the braneworld is sitting between two patches of AdS$_5$, one on either side, and is therefore dubbed "two-sided" [15, 17]. In contrast, in the "one-sided" RSII model the region $0 \leq z \leq z_{\rm br}$ is simply cut off so the bulk is the section of spacetime $z_{\rm br} \leq z < \infty$.

The Planck mass scale is determined by the curvature of the five-dimensional space-time

$$\frac{1}{G_{\rm N}}=\frac{\gamma}{G_5}\int_{y_{\rm br}}^{\infty}dy\psi^2=\frac{\gamma\ell}{2G_5}. \tag{13}$$

where we have introduced the "sidedness" parameter γ [6] to facilitate a joint description of the two versions of RSII model: the one-sided ($\gamma=1$) and two-sided ($\gamma=2$). One usually imposes a fine tuning condition on the brane tension

$$\sigma=\sigma_0\equiv\frac{3\gamma}{8\pi G_5\ell}=\frac{3}{4\pi G_{\rm N}\ell^2}. \tag{14}$$

which eliminates the 4-dim cosmological constant. Note that the RSII fine tuning condition does not depend on the sidedness γ if σ_0 is expressed in terms of the four-dimensional Newton constant.

Table top measurements of the Newton gravitational law impose a bound on the AdS_5 curvature radius. The classical 3+1 dimensional gravity is altered on the RSII brane due to the extra dimension. For $r\gg\ell$ the weak gravitational potential created by an isolated matter source on the brane is given by [21]

$$\Phi(r)=\frac{G_{\rm N}M}{r}\left(1+\frac{2\ell^2}{3r^2}\right). \tag{15}$$

Table-top tests of Long et al [20] find no deviations of Newton's potential at distances greater than 0.1 mm and place the limit curvature

$$\ell<0.1\text{mm},\quad \text{or}\quad \ell^{-1}>10^{-12}\text{GeV}. \tag{16}$$

2.3. RSII cosmology – Dynamical brane

Branewarld cosmology is based on the scenario in which matter is confined on a brane moving in the higher dimensional bulk with only gravity allowed to propagate in the bulk [1, 2, 3, 4]. In this section we give a simple derivation of the RSII braneworld cosmology following J. Soda [22]. Cosmology on the brane is obtained by allowing the brane to move in the bulk. Equivalently, one could keep the brane fixed at $y=0$ while making the metric in the bulk time dependent.

Consider a time dependent brane hypersurface defined by

$$r-a(t)=0, \tag{17}$$

in AdS-Schwarzschild background [23, 24] where $a=a(t)$ is an arbitrary positive function. The induced line element on the brane is

$$ds^2_{\rm ind}=n^2(t)dt^2-a(t)^2d\Omega^2_\kappa, \tag{18}$$

where

$$n^2 = f(a) - \frac{(\partial_t a)^2}{f(a)}, \tag{19}$$

and f is defined by (8). The junction conditions on the brane with matter

$$K_{\mu\nu}|_{r=a-\epsilon} = \frac{8\pi G_5}{3\gamma}(\sigma g_{\mu\nu} + 3T_{\mu\nu}) \tag{20}$$

yield

$$\frac{(\partial_t a)^2}{n^2 a^2} + \frac{f}{a^2} = \frac{1}{\ell^2 \sigma_0^2}(\sigma + \rho)^2. \tag{21}$$

Now, imposing the fine tuning condition (14) one finds a modified Friedmann equation [25, 26, 27, 28].

$$\mathcal{H}^2 = \frac{8\pi G_{\rm N}}{3}\rho + \left(\frac{4\pi G_{\rm N}\ell}{3}\right)^2 \rho^2 + \frac{\mu\ell^2}{a^4}, \tag{22}$$

where

$$\mathcal{H}^2 = H^2 + \frac{\kappa}{a^2}, \quad H = \frac{\partial_t a}{na}. \tag{23}$$

Equation (22) differs from the standard Friedmann equation by the last two terms on the right-hand side. RSII cosmology is thus subject to astrophysical and cosmological tests (see, e.g., Refs. [29, 30]). The deviation proportional to ρ^2 poses no problem as it decays as a^{-8} in the radiation epoch and will rapidly become negligible after the end of the high-energy regime $\rho \simeq \sigma_0$. The last term on the right-hand side, the so called "dark radiation", for positive μ should not exceed 10% of the total radiation content in the epoch of BB nucleosynthesis whereas for negative μ could be as large as the rest of the radiation content [31, 32]. As expected, both one-sided and two-sided versions of the RSII model yield identical braneworld cosmologies.

The second Friedmann equation may be easily obtained by combining the time derivative of (22) with the energy conservation equation

$$\partial_t \rho + 3(\rho + p)\frac{\partial_t a}{a} = 0. \tag{24}$$

3. Connection with AdS/CFT

AdS/CFT correspondence is a holographic duality between gravity in d+1-dimensional space-time and quantum conformal field theory (CFT) on the d-dim boundary. Original formulation stems from string theory: the original AdS/CFT conjecture establishes an equivalence of a four dimensional $\mathcal{N} = 4$ supersymmetric Yang-Mills theory and string theory in a ten dimensional $\mathrm{AdS}_5 \times \mathrm{S}_5$ bulk [5, 33, 34].

3.1. RSII braneworld as a cutoff in AdS_5

In the RSII model by introducing the boundary in AdS_5 at $z = z_{\rm br}$ instead of $z = 0$, the model is conjectured to be dual to a cutoff CFT coupled to gravity, with $z = z_{\rm br}$ providing the IR cutoff (corresponding to the UV cutoff of the boundary CFT) [15]. In the one-sided RSII model, the model involves a single CFT at the boundary of a single patch of AdS_5. In the two-sided RSII model one would instead require two copies of the CFT, one for each of the AdS_5 patches.

The on-shell bulk action

$$S_{\rm bulk} = \frac{1}{8\pi G_5}\int d^5x\sqrt{G}\left[-\frac{R^{(5)}}{2} - \Lambda_5\right], \tag{25}$$

is infrared divergent because physical distances diverge at $z = 0$. The asymptotically AdS metric near $z = 0$ can be expanded as

$$ds^2_{(5)} = \frac{\ell^2}{z^2}\left(g_{\mu\nu}dx^\mu dx^\nu - dz^2\right), \tag{26}$$

$$g_{\mu\nu} = g^{(0)}_{\mu\nu} + z^2 g^{(2)}_{\mu\nu} + z^4 g^{(4)}_{\mu\nu} + z^6 g^{(6)}_{\mu\nu} + \dots . \tag{27}$$

Explicit expressions for $g^{(2n)}_{\mu\nu}$, $n = 1, 2, 3$ in terms of arbitrary $g^{(0)}_{\mu\nu}$ may be found in Ref. [35]. In particular, we will need

$$g^{(2)}_{\mu\nu} = \frac{1}{2}\left(R_{\mu\nu} - \frac{1}{6}Rg^{(0)}_{\mu\nu}\right) \tag{28}$$

and the relation

$$\mathrm{Tr}g^{(4)} = -\frac{1}{4}\mathrm{Tr}(g^{(2)})^2, \tag{29}$$

where the trace of a tensor $A_{\mu\nu}$ is defined as

$$\mathrm{Tr}A = A^\mu_\mu = g^{(0)\mu\nu}A_{\mu\nu}. \tag{30}$$

Now, we regularize the action by placing the RSII brane near the AdS boundary, i.e., at $z = \epsilon\ell$, $\epsilon \ll 1$, so that the induced metric is

$$h_{\mu\nu} = \frac{1}{\epsilon^2}(g^{(0)}_{\mu\nu} + \epsilon^2\ell^2 g^{(2)}_{\mu\nu} + \dots). \tag{31}$$

The bulk splits in two regions: $0 \le z < \epsilon\ell$, and $\epsilon\ell \le z \le \infty$. We can either discard the region $0 \le z < \epsilon\ell$ (one-sided regularization, $\gamma = 1$) or invoke the Z_2 symmetry and identify two regions (two-sided regularization, $\gamma = 2$). Then, the regularized bulk action is

$$S^{\rm reg}_{\rm bulk} = \gamma S^{\rm reg}_0 = \frac{\gamma}{8\pi G_5}\int_{z\ge\epsilon\ell} d^5x\sqrt{G}\left[-\frac{R^{(5)}}{2} - \Lambda_{(5)}\right] + S_{\rm GH}[h] \tag{32}$$

The renormalized action is obtained by adding counterterms to $S_0^{\rm reg}$ and taking the limit $\epsilon \to 0$ [35, 36]

$$S_0^{\rm ren}[g^{(0)}] = \lim_{\epsilon\to 0}(S_0^{\rm reg}[G] + S_1[h] + S_2[h] + S_3[h]), \tag{33}$$

The necessary counterterms are [35]

$$S_1[h] = -\frac{6}{16\pi G_5 \ell}\int d^4x\sqrt{-h}, \tag{34}$$

$$S_2[h] = -\frac{\ell}{16\pi G_5}\int d^4x\sqrt{-h}\left(-\frac{R[h]}{2}\right), \tag{35}$$

$$S_3[h] = -\frac{\ell^3}{16\pi G_5}\int d^4x\sqrt{-h}\frac{\log\epsilon}{4}\left(R^{\mu\nu}[h]R_{\mu\nu}[h] - \frac{1}{3}R^2[h]\right). \tag{36}$$

Now we demand that the variation with respect to the induced metric $h_{\mu\nu}$ of the total RSII action (the sum of the regularized on shell bulk action and the brane action (4)) vanishes, i.e., we require

$$\delta(S_{\rm bulk}^{\rm reg}[h] + S_{\rm br}[h]) = 0, \tag{37}$$

which may be expressed as

$$\delta\left[\gamma S_0^{\rm ren} - \gamma S_3 - \left(\sigma - \frac{3\gamma}{8\pi\ell G_5}\right)\int d^4x\sqrt{-h} + \int d^4x\sqrt{-h}\mathcal{L}_{\rm matt} \right.$$
$$\left. + \frac{\gamma\ell}{16\pi G_5}\int d^4x\sqrt{-h}\frac{R[h]}{2}\right] = 0. \tag{38}$$

The third term gives the contribution to the cosmological constant and may be eliminated by imposing the RSII fine tuning condition (14). The variation of the scheme dependent S_3 may be combined with the first term. Then, according to the AdS/CFT prescription by functionally differentiating the renormalized on-shell bulk gravitational action with respect to the boundary metric $g_{\mu\nu}^{(0)}$ one obtains the expectation value $\langle T_{\mu\nu}^{\rm CFT}\rangle$ and hence

$$\delta(S_0^{\rm ren} - S_3) = \frac{1}{2}\int d^4x\sqrt{-h}\langle T_{\mu\nu}^{\rm CFT}\rangle\delta h^{\mu\nu}, \tag{39}$$

With this the variation of the action yields Einsteins equations on the boundary

$$R_{\mu\nu} - \frac{1}{2}Rg_{\mu\nu} = 8\pi G_{\rm N}(\gamma\langle T_{\mu\nu}^{\rm CFT}\rangle + T_{\mu\nu}^{\rm matt}), \tag{40}$$

where the energy-momentum tensor $T^{\rm matt}_{\mu\nu}$ describes matter on the holographic brane in addition to the holographic conformal part given by [35]

$$\langle T^{\rm CFT}_{\mu\nu}\rangle = -\frac{\ell^3}{4\pi G_5}\left\{ g^{(4)}_{\mu\nu} - \frac{1}{8}\left[({\rm Tr} g^{(2)})^2 - {\rm Tr}(g^{(2)})^2\right] g^{(0)}_{\mu\nu} \right.$$
$$\left. -\frac{1}{2}(g^{(2)})^2_{\mu\nu} + \frac{1}{4}{\rm Tr} g^{(2)} g^{(2)}_{\mu\nu} \right\}. \tag{41}$$

This is an explicit realization of the AdS/CFT correspondence: the vacuum expectation value of a boundary CFT operator is obtained solely in terms of geometrical quantities of the bulk.

3.2. Conformal anomaly

It is of particular interest to check whether the stress tensor $T^{\rm CFT}$ obtained using AdS/CFT prescription correctly reproduces the conformal anomaly. From (41) with the help of (28) and (29) we find

$$\langle T^{\rm CFT}{}^{\mu}{}_{\mu}\rangle = \frac{\ell^3}{128\pi G_5}\left(G_{\rm GB} - C^2\right), \tag{42}$$

where

$$G_{\rm GB} = R^{\mu\nu\rho\sigma}R_{\mu\nu\rho\sigma} - 4R^{\mu\nu}R_{\mu\nu} + R^2 \tag{43}$$

is the Gauss-Bonnet invariant and

$$C^2 \equiv C^{\mu\nu\rho\sigma}C_{\mu\nu\rho\sigma} = R^{\mu\nu\rho\sigma}R_{\mu\nu\rho\sigma} - 2R^{\mu\nu}R_{\mu\nu} + \frac{1}{3}R^2 \tag{44}$$

is the square of the Weyl tensor $C_{\mu\nu\rho\sigma}$. This result should be compared with the standard conformal anomaly calculated in field theory [37]

$$\langle T^{\rm CFT}{}^{\mu}{}_{\mu}\rangle = bG_{\rm GB} - cC^2 + b'\Box R. \tag{45}$$

The two results agree if we ignore the last term in (45) and identify

$$b = c = \frac{G_5}{128\pi\ell^3}. \tag{46}$$

The standard field theory calculations give [37, 38]

$$b = \frac{n_{\rm s} + (11/2)n_{\rm f} + 62n_{\rm v}}{360(4\pi)^2}, \qquad c = \frac{n_{\rm s} + 3n_{\rm f} + 12n_{\rm v}}{120(4\pi)^2}. \tag{47}$$

where $n_{\rm s}$, $n_{\rm f}$, $n_{\rm v}$ are the numbers of massless scalar bosons, Weyl fermions and vector bosons, respectively. Hence, generally $b \neq c$. However, in the $\mathcal{N} = 4$ U(N) super-Yang-Mills theory, $n_{\rm s} = 6N^2$, $n_{\rm f} = 4N^2$, and $n_{\rm v} = N^2$, in which case the equality $b = c$ holds and the conformal anomaly is correctly reproduced by the holographic expression (42) if we identify [39]

$$\frac{\ell^3}{G_5} = \frac{2N^2}{\pi}. \tag{48}$$

4. Holographic cosmology

As we have shown in the previous section, in the limit in which the RSII brane approaches the AdS_5 boundary, the geometry on the boundary brane, referred to as the *holographic brane*, satisfies a special form of Einstein's equations (40). To derive the corresponding cosmology we start from AdS-Schwarzschild static coordinates in the bulk and make the coordinate transformation

$$t = t(\tau, z), \quad r = r(\tau, z). \tag{49}$$

The line element will take a general form

$$ds^2_{(5)} = \frac{\ell^2}{z^2}\left(n^2(\tau, z)d\tau^2 - a^2(\tau, z)d\Omega^2_\kappa - dz^2\right), \tag{50}$$

Imposing the boundary conditions at $z = 0$:

$$n(\tau, 0) = 1, \quad a(\tau, 0) = a_{\rm h}(\tau), \tag{51}$$

we obtain the induced metric at the boundary in the general FRW form

$$ds^2_{(0)} = g^{(0)}_{\mu\nu}dx^\mu dx^\nu = d\tau^2 - a^2_{\rm h}(\tau)d\Omega^2_\kappa. \tag{52}$$

Solving Einsteins equations in the bulk one finds [8]

$$a^2 = a^2_{\rm h}\left[\left(1 - \frac{\mathcal{H}^2_{\rm h} z^2}{4}\right)^2 + \frac{1}{4}\frac{\mu z^4}{a^4_{\rm h}}\right], \quad \mathcal{N} = \frac{\dot{a}}{\dot{a}_{\rm h}}. \tag{53}$$

where

$$\mathcal{H}^2_{\rm h} = H^2_{\rm h} + \frac{\kappa}{a^2_{\rm h}}, \tag{54}$$

and $H_{\rm h} = \dot{a}_{\rm h}/a_{\rm h}$ is the Hubble expansion rate on boundary. Comparing the exact $g_{\mu\nu}(\tau, z)$ in (50) with the expansion (26) we can extract $g^{(2)}_{\mu\nu}$ and $g^{(4)}_{\mu\nu}$. Then, using the expression (41) we obtain

$$\langle T^{\rm CFT}_{\mu\nu}\rangle = t_{\mu\nu} + \frac{1}{4}\langle T^{\rm CFT}{}^\alpha{}_\alpha\rangle g^{(0)}_{\mu\nu}, \tag{55}$$

where the second term on the right-hand side corresponds to the conformal anomaly

$$\langle T^{\rm CFT}{}^\alpha{}_\alpha\rangle = \frac{3\ell^3}{16\pi G_5}\frac{\ddot{a}_{\rm h}}{a_{\rm h}}\mathcal{H}^2_{\rm h}, \tag{56}$$

and the first term is a traceless tensor with non-zero components

$$t_{00} = -3t^i_i = \frac{3\ell^3}{64\pi G_5}\left(\mathcal{H}^4_{\rm h} + \frac{4\mu}{a^4_{\rm h}} - \frac{\ddot{a}_{\rm h}}{\dot{a}_{\rm h}}\mathcal{H}^2_{\rm h}\right). \tag{57}$$

Hence, apart from the conformal anomaly, the CFT dual to the time dependent asymptotically AdS_5 bulk metric is a conformal fluid with the equation of state $p_{CFT} = \rho_{CFT}/3$, where $\rho_{CFT} = t_{00}$, $p_{CFT} = -t^i_i$.

Using this, from the boundary Einstein equations we obtain the holographic Friedmann equation [8, 7]

$$\mathcal{H}_{\rm h}^2 = \frac{\ell^2}{4}\left(\mathcal{H}_{\rm h}^4 + \frac{4\mu}{a_{\rm h}^4}\right) + \frac{8\pi G_{\rm N}}{3}\rho_{\rm h}. \tag{58}$$

Here we have used the energy-momentum tensor with nonvanishing components

$$T_{00}^{\rm matt} = \rho_{\rm h}, \quad T_{ij}^{\rm matt} = p_{\rm h} g_{ij}^{(0)}, \tag{59}$$

where $\rho_{\rm h}$ and $p_{\rm h}$ are the matter energy density and pressure, respectively. The second Friedmann equation can be derived by combining the time derivative of (58) with the energy conservation equation

$$\dot{\rho}_{\rm h} + 3(\rho_{\rm h} + p_{\rm h})H_{\rm h} = 0. \tag{60}$$

One finds

$$\frac{\ddot{a}_{\rm h}}{a_{\rm h}}\left(1 - \frac{\ell^2}{2}\mathcal{H}_{\rm h}^2\right) + \mathcal{H}_{\rm h}^2 = \frac{4\pi G_{\rm N}}{3}(\rho_{\rm h} - 3p_{\rm h}). \tag{61}$$

5. Holographic map

The time dependent bulk spacetime with metric (50) may be regarded as a z-foliation of the bulk with FRW cosmology on each z-slice [6]. In particular, at $z = z_{\rm br}$ one has the RSII cosmology and at at $z = 0$ the holographic cosmology. A map between a z-cosmology and $z = 0$-cosmology can be constructed using (53) and the inverse relation

$$a_{\rm h}^2 = \frac{a^2}{2}\left(1 + \frac{\mathcal{H}^2 z^2}{2} + \mathcal{E}\sqrt{1 + \mathcal{H}^2 z^2 - \frac{\mu z^4}{a^4}}\right), \tag{62}$$

where

$$\mathcal{E} = \begin{cases} -1, & \text{for two-sided version,} \\ \pm 1, & \text{for one-sided version.} \end{cases} \tag{63}$$

A functional relationship between Hubble rates can be obtained by making use of (53) and (54). One finds

$$\mathcal{H}^2 = \mathcal{H}_{\rm h}^2\left[1 - \frac{\mathcal{H}_{\rm h}^2 z^2}{2} + \frac{1}{16}\left(\mathcal{H}_{\rm h}^4 + \frac{4\mu}{a_{\rm h}^4}\right) z^4\right]^{-1}. \tag{64}$$

The map is schematically illustrated as

$$\begin{array}{ccc} d\tau^2 - a_{\mathrm{h}}^2 d\Omega_\kappa^2 & \xrightarrow{\tau\to\tilde{\tau}} & (1/n^2)d\tilde{\tau}^2 - a_{\mathrm{h}}^2 d\Omega_\kappa^2 \\ {\scriptstyle z}\big\downarrow & & \big\downarrow{\scriptstyle z} \\ n^2 d\tau^2 - a^2 d\Omega_\kappa^2 & \xrightarrow[\tau\to\tilde{\tau}]{} & d\tilde{\tau}^2 - a^2 d\Omega_\kappa^2 \end{array}$$

where τ and $\tilde{\tau}$ are the holographic and RSII synchronous times, respectively.

As an example, consider the RSII braneworld at $z_{\mathrm{br}} = \sqrt{2}\ell$ and the holographic braneworld at $z = 0$ with the corresponding Hubble rates $\mathcal{H}_{\mathrm{br}}^2$ and $\mathcal{H}_{\mathrm{h}}^2$. In Fig. 2 we plot $\mathcal{H}_{\mathrm{br}}^2$ versus $\mathcal{H}_{\mathrm{h}}^2$ for two values of the black hole mass parameter $\mu = 0$ (left panel) and $\mu\ell^4/a_{\mathrm{h}}^4 = 1/2$ with $z_{\mathrm{br}}^2/\ell^4 = 2$ (right panel). In both panels the shaded area, corresponding to the physical region $\rho_{\mathrm{h}} > 0$, is determined by the condition

$$2 - 2\sqrt{1 - \mu\ell^4/a_{\mathrm{h}}^4} \le \mathcal{H}_{\mathrm{h}}^2\ell^2 \le 2 + 2\sqrt{1 - \mu\ell^4/a_{\mathrm{h}}^4}. \tag{65}$$

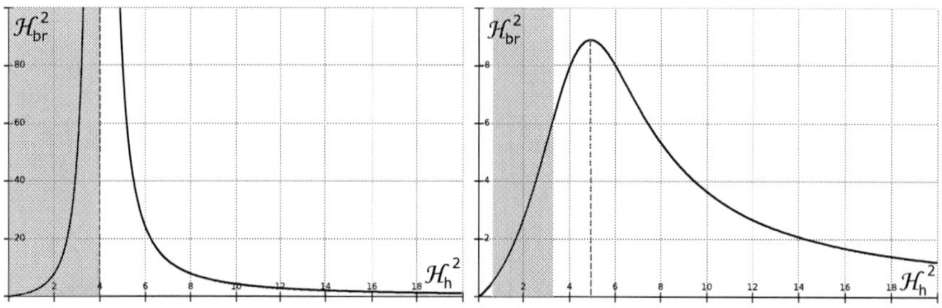

Figure 2: $\mathcal{H}_{\mathrm{br}}^2$ as a function of $\mathcal{H}_{\mathrm{h}}^2$ (both in units of z_{br}^{-2}) defined by (64) for $\mu = 0$ (left panel) and $\mu\ell^4/a_{\mathrm{h}}^4 = 2$ with $z_{\mathrm{br}}^2/\ell^2 = 2$ (right panel). The region left from the vertical dashed red line is relevant for the one-sided version only. The shaded area corresponds to the physical region $\rho_{\mathrm{h}} > 0$.

6. Effective energy density

Next, we analyze a few special cases in two scenarios.

1. The RSII scenario with the primary braneworld at $z = z_{\mathrm{br}}$.
2. The holographic scenario with the primary cosmology on the AdS boundary at $z = 0$.

In each of the two scenarios we assume the presence of matter on the primary brane only and no matter in the bulk.

RSII scenario

In the RSII scenario the primary braneworld is the RSII brane at $z = z_{\rm br}$. The cosmology on the $z = 0$ brane emerges as a reflection of the RSII cosmology. For simplicity we take $z_{\rm br} = \ell$ and we fine tune the tension σ as in (14). Then, assuming the modified Friedmann equations (58) and (61) hold on the holographic brane, the effective energy density is given by

$$\frac{\rho_{\rm h}}{\sigma_0} = \frac{4\mathcal{E}(\rho/\sigma_0 + 1 - \mathcal{E})}{(\rho/\sigma_0 + 1 + \mathcal{E})^2 + \mu\ell^4/a^4}. \tag{66}$$

where $\mathcal{E}$ is defined by (63). Given the equation of state $p = p(\rho)$ on the RSII brane, the cosmological scale a is derived by integrating (22) and (24).

Thus, the two-sided model with positive energy density and positive μ maps into a holographic cosmology with negative effective energy density $\rho_{\rm h}$. For $\mu = 0$ the density$\rho_{\rm h}$ diverges with ρ as $1/\rho$. The one-sided model maps into two branches: $\mathcal{E} = -1$ branch identical with the two-sided map and the $\mathcal{E} = +1$ branch with a smooth positive function $\rho_{\rm h} = \rho_{\rm h}(\rho)$.

Holographic scenario

Suppose the cosmology on the $z = 0$ brane is known, i.e., the density $\rho_{\rm h}$, the pressure $p_{\rm h}$, and the cosmological scale $a_{\rm h}$ are known. If there is no matter in the bulk the induced cosmology on an arbitrary z-slice will be completely determined. Observers on the RSII brane on an arbitrary z-slice experience an emergent cosmology which is a reflection of the boundary cosmology. The general expression for the effective energy density ρ on the RSII brane is rather complicated but simplifies considerably for $z_{\rm br} = \ell$. In this case

$$\frac{\rho}{\sigma_0} = \left| \frac{1 + \rho_{\rm h}/\sigma_0 - \epsilon\sqrt{1 - 2\rho_{\rm h}/\sigma_0 - \mu\ell^4/a_{\rm h}^4}}{1 - \rho_{\rm h}/\sigma_0 - \epsilon\sqrt{1 - 2\rho_{\rm h}/\sigma_0 - \mu\ell^4/a_{\rm h}^4}} \right| - \frac{\sigma}{\sigma_0}, \tag{67}$$

where ϵ may take the values $+1$ or -1 . Hence, the function $\rho = \rho(\rho_{\rm h}, a_{\rm h})$ is not uniquely defined, although the mapping $a_{\rm h} \to a$ is unique (Fig. 3)

For an arbitrary $z_{\rm br} \neq \ell$ in the low density regime (relevant for the one sided version only), i.e., $\rho_{\rm h}^2 \ll \sigma_0^2$ and $\mu\ell^4 \ll a_{\rm h}^4$ we find:
a) For $\epsilon = -1$ at linear order in μ and quadratic order in $\rho_{\rm h}$ the effective energy density

$$\begin{aligned}\frac{\rho}{\sigma_0} &= 1 - \frac{\sigma}{\sigma_0} + \frac{z_{\rm br}^2}{\ell^2}\frac{\rho_{\rm h}}{\sigma_0} + \frac{1}{2}\frac{z_{\rm br}^2}{\ell^2}\left(\frac{z_{\rm br}^2}{\ell^2} + 1\right)\frac{\rho_{\rm h}^2}{\sigma_0^2} \\ &\quad - \frac{1}{2}\frac{z_{\rm br}^2}{\ell^2}\left(\frac{z_{\rm br}^2}{\ell^2} - 1\right)\frac{\mu\ell^4}{a_{\rm h}^4} + \dots . \end{aligned} \tag{68}$$

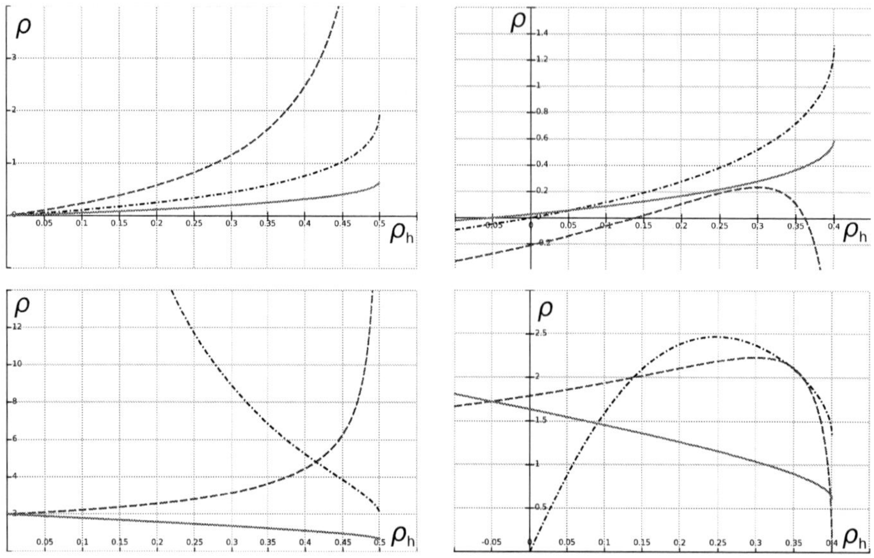

Figure 3: The effective density ρ on the RSII brane as a function of the density $\rho_{\rm h}$ on the holographic brane (both in units of σ_0) for $\sigma = \sigma_0$, $\epsilon = -1$ (top panels), $\epsilon = +1$ (bottom panels), and $\mu\ell^4/a_{\rm h}^4 = 0$ (left panels), 0.2 (right panels). The full red, dash-dotted black, and dashed blue lines represent $z_{\rm br}^2/\ell^2 = 0.5$, 1, and 2, respectively.

and pressure

$$p = -(\sigma_0 - \sigma) + \frac{z_{\rm br}^2}{\ell^2} p_{\rm h} + \dots . \tag{69}$$

Hence, at linear order the effective fluid on the RSII brane satisfies the same equation of state as the fluid on the holographic brane. The cosmological constant term will vanish on both branes if the RSII fine tuning condition is imposed.

b) For $\epsilon = +1$ at linear order

$$\frac{\rho}{\sigma_0} = \frac{z_{\rm br}^2/\ell^2 + 1}{z_{\rm br}^2/\ell^2 - 1} - \frac{\sigma}{\sigma_0} + \frac{z_{\rm br}^2/\ell^2}{(z_{\rm br}^2/\ell^2 - 1)^2} \frac{\rho_{\rm h}}{\sigma_0} - \frac{z_{\rm br}^2/\ell^2}{2(z_{\rm br}^2/\ell^2 - 1)^3} \frac{\mu\ell^4}{a_{\rm h}^4} + \dots . \tag{70}$$

Hence, in this case, the effective energy density ρ on the RSII brane differs from $\rho_{\rm h}$ on the holographic brane by a multiplicative constant and diverges in the limit $z_{\rm br} \to \ell$. The effective cosmological constant on the RSII brane does not vanish even if $\sigma = \sigma_0$ in which case

$$\Lambda_{\rm br} = \frac{6}{\ell^2} \frac{z_{\rm br}^2/\ell^2 + 1}{z_{\rm br}^2/\ell^2 - 1} - \frac{6}{\ell^2}. \tag{71}$$

7. Conclusions & Outlook

Our study can be summarized as follows:

- We have explicitly constructed the mapping between two cosmological braneworlds: holographic and RSII .
- The cosmologies are governed by the corresponding modified Friedmann equations.
- There is a clear distinction between 1-sided and 2-sided holographic map with respective 1-sided and 2-sided versions of RSII model.
- In the 2-sided map the low-density regime on the two-sided RSII brane corresponds to the high negative energy density on the holographic brane.
- The low density regime is maintained on both branes only in the one-sided RSII
- We have analyzed the effective energy density in two scenarios: the RSII scenario with the primary braneworld at $z = z_{\rm br}$ and the holographic scenario with the primary cosmology on the AdS boundary at $z = 0$. Then, in the holographic and RSII scenarios we will have emergent cosmologies on the RSII and holographic branes, respectively.

It is conceivable that we live in a braneworld with emergent cosmology. That is, dark energy and dark matter could be emergent phenomena induced by what happens on the primary braneworld. In this regards, the holographic scenario offers a few interesting possibilities. For example, suppose our universe is a one-sided RSII braneworld the cosmology of which is emergent in parallel with the primary holographic cosmology. If the energy density $\rho_{\rm h}$ on the holographic brane describes matter with the equation of state satisfying $3p_{\rm h} + \rho_{\rm h} > 0$, as for, e.g., cold dark matter, according to (70) and (71) we will have an asymptotically de Sitter universe on the RSII brane. If we choose the curvature radius ℓ so that the cosmological constant $\Lambda_{\rm br}$ fits the observed value, the quadratic term will be comparable with the linear term today but will strongly dominate in the past and hence will spoil the standard cosmology. However, the standard ΛCDM cosmology could be recovered by including a negative cosmological constant term in $\rho_{\rm h}$ and $p_{\rm h}$ and fine tune it to cancel $\Lambda_{\rm br}$ up to a small phenomenologically acceptable contribution.

Acknowledgments

This work has been supported by the Croatian Science Foundation under the project IP-2014-09-9582 and partially supported by the H2020 CSA Twinning project No. 692194, "RBI-T-WINNING" and by ICTP - SEENET-MTP project NT-03 Cosmology - Classical and Quantum Challenges.

References

[1] L. Randall and R. Sundrum, *Phys. Rev. Lett.* **83** (1999) 3370.

[2] L. Randall and R. Sundrum, *Phys. Rev. Lett.* **83** (1999) 4690.

[3] N. Arkani-Hamed, S. Dimopoulos, and G. Dvali, *Phys. Lett. B* **429**, 263 (1998).

[4] I. Antoniadis, N. Arkani-Hamed, S. Dimopoulos and G. R. Dvali, *Phys. Lett. B* **436**(1998) 257.

[5] J.M. Maldacena, *Adv. Theor. Math. Phys.* **2** (1998) 231; *Int. J. Theor. Phys.* **38** (1999) 1113 [arXiv:hepth/9711200].

[6] N. Bilic, *Phys. Rev. D* **93** (2016) 066010 [arXiv:1511.07323 [gr-qc]].

[7] E. Kiritsis, *JCAP* **0510** (2005) 014 [hep-th/0504219].

[8] P. S. Apostolopoulos, G. Siopsis and N. Tetradis, *Phys. Rev. Lett.* **102** (2009) 151301 [arXiv:0809.3505 [hep-th]].

[9] N. Tetradis, *JHEP* **1003** (2010) 040 [arXiv:0905.2763 [hep-th]].

[10] P. Brax and R. Peschanski, *Acta Phys. Polon.* B **41** (2010) 2645 [arXiv:1006.3054 [hep-th]].

[11] S. S. Gubser, *Phys. Rev. D* **63** (2001) 084017 [hep-th/9912001].

[12] S. Nojiri, S. D. Odintsov and S. Zerbini, *Phys. Rev. D* **62** (2000) 064006 [hep-th/0001192].

[13] S. B. Giddings, E. Katz and L. Randall, *JHEP* **0003** (2000) 023 [hep-th/0002091].

[14] S. W. Hawking, T. Hertog and H. S. Reall, *Phys. Rev. D* **62** (2000) 043501 [hep-th/0003052].

[15] M. J. Duff and J. T. Liu, *Class. Quant. Grav.* **18** (2001) 3207 [*Phys. Rev. Lett.* **85** (2000) 2052] [hep-th/0003237].

[16] S. Nojiri and S. D. Odintsov, *Phys. Lett. B* **484** (2000) 119 [hep-th/0004097].

[17] S. de Haro, K. Skenderis and S. N. Solodukhin, *Class. Quant. Grav.* **18** (2001) 3171 [hep-th/0011230].

[18] R. C. Myers and M. J. Perry, *Annals Phys.* **172** (1986) 304.

[19] E. Witten, *Adv. Theor. Math. Phys.* **2** (1998) 505 [hep-th/9803131].

[20] J. C. Long, H. W. Chan, A. B. Churnside, E. A. Gulbis, M. C. M. Varney and J. C. Price, Letter version published in *Nature* **421** 922-925 (2003) 922 [hep-ph/0210004].

[21] J. Garriga and T. Tanaka, *Phys. Rev. Lett.* **84** (2000) 2778 [hep-th/9911055].

[22] J. Soda, *Lect. Notes Phys.* **828** (2011) 235 [arXiv:1001.1011 [hep-th]].

[23] C. Gomez, B. Janssen and P. J. Silva, *JHEP* **0004** (2000) 027 [hep-th/0003002].

[24] D. Birmingham and M. Rinaldi, *Mod. Phys. Lett. A* **16** (2001) 1887 [hep-th/0106237].

[25] P. Binetruy, C. Deffayet and D. Langlois, *Nucl. Phys. B* **565** (2000) 269 [hep-th/9905012].

[26] P. Kraus, *JHEP* **9912** (1999) 011 [hep-th/9910149].

[27] E. E. Flanagan, S. H. H. Tye and I. Wasserman, *Phys. Rev. D* **62** (2000) 044039 [hep-ph/9910498].

[28] S. Mukohyama, *Phys. Lett. B* **473** (2000) 241 [hep-th/9911165].

[29] R. Maartens and K. Koyama, *Living Rev. Rel.* **13** (2010) 5 [arXiv:1004.3962 [hep-th]].

[30] W. Godlowski and M. Szydlowski, *Gen. Rel. Grav.* **36** (2004) 767 [astro-ph/0404299].

[31] K. Ichiki, M. Yahiro, T. Kajino, M. Orito and G. J. Mathews, *Phys. Rev. D* **66** (2002) 043521 [astro-ph/0203272].

[32] J. D. Bratt, A. C. Gault, R. J. Scherrer and T. P. Walker, *Phys. Lett. B* **546** (2002) 19 [astro-ph/0208133].

[33] S.S. Gubser, I.R. Klebanov, and A.M. Polyakov, *Phys. Lett. B* **428** (1998) 105 [arXiv:hep-th/9802109].

[34] E. Witten, *Adv. Theor. Math. Phys.* **2** (1998) 253 [arXiv:hep-th/9802150].

[35] S. de Haro, S. N. Solodukhin and K. Skenderis, *Commun. Math. Phys.* **217** (2001) 595 [hep-th/0002230].

[36] S. W. Hawking, T. Hertog and H. S. Reall, *Phys. Rev. D* **62** (2000) 043501 [hep-th/0003052].

[37] M. J. Duff, *Nucl. Phys. B* **125** (1977) 334

[38] M. J. Duff, *Class. Quant. Grav.* **11** (1994) 1387 [hep-th/9308075].

[39] M. Henningson and K. Skenderis, *JHEP* **9807**(1998) 023 [hep-th/9806087].

Cycles cohomology by integral transforms in derived geometry to ramified field theory*

Prof. Dr. Francisco Bulnes†
Head of Research Department in Mathematics and Engineering,
TESCHA, IINAMEI Director

Abstract

Several geometrical Langlands correspondences are considered to determine equivalences necessary to the obtaining in the quantized context from differential operators algebra (actions of the algebra on modules) and the holomorphic bundles in the lines bundle stacks required to the modeling of the elements of the different physical stacks and the extension of their field ramifications to the meromorphic case. In this point, is obtained a result that establish a commutative diagram of rings and their spectrum functor involving the non-commutative Hodge theory, and using integral transforms to establish the decedent isomorphisms in the context of the geometrical stacks to a good Opers, level. The co-cycles obtained through integral transforms are elements of the corresponding deformed category to mentioned different physical stacks (where are had, even field singularities). In this point, a justification on the nature of our twisted derived categories and their elements as ramifications of a field (to the field equations) is the followed through the Yoneda algebra where is searched extends the action of certain endomorphisms of Verma modules of critical level through the Lie algebra of ramifications, whose cohomological space has zero dimension. This establishes solution classes to the QFT-equations in field theory through the spectrum of their corresponding differential operators.

Key words: Deformed Derived Categories, Field Ramifications, Geometrical Langlands Correspondences, Generalized Verma Modules, Grothendieck Schemes, Moduli Stacks, Integral Transforms, Spec Functor, Twisted Derived Categories, Yoneda Algebra, QFT-equations, "Quantized" Differential Operators Algebra.

AMS 2010 Classification. 53D37; 11R39; 14D24; 83C60; 11S15

---

* Post-Doctoral Research supported by the State of Mexico Government. Presented talk in the 9th MATHEMATICAL PHYSICS MEETING: School and Conference on Modern Mathematical Physics, in the Physics Institute of Belgrade City, Serbia, 2017.

† e-mail address: francisco.bulnes@tesch.edu.mx

1. Introduction and General Discussion.

Equivalence classes on field solutions to establish an Universe theory involving the QFT, TFT, SUSY and the inclusion of the strings theory in all their modalities (for examples the heterotic strings) are obtained. Then through corresponding sheaves of differential operators of the field equations; sheaves of coherent D-modules are identified [1].

This identification will serve to build, on the equivalences basis of the corresponding Zuckerman functor, the universal functors on derived categories of Harish-Chandra modules to the geometrical Langlands programme in mirror theory [2], [3]. Likewise, the incorporation of the geometrical Langlands ramifications will establish extensions of a connection beyond the holomorphicity to much of the vector complex bundles that can be constructed to a large stack of physical phenomena in the search to obtain solutions of field through the curvature and torsion tensors of the space-time.

The obtained development includes D-modules complex of infinite dimension, generalizing for this via, to the BRST-cohomology. Then the integrable systems class can be extended in mathematical-physics and with it the possibility of obtaining of a general theory of integral transforms to the study of the space-time (cohomology of integral operators [4]) considering the kernels of the germs of the corresponding sheaves to the vector complex bundles, and then the measure of much of their field observables [3].

Having these Langlands correspondences we can to tend a bridge to complete a classification of the different operators to the field equations using on the base the Verma modules that are classifying spaces, the differential operators of $SO(1, n+1)$, where elements of the Lie algebra $\mathfrak{sl}(1, n+1)$, are differential operators of the modern mathematical physics [1]. The cosmological problem that exist is to reduce the number of the field equations that are resoluble under the same gauge field (Verma modules) and extend the gauge solutions to other fields using the symmetries of topological groups that define their interactions.

This extension can be given for a global Langlands correspondence between categories of Hecke sheaves on an adequate moduli stack and the category of holomorphic ${}^{L}G$- bundles with a special connection (Deligne connection). The corresponding D-modules can be viewed as sheaves of conformal blocks (or co-invariants) being images under a version of the generalized Penrose transform [1], [6], naturally arising in the frame of the conformal field theory.

Finally we can to determine "quantized versions" of the derived categories and their cohomology.

2. Background Results.

We need a scheme theory where we can exhibit the image in the context of commutative rings of the corresponding spectrum functor of the derived category of the sheaves of the field equations germs to Lie algebra $\hat{\mathfrak{g}}$, of the connection of the corresponding holomorphic bundle.

Theorem 2. 1 (F. Bulnes) [7]. If we consider the category $M_{\mathcal{K}_{\mathcal{F}}}(\hat{\mathfrak{g}}, Y)$, then a scheme of their spectrum $\mathrm{V}_{critical}^{Def}$, where Y, is a Calabi-Yau manifold comes given as:

$$Hom_{\hat{\mathfrak{g}}}(X, V_{critical}^{Def}) \cong Hom_{Loc_{L_G}}(V_{critical}, M_{\mathcal{K}_F}(\hat{\mathfrak{g}}, Y)), \qquad (1)$$

Proof. [7].▪

Then we can to establish the following results considering the moduli problems between objects of an algebra A_{∞}, and objects of a vector bundle.

Studies realized using commutative rings extended by the Yoneda algebra establish the existence inside the moduli problems of an isomorphism between functors Ext with deformations in the moduli problem defined for Loc_{L_G}, and spectrum, the corresponding deformed derived category $\mathcal{D}^s(Bun_G)$. Likewise we have the following result.

Theorem. 2.2. The Yoneda algebra $\mathrm{Ext}_{\mathcal{D}^s(Bun_G)}(\mathcal{D}^s, \mathcal{D}^s)$, is abstractly A_{∞}, isomorphic to $\mathrm{Ext}^{\bullet}_{Loc_{L_G}}(\mathcal{O}_{Op_{L_G}}, \mathcal{O}_{Op_{L_G}})$.

Proof. [8],[9].▪

This result bring in particular that formal deformations of the sheaf $\mathcal{D}^s$, can be consigned in $\mathcal{D}^s_{Bun_G}$ -mod, which in the moduli stack language can be re-written using the theorem 2.1, as

$$Spec_{Bun_G} TBun_G = T^{\vee} Bun_G, \qquad (2)$$

We can establish a long sequence where the correspondence between moduli stacks and cohomological classes as products of the generalized Verma modules (see table [10]) can be given until critical level of the vector holomorphic bundle. But is here where precisely the cohomological space $H^{\bullet}(T^{\vee}Bun_G, \mathcal{O}))$, has their corresponding version with coefficients in the Verma module at critical level [1] as $H^{\bullet}(\mathfrak{g}[[t]], \mathfrak{g}; \mathbb{V}_{crit}))$. Of fact, this appears inside moduli identity given in [11], using the critical level bundle

[1] $\mathbb{V}_{crit} = U_{crit}\hat{\mathfrak{g}} \otimes_{\mathfrak{g}[[z]]} \mathbb{C}$.

$K^{1/2}$.

The integration process is given or described for the main theorem in [12], which was demonstrated with detail.

3. Main Result.

We establish the following hypothesis using the before results as platform and as part of this hypothesis. Likewise, to define an adequate moduli problem we use the scheme of the Grothendieck type given by (1), and to extension of commutative rings, we use the theorem 2. 2.

The important prospective of the theorem 2. 2., that we need, is the fact that formal deformations of sheaves can be extended to deformation of categories in QFT.

But also, as was mentioned in the before section, the twisted nature of the derived categories $\mathcal{D}^{\times}$, must be done on an appropriated stack (which as we know must be on an appropriated vector bundle).

Theorem 3. 1. One meromorphic extension of one flat connection given through a Hitchin construction we can give the following commutative cocycles diagram to the category $M_{\mathcal{K}_{\mathcal{F}}}(\hat{\mathfrak{g}}, Y)$,

$$\begin{array}{ccccc} \mathbf{h} \epsilon H^0(T^{\vee} Bun_G, \mathcal{D}^s) & \xrightarrow{d} & H^1(T^{\vee} Bun_G, \mathcal{O}) & \xrightarrow{\cong} & \Omega^1[\mathbf{H}] \\ \cong\downarrow & & \cong_{\Phi\mu}\downarrow & & \downarrow \pi \\ a \epsilon \mathbb{C}[Op_{{}^LG}] & \xrightarrow{d} & \Omega^1[\mathcal{O}_{Op_{{}^LG}}] & \xrightarrow{d} & C \times B \end{array} \qquad (3)$$

Their demonstration was given with detail in [13]. However, we will give some sketch and we discuss some key steps and conclusions from their demonstration.

From the lemma 3. 1 of [11], and using some arguments inside the demonstration of the main theorem given in [12], the cohomology space $H^{\bullet}(?, \Omega^{\bullet})$, is exhibited as the space $H^{\bullet}(H^{\vee}, \Omega^{\bullet})$, inside the quasi-coherent category given by $M_{\mathcal{K}_{\mathcal{F}}}(\hat{\mathfrak{g}}, Y)$, which carry us to the ramification problem. But we consider QFT and TFT, in the derived categories framework to define cocycles of $M_{\mathcal{K}_{\mathcal{F}}}(\hat{\mathfrak{g}}, Y)$

To the demonstration of the theorem 3. 1, firstly was demonstrated the equality between $R^1\chi_*(\mathcal{D}^s)$ and $R^1\chi_*(\mathbf{h}) \in \Omega^1[\mathbf{H}]$ [2] in the derived class $\mathcal{D}^\times$. Then we consider a Langlands correspondence such that:

$$\Phi^i(\mathfrak{c}(\mathcal{O}_{\mathbb{V}})) = \mathfrak{c}(\mathcal{O}_{\mathbb{V}}) \boxtimes \wedge^i \mathbb{V}, \tag{4}$$

arriving to $\Omega^i[Op_{L_G}]$, which gives the equivalence of complexes

$$\{d\mathbf{h} = 0\} \cong {}^L\Phi^\mu\{da = 0\}, \tag{5}$$

which requires the correspondence

$$\tilde{\mathfrak{c}} : D_{Coh}(T^\vee Bun_G, \mathcal{O}) \cong D_{Coh}({}^LT^\vee Bun_G, \mathcal{O}), \tag{6}$$

Then is proved the first descendent isomorphism of (3).

By the Frenkel equivalence $D^b(\mathfrak{g}_{\mathcal{K}_C} - mod_{nilp})^{I_0} \cong D^b(\mathcal{Q}Coh(MOp^{nilp}_{L_G}))$, each quasi-coherent sheaf on the kernel of the right side of this equivalence corresponds an object of $D^b(M_{\mathcal{K}_\mathcal{F}}(\hat{\mathfrak{g}}, Y))^{I_0}$. Then the functor $\Phi_{\mathcal{O}^{DG}_{Spu}}$, is a Hecke functor. And also, their integral transform is such that $h : Bun_{Higgs} \to B$ (to a quantized) and their image is equivalent to $D_{coh}({}^LBun, \mathcal{D})$. Then the geometrical Langlands conjecture in terms of Higgs bundles consider a functor between the categories $D_{Coh}({}^LLoc, \mathcal{O})$, with the action of the Hecke functors on $D_{Coh}({}^LBun, \mathcal{D})$. But $MOp^{nilp}_{L_G} \cong Op^{nilp}_{L_G} \times \tilde{\tilde{\mathcal{N}}} \ /^L \ G$, by the Steinberg manifold structure to the Langlands correspondence given by $\mathfrak{c}$, such that $\tilde{\mathfrak{c}} = \mathfrak{c}(\mathcal{O}_{Sp_uC}(\mathbb{V}) \times \mathbb{C}^\times)$, where $\mathcal{O}_{Sp_uC} = \mathcal{O}_{\tilde{\mathcal{N}}} \otimes_{\mathcal{O}\mathcal{N}} \mathbb{C}$, and $\mathcal{N} \subset^{\mathcal{L}} \mathfrak{g}$. By K-theory, the Steinberg manifold have elements $C \times B$, that satisfy $Isom(d\mathbf{h}) = d(da), \forall a \in \mathbb{C}[Op_{L_G}(D)]$. Then is had that

$$d(da) \overset{{}^L\Phi^\mu}{\longleftrightarrow} Isom d\mathbf{h}, \tag{7}$$

That is to say, the kernels of $\Omega^i, i = 1, 2, \ldots,$ are the that are in the sheaf

[2] The first differential in a spectral sequence $H^*(\mathcal{D})[[s]]$ implies $H^*(\mathcal{D}^s)$, for the deformation $\mathcal{D} \to \mathcal{D}^s$.

$\mathcal{O}_{Op_{LG}}$, [3] that is to say, there is an extended Penrose transform such that their kernel set has as elements the fields $\mathbf{h}$, with $\mathrm{Isom}(d\mathbf{h}) = 0$, in the hyper-cohomology. Then in hyper-cohomology, in the down line and using the Hitchin mapping the fields a, are those that satisfy $d(da) = 0$, in the hyper-cohomology $\mathbb{H}(\Omega^1 \xrightarrow{d} \Omega^2 \xrightarrow{d} \ldots)$.

In the context of the differential operators algebra we can give the commutative rings diagram:

$$\begin{array}{ccccc} H^\bullet(\mathfrak{g}[[z]], \mathbb{V}_{critical}) & \to H^\bullet(\mathfrak{g}[[z]], \mathfrak{g}, \mathbb{V}_{critical}) & \xrightarrow{\cong} & \Omega^\bullet[\mathbf{H}] \\ \cong\downarrow & \cong_\Phi\downarrow & & \downarrow \pi \\ \mathbb{C}[Op_{LG}] \xrightarrow{d} & \Omega^\bullet[Op_{LG}] \xrightarrow{d} & & \mathbf{H}^\vee \end{array} \tag{8}$$

where $\mathbf{H}^\vee = Spec_H SymT[Op_{LG}(D)]$. Then we have the Penrose transform in the descendent isomorphism whose field solutions are to the equations $d(da) = 0$. An extended Penrose transform to the deformed modules version $H^\bullet(\mathfrak{g}[[z]], \mathfrak{g}, \mathbb{V}_{critical})$, consider the Fourier-Mukai transform. Then the Yoneda algebra given by $\mathrm{Ext}_{\mathcal{D}^s(Bun_G)}(\mathcal{D}^s, \mathcal{D}^s)$, can to establish the endomorphism of critical level modules. Then is completed the sequence of critical Verma modules (projective Harish Chandra module to whole the sequence). Likewise the global functor to the diagram in question until $\Omega^\bullet[Op_{LG}]$, is ${}^L\Phi^\mu(\mathcal{M}) = \mathcal{M} \boxtimes \rho^\mu(\mathbb{V})$ with ${}^L\Phi^\mu$ is the Hecke functor. Finally to obtain the spectrum (in a Hamiltonian variety) $\mathbf{H}$, we use the quantized version of the corresponding cohomology space $H^q(Bun_G, \mathcal{D}^s) = \mathbb{H}^q_{G[[z]]}(\mathbf{G}, (\wedge^\bullet[\Sigma^0] \otimes \mathbb{V}_{critical}; \partial))$. Finally and after of a good work in the graded vector spaces in the deformed derived category $\mathrm{D}_{coh}({}^L Bun, \mathcal{D}^\times)$, which are included in the quasi-coherent category $M_{\mathcal{K}_F}(\hat{\mathfrak{g}}, Y)$, we have $Spec_G^{\mathfrak{g}[[z]]/G}(\Omega^1(\mathbf{H})) = Y$.

4. Application (Gravitational waves as oscillations in the space-time curvature/spin).

We consider now $G = GL(1, \mathbb{C}) \cong^L G$. We consider the bundle stack given for [4]

[3] $H^\bullet Bun_G, \mathcal{O}) \cong \mathcal{O}_{Op_{LG}} = Ker(U, \nabla)$, where ∇, induces a holomorphic connection on lines bundles.

[4] Here $Pic(C)$, is the moduli stack Bun.

$$\mathbb{M} = Pic(C), \tag{9}$$

which is known as Picard variety of C. Then the Hecke functor is the mapping

$$\Phi^1 : D_{coh}(Pic(C), \mathcal{D}) \to D_{coh}(C \times Pic(C), \mathcal{D}), \tag{10}$$

which is pull-back of the Abel-Jacobi mapping:

$$aj : C \times Pic^d(C) \to Pic^{d+1}(C), \tag{11}$$

with correspondence rule

$$(L, \chi) \mapsto L(\chi), \tag{12}$$

In this case, the geometrical Langlands correspondence comes give as:

$$\mathfrak{c}(\mathbb{L}) = \left\{ \begin{array}{c} \textit{The unique translation invariant rank on local system} \\ \textit{on variety } Pic(C)\textit{, whose restriction on each component} \\ Pic^d(C)\textit{, has the same monodromy as } \mathbb{L}. \end{array} \right\} \tag{13}$$

where $\mathbb{L}$, is the space of the Langlands data (bundle and connection) (L, ∇), that is a rank one local system on C. Due to that $\pi_1(Pic^d(C))$, is the abelianization of $\pi_1(C)$, and the monodromy of the space $\mathbb{L}$, is Abelian, we can view this space $\mathbb{L}$, as a local system on each component $Pic^d(C)$, of $Pic(C)$.

Them likewise, considering the pull-back of the local system $\mathbb{L}$, to the various factors of the $d-th$ Cartesian power $C^{\times d}$, of C, and tensor of these pull-backs to get rank one local system $\mathbb{L}^{\boxtimes d}$, where $\boxtimes$, is a micro-local tensor product.From a point of view of the field equations, each component of the correspondences space $\mathfrak{c}(\mathbb{L})$, on $Pic^d(C)$, a trace of particles in the symplectic geometry that can be characterized in a Hamiltonian manifold, with the due quantization of the coherent sheaves of the differential operators of the field equations.

Likewise, using a Hitchin′s abelianization we can induce the geometrical Langlands correspondence $\mathfrak{c}$, as was planted to the case of the group $G = GL(n, \mathbb{C})$, considering the correspondence $\mathfrak{c}$, as:

$$\mathfrak{c} = quant_{Bun} \circ \Phi \circ quant_C^{-1}, \tag{14}$$

where Φ, is the Fourier-Mukai transform defined to this case as:

$$\Phi : D_{coh}(quant_{Bun}, \mathcal{D}) \to D_{coh}(C \times quant_{Bun}, \mathcal{D}), \tag{15}$$

Here, the quantization procedures $quant_{Bun}$, and $quant_C$, are appropriately understood non-Abelian Hodge correspondences. An adequate Hitchin mapping can give solution to the equations through Hamiltonian states, in the non-Abelian context of the Hodge theory [13] in hypercohomology:

$$d(da) = 0, \tag{16}$$

$\forall a \in \mathbb{C}[Op_{L_G}(D)]$, having as integral the integral transforms composition:

$$\mathfrak{c} \circ \Phi^{\mu} =^{L} \Phi^{\mu} \tag{17}$$

where the states of the quantum field are the cotangent vector (Higgs fields) $\mathbf{h}$, such that

$$Isomd\mathbf{h} = 0, \tag{18}$$

Then by superposing of these states considering the field corresponding ramifications, we have:

$$\mathcal{H} = H^0(\omega_C) \oplus H^0(\omega_C^{\otimes 2}) \oplus \cdots \oplus H^0(\omega_C^{\otimes n}), \tag{19}$$

which has their re-interpretation as the curvature energy expressed through the H-states which can be written using the superposing principle to each connection $\omega_C^{\otimes j}$, (with C, a curve) that describes the corresponding dilaton (photon). Likewise, in a Hamiltonian densities space [14] we have the figure 1A, considering a Hitchin base.

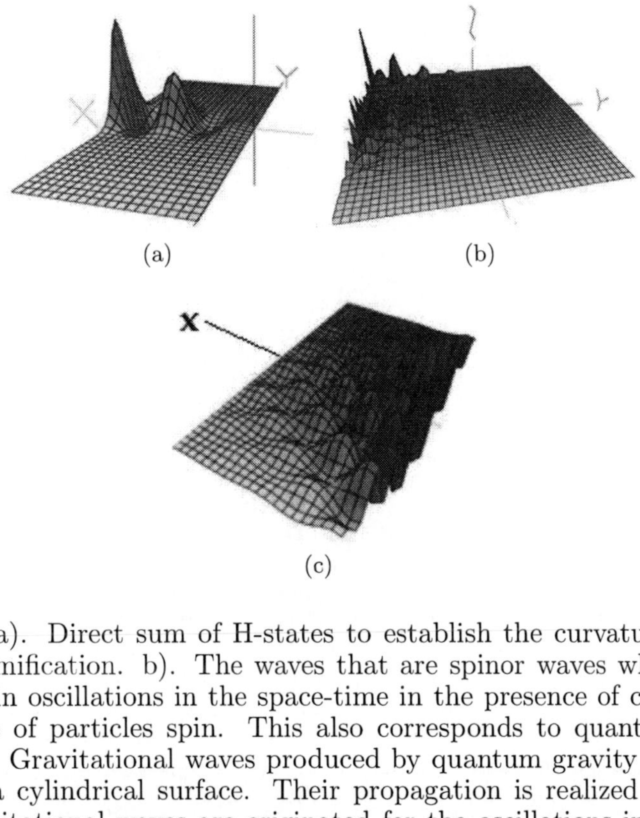

(a) (b)

(c)

Figure 1: a). Direct sum of H-states to establish the curvature measure by field ramification. b). The waves that are spinor waves which can be consigned in oscillations in the space-time in the presence of curvature to the change of particles spin. This also corresponds to quantum gravity model. c). Gravitational waves produced by quantum gravity due the H-states on a cylindrical surface. Their propagation is realized on axis X. These gravitational waves are originated for the oscillations in the space-time-curvature/spin (that is to say using causal fermions systems).

In the case of spinor representation the corresponding H-states can be given as spinor waves (see figure B) which can be consigned in oscillations in the space-time-curvature/spin, to a microscopic deformation measured in $\mathcal{H}$.

Specialized Notation

$\mathcal{D}^{\times}$ -Twisted sheaf of differential operators to our Oper, given by $\mathrm{Loc}_{^LG}(\mathcal{D}^{\times})$.

$K^{1/2}$ - Root square of the canonical line bundle on Bun_G, corresponding to the critical level. This is a divisor vector bundle.

$\mathrm{Bun}_G(X)$ -Category of principal G - bundles over $C \times X$. Also is the moduli stack of principal G -bundles over C.

$\mathrm{Loc}_{{}^LG}(\mathcal{D}^{\times})$ -Set of equivalence classes of ${}^LG$ -bundles with a connection on $\mathcal{D}^{\times}$. This space shape a bijection with the set of gauge equivalence classes of the ramified operators, as defined in [14],[15].

$Pic(C)$- Picard variety of C.

H− Hamiltonian variety. Space of the function of H-states. Their manifestation are energy waves corresponding to the connection of the level field equation.

References

[1] F. Bulnes, Cohomology of Moduli Spaces in Differential Operators Classification to the Field Theory (II), in: Proceedings of Function Spaces, Differential Operators and Non-linear Analysis., 2011, Tabarz Thur, Germany, Vol. 1 (12) pp001-022.

[2] F. Bulnes, Penrose Transform on Induced DG/H-Modules and Their Moduli Stacks in the Field Theory, Advances in Pure Mathematics 3 (2) (2013) 246-253. doi: 10.4236/apm.2013.32035.

[3] A. Kapustin, M. Kreuser and K. G. Schlesinger, Homological mirror symmetry: New Developments and Perspectives, Springer. Berlin, Heidelberg, 2009.

[4] F. Bulnes, Integral geometry and complex integral operators cohomology in field theory on space-time, in: Proceedings of 1st International Congress of Applied Mathematics-UPVT (Mexico)., 2009, vol. 1, Government of State of Mexico, pp. 42-51.

[5] F. Bulnes, M. Shapiro, On general theory of integral operators to analysis and geometry, IM-UNAM, SEPI-IPN, Monograph in Mathematics, 1st ed., J. P. Cladwell, Ed. Mexico: 2007.

[6] F. Bulnes, R. Goborov, Integral Geometry and Complex Space-Time Cohomology in Field Theory, Pure and Applied Mathematics Journal. Special Issue:Integral Geometry Methods on Derived Categories in the Geometrical Langlands Program. Vol. 3, No. 6-2, 2014, pp. 30-37. doi: 10.11648/j.pamj.s.2014030602.16.

[7] F. Bulnes, Integral Geometry Methods on Deformed Categories in Field Theory II, Pure and Applied Mathematics Journal. Special Issue: Integral Geometry Methods on Derived Categories in the Geometrical Langlands Program, 3(6-12), (2014), 1-5, doi: 10.11648/j.pamj.s.2014030602.11.

[8] N. Yoneda, On the homology theory of modules, J. Fac. Sci. Univ. Tokyo. Sect. I, 7, (1954), 193227.

[9] E. Frenkel, C. Teleman, Self extensions of Verma modules and differential forms on opers, Comps. Math., 142, (2006), 477-500.

[10] F. Bulnes, Geometrical Langlands Ramifications and Differential Operators Classification by Coherent D-Modules in Field Theory, Journal of Mathematics and System Sciences, 3(10), (2013), 491-507.

[11] F. Bulnes, Integral geometry methods on deformed categories to geometrical Langlands ramifications in field theory, Ilirias Journal of Mathematics, 3(1), 1-13.

[12] F. Bulnes, Moduli Identities and Cycles Cohomologies by Integral Transforms in Derived Geometry, Theoretical Mathematics and Applications, 6(4), (2016), 1-12.

[13] F. Bulnes, Extended d Cohomology and Integral Transforms in Derived Geometry to QFT-equations Solutions using Langlands Correspondences, Theoretical Mathematics and Applications, Vol. 7 (2), pp51-62.

[14] F. Bulnes, Integral Geometry Methods in the Geometrical Langlands Program, SCIRP, USA, 2016.

[15] A. Abbondandolo, M. Schwarz, Floer homology of cotangent bundle and the loop product, Geom. Top., 14(3), (2010), 1569-1722.

Supersymmetric path to unification of gravity with particle physics*

Alexander Burinskii†
Nuclear Safety Institute, Russian Academy of Sciences

ABSTRACT

It has been showed recently that spin curves space along with mass and ultra-high spin/mass ratio of the elementary particles makes Einstein's gravity strong, shifting gravitational influence from Planck to the Compton scale. Although this increases conflict between gravity and quantum theory, the compatibility of the spinning Kerr-Newman (KN) gravity with electroweak quantum particles can be achieved by using the supersymmetric Higgs model, which creates a free from gravity supersymmetric and superconducting core of the particle. The corresponding BPS-saturated solution to generalized LG model takes the form of a supersymmetric bag model, which provides a flat supersymmetric vacuum state inside the bag. The bag is deformable, and its shape is controlled by supersymmetry providing compatibility of the core with external gravitational and electromagnetic (EM) field. In particular, for the spinning KN gravity bag takes the form of oblate disk with a circular string placed on the disk border. Excitations of the KN EM field create circular traveling waves. The super-bag solution is upgraded to the Wess-Zumino supersymmetric QED model, indicating a bridge from the super-bag model to perturbative formalism of the conventional QED.

PACS:11.27.+d, 04.20.Jb, 04.70.Bw

1. Introduction

Modern physics is based on Quantum theory and Gravity. The both theories are confirmed experimentally with great precision. Nevertheless, they are contradicting and cannot be combined in a unified theory. This contradiction is mutual. On the one hand, gravitational field cannot be quantized, on the other hand, the stochastic representation of the quantum particles as point-like objects controlled by the wave function is not suitable for gravity, which requires representation in terms of the real physical fields generating the right hand side of the Einstein equations, $G_{\mu\nu} = 8\pi T_{\mu\nu}$.

* This work has been supported by Program of Fundamental Scientific Research of the State Academies of Science for 2013-2020, SR-11407/2017/23

† e-mail address:Burinskii@mail.ru

Treatment of the quantum particles as the extended semiclassical objects was started in 1970's in the models of solitons, strings and bag models, and superstring theory was considered to be the most promising among them. However, as mentioned John Schwarz, *"...Since 1974 superstring theory stopped to be considered as particle physics... "* and *"... a realistic model of elementary particles still appears to be a distant dream ... "* [1]). One of the reasons of this is that extra dimensions are compactified with extra tiny radii of order the Planck length 10^{-33} cm, which does not correlate with characteristic lengths of quantum physics and makes impossible to test extra dimensions with currently available energies. The idea to bring fundamental gravitational scale close to the weak scale was considered in different approaches, and in particular, in the brane world scenario, where the weakness of the localized 4d gravity is explained by its "leaks" into the higher-dimensional bulk, and the brane world mechanism allowed to realize ideas of the superstring theory for any numbers of the extra dimensions [2].

Alternative ideas were related with nonperturbative 4D solutions of the non-linear field models – solitons, in particular, solitonic solutions to low energy string theory [3, 4, 5, 6]. This approach, being nonperturbative, is akin to the Higgs mechanism of symmetry breaking and linked with nonperturbative approach to electroweak sector of the Standard Model. The most known is the Nielsen-Olesen model of dual string based on the Landau-Ginzburg field model for a phase transition in superconducting media, and also the famous MIT and SLAC bag models [7, 8, 9] which are similar to solitons, but being soft, deformable and oscillating, acquire many properties of the dual string models. Besides, being suggested for confinement of quarks, the bag models assume consistent implementation of the Dirac equation. The question on consistency with gravity is not discussed usually for the solitonic models, as it is conventionally assumed that gravity is very weak and is not essential on the scale of electroweak interactions. For example, in [10] we read *"... quantum gravity effects are usually very small, due to the weakness of gravity relative to other forces. Because the effects of gravity are proportional to the mass, or the energy of the particle, they grow at high energies. At energies of the order of E 10^{19} GeV, gravity would have a strength comparable with that of the other Standard Model interactions."*

Meanwhile, in a relativistic theory, spin is not separable from rotation, and its influence on metric should also be taken into account. Analyzing metric of the Kerr-Newman (KN) rotating black hole solution with parameters of an electron, we obtain strong influence of spin and recognize that the usual assumption on the weakness of gravity is not correct. Indeed, nobody says that gravity is weak in Cosmology where physics is determined by giant masses. Similarly, the giant spin/mass ratio of the elementary particles, $J/m = 10^{20} - 10^{22}$ (in the dimensionless Planck's units $G = c = \hbar = 1$) shows that the energy equivalent to unit of quantum spin on the Planck scale must be the unit of Planck mass $E_P = M_P = 1$, leading to the ratio $J/m \approx M_P/m \sim 10^{20}$. It shows that giant spin of particles must have a giant impact on the metric. The commonly accepted view that gravity is weak and not essential in particle physics becomes not compatible with Planck scale of gravitational interaction, and should be replaced by principally new point of view that scale of gravitational interaction is to be shifted about 20 orders of magnitude to the Compton lengths [11, 12, 13]. Influence of spin on metric becomes extreme strong, and even crucial for the electroweak interactions.

Analysis of the Kerr-Newman (KN) solution with parameters of an electron confirms this statement, showing that space-time is strongly deformed in the Compton zone, contrary to the usually accepted Planck length, and reason of this deformation is that spin deforms space along with mass. In other words, *spin is gravitating*, and ultra-high spin/mass ration of the electron, $a = J/m \sim 10^{22}$, breaks space topologically, creating the naked singular ring and two-sheeted space, which differs from Minkowski space so strongly that neither Dirac theory nor perturbative QED can be applied.

This great influence of spin, leading to drastic shifts of the scale of gravitational interaction was not taken into account before, and apparently, it was the main reason of the failure previous attempts of unification of gravity with particle physics. So large shifts of the scale of interaction creates a new paradigm and requires a principally new approach, in which gravity is not subordinated to quantum theory, and both these theories impact on process of unification on an equal footing.

In this article we suggest a solution of the problem of unification with gravity based on this new paradigm. We show that conflict between gravity and quantum theory can be resolved *without modification of the Einstein-Maxwell gravity – the spoiled by gravity space can be cured by a supersymmetric bag model*, in which the singular region of KN solution is replaced by the flat internal space of the Compton size. We find the corresponding non-perturbative BPS-saturated solution in frame of the supersymmetric generalized Landau-Ginzburg model, in which boundary of the bag is formed by a domain wall interpolating between the external KN gravity and the supersymmetric vacuum state inside the bag. Similar to the usual bag models, the super-bag model is deformable and displays a super-consistency with the external gravitational and electromagnetic KN field, in the sense that its shape and dynamics are fully defined by matching its boundary with a special surface (which can be called as "zero gravity surface"), where the external gravitational field is compensated by electromagnetic field. This surface determines position of the domain wall, and therefore, it determines shape of the bag, which gets a disk-like configuration with a closed string lying along sharp border of the disk [14, 15, 16, 17]. Therefore, this new concept leads us again to a string theory, which is however, four-dimensional, and differs essentially from the famous superstring theory based on compactification of higher dimensions at Planck scale.

One of the main and the most studied quantum particles is an electron, and assuming that dressed electron has such a SuperBag structure we have to explain first of all, while this structure was not observed experimentally, and second, to put out the possible relationship with the Dirac theory of electron and with perturbative QED. The first signal on the relation of the KN solution with Dirac electron was obtained by Carter [18, 19], who obtained that KN solution has gyromagnetic ratio $g = 2$ as that of the Dirac electron. Besides, one of the basic features of the Bag models is compatibility with the Dirac equation [7, 8], which in the Kerr geometry was supported by two aligned spinor solutions given by the Kerr theorem, [20, 14]. The link with QED is much more problematic, and we show here that the supersymmetric Landau-Ginzburg model can naturally be upgraded to the Wess-Zumino SuperQED model [21], revealing connections of the non-perturbative SuperBag solution with the conventional perturbative technics used in QED.

2. Super-bag model as BPS solution to generalized LG model

2.1. Basic features of the ultra-rotating Kerr-Newman solution

It has been recently obtained [22, 23], that the source of ultra-spinning Kerr-Newman (KN) solution can be considered as a superconducting soliton having many features of the bag model [14, 15, 24], but with the essential advantage of compatibility with Einstein-Maxwell gravity in four dimensions. As is known, the bag models take intermediate position between strings and solitons [25, 26, 27]. Although, the bags were initially offered as the extended models of hadrons, [7, 8, 9], being based on the Abelian Higgs model of symmetry breaking their indicated rather applicability to the Salam-Weinberg model of leptons, which was one of the reasons to consider the gravitating KN bag as the model for consistent with gravity leptons.

The spinning KN solution is of particular interest in this regard, since, as it was obtained by Carter [19, 18], that gyromagnetic ratio of the KN solution is $g = 2$, and therefore corresponds to the external field of the electron. The spin/mass ratio of the electron is about 10^{22}, and structure of source of the KN solution for such a huge spin should shed the light on origin of the conflict between gravity and quantum theory. One can see that the KN field with parameters of electron becomes extremely strong on the Compton distances, so that the BH horizons disappear and the Kerr singular ring of the Compton radius $a = \hbar/m$ becomes open, which breaks topology of space-time and creates two-sheeted metric.
In the Kerr-Schild approach, metric of the KN solutions is [19]

$$g_{\mu\nu} = \eta_{\mu\nu} + 2Hk_\mu k_\nu, \tag{1}$$

where $\eta_{\mu\nu}$ is metric of an auxiliary Minkowski space M^4, (signature $(-+++)$), and H is the scalar function which for the KN solution takes the form

$$H_{KN} = \frac{mr - e^2/2}{r^2 + a^2\cos^2\theta}, \tag{2}$$

where r and θ are oblate spheroidal coordinates, and k_μ is a null vector field $k_\mu k^\mu = 0$, forming a Kerr congruence – the vortex of polarization of gravitational and electromagnetic field in the Kerr space-time, see Fig.1.

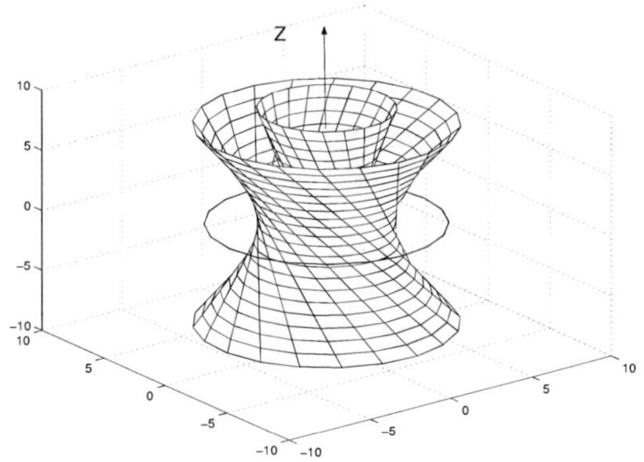

Figure 1: Vortex of the Kerr light-like (null) congruence k^μ propagates analytically from negative sheet of Kerr metric, $r < 0$, to positive one, $r > 0$. In the equatorial plane, $\cos\theta = 0$, the Kerr congruence is focused on the Kerr singular ring, $r = \cos\theta = 0$.

The Kerr singular ring corresponds to border of the disk $r = 0$, in the equatorial plane $\cos\theta = 0$.

Similarly, vector potential of KN solution is also collinear with the null direction k_μ,

$$A_\mu = -\frac{er}{(r^2 + a^2\cos^2\theta)} k_\mu \tag{3}$$

The KN metric becomes two-sheeted, since the Kerr congruence

$$k_\mu dx^\mu = dr - dt - a\sin^2\theta d\phi, \tag{4}$$

is out-going at the 'positive' sheet of the metric, $r > 0$, and passes analytically to 'negative' sheet, $r < 0$, being extended via ring $r = 0$, where it becomes in-going. The two null vector fields $k_\mu(x)^\pm$ become different at $r > 0$ and $r < 0$, leading to two different metrics $g^\pm_{\mu\nu} = \eta_{\mu\nu} + 2Hk^\pm_\mu k^\pm_\nu$ on the positive and negative sheet of the same Minkowski background. Similarly, it leads also to two-sheeted vector-potential $A^\pm_\mu$, that makes space inappropriate for quantum theory, and therefore, conflict between quantum theory and gravity is shifted by 22 orders earlier then it is usually expected, from the Planck to the Compton scale. As usually, singularity is signal to new physics – theory of more high level. The KN gravitational field is strong near the Kerr singular ring and creates vortex of the space-time polarization in the Compton zone of the dressed electron, which should be flat for work of quantum theory. It is usually assumed that in vicinity of strong field, gravity should be modified to a new Quantum Gravity. Taking into account sharp incompatibility of Quantum and Gravity, natural requirement for such new theory would be separation of their zones of influence: formation of the internal zone

(I) – flat core for quantum theory, and external zone

(E) – for undisturbed gravitational and electromagnetic fields.

There should also be selected intermediate zone

(R) – interpolating between **(I)** and **(E)**.

In the case of strong KN field, these demands become so restrictive that determine structure of the new theory almost uniquely. It turns out that the flat Compton zone free from gravity may be achieved *without modification of the Einstein-Maxwell equations,* through SUPERSYMMETRY, which eats up the strong gravitational field in the core of particle. Expelling gravity from the core of the KN spinning particle is similar to expelling the EM field from superconducting core, and both of these super-phenomena are realized in core of the KN solution by the supersymmertric Landau-Ginzburg field model [28, 29, 30, 31, 32, 33, 34, 21] in the form of a BPS-saturated Super-Bag solution, for which just the strong contradiction between Quantum and Gravity determines extreme sensitivity of the model to the choice of the separating surface **(R)**.

The natural choice of this surface for the KN solution was suggested by C. López [35]. According (1) and (2) it should be the "zero gravity" (ZG) surface

$$r = R = \frac{e^2}{2m}, \tag{5}$$

where function H vanishes

$$H_{KN}(R) = 0, \tag{6}$$

metric becomes flat, and can be matched with flat Minkowski space for $r < R$. It turns the López source of the KN solution in a shell-like bubble.

The corresponding metrics were suggested by Gürsay and Gürses [36]. They have the Kerr-Schild form (1) and retain the form of the Kerr congruence (4), while the function H is changed as follows,

$$H = \frac{f(r)}{r^2 + a^2 \cos^2\theta}, \tag{7}$$

where $f(r)$ is arbitrary smooth function of the Kerr radial coordinate r, taking at the far distances the KN form (2).

So far as in the Kerr solution r is oblate spheroidal coordinate [19], which is related with Cartesian coordinates by transformations

$$x + iy = (r + ia)\exp\{i\phi_K\}\sin\theta, \quad z = r\cos\theta, \quad \rho = r - t, \tag{8}$$

the bubble surface $r = R$ takes the oblate ellipsoidal form – the disk of the thickness R and radius $r_c = \sqrt{R^2 + a^2}$, where $a = J/m$.

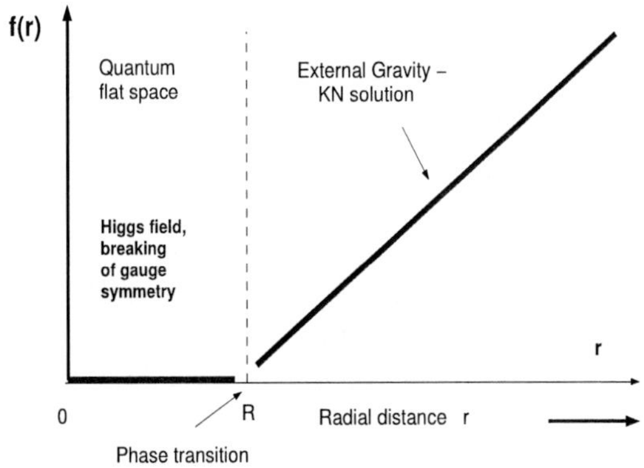

Figure 2: Behavior of the function $f(r)$ in bag model. Domain wall placed at $r = R$ separates flat Quantum interior from external KN Gravity .

For solutions without rotation, $a = 0$, and bubble turns into a sphere of the classical radius r_e. Such spherical shape was suggested by Dirac in [37] as an "extensible electron model" – prototype of the bag models, displaying one of their basic features of the bags – their deformability.

We see that deformations of the KN Super-Bag appear as consequence of the requirement on sharp separation of the zones **(I)**,**(E)**,**(R)**.

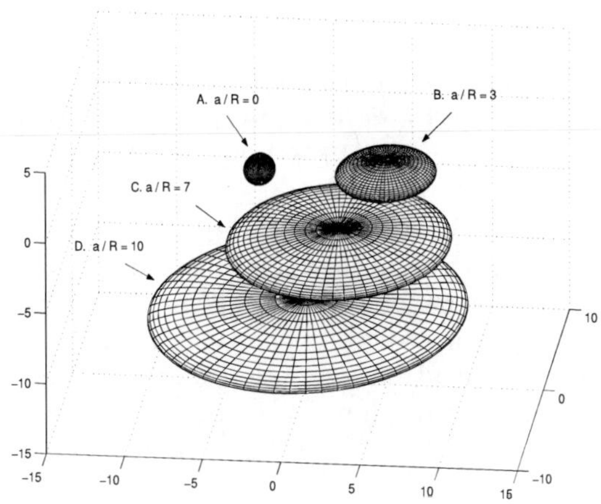

Figure 3: (A): Spherical bag without rotation $a/R = 0$, and disk-like bags for different values of the rotation parameter: (B)- $a/R = 3$; (C) - $a/R = 7$; and (D) - $a/R = 10$.

3. Supersymmetry ensures consistency with gravity

3.1. Generalized Landau-Ginzburg field model and domain wall (DW) phase transition

The Landau-Ginzburg field model of superconductivity is used in many solitonic models, in particular, in the Nielsen-Olesen dual string model, as a field model in the MIT and SLAC bag models, and really, it is also the the Higgs model of symmetry breaking, because the Higgs vacuum itself "... is analog to a superconducting metal", [9]. The Landau-Ginzburg Lagrangian used in the Nielsen-Olesen model (minimal Landau-Ginzburg model) is

$$\mathcal{L}_{NO} = -\frac{1}{4}F_{\mu\nu}F^{\mu\nu} - \frac{1}{2}(\mathcal{D}_\mu\Phi)(\mathcal{D}^\mu\Phi)^* - V(|\Phi|), \tag{9}$$

where $\mathcal{D}_\mu = \nabla_\mu + ieA_\mu$ are the $U(1)$ covariant derivatives, and $F_{\mu\nu} = A_{\mu,\nu} - A_{\nu,\mu}$ is the corresponding field strength, and potential V has the quartic form

$$V = \lambda(\Phi^\dagger\Phi - \eta^2)^2, \tag{10}$$

where η is condensate of the Higgs field Φ, its vacuum expectation value (vev) $\eta < |\Phi| >$, [43].

The minimal Landau-Ginzburg model can be used to describe superconductivity inside the bag – interplay of the KN vector-potential with the Higgs condensate. Since requirements **(I)**,**(E)**,**(R)** define inside the bag a flat space, the corresponding covariant derivatives can be taken as flat,

$$\mathcal{D}_\mu = \nabla_\mu + ieA_\mu \to \mathcal{D}_\mu = \partial_\mu + ieA_\mu. \tag{11}$$

However, the NO and KN models have opposite spacial configurations: the KN bag model should describe a superconducting disk surrounded by the long-range EM and gravitational field, while the NO model describes vortex of the EM field inside the superconducting Higgs condensate which breaks the external long-range EM and gravitational field. Note, that this is a typical drawback of the most of soliton models and, in particular, the usual bag models which are formed as a "cavity in superconductor" [9]. The reason of this disadvantage lies in the use of the potential (10).

The correct opposite configuration – condensation of the Higgs field inside the core – requires more complex scalar potential V formed from several complex fields Φ_i, $i = 1, 2, 3$, [22]. Kinetic part of the corresponding generalized LG model differs from those of the minimal Landau-Ginzburg model (9) only by summation over the fields Φ_i,

$$\mathcal{L}_{GLGkin} = -\frac{1}{4}F_{\mu\nu}F^{\mu\nu} - \frac{1}{2}\sum_i(\mathcal{D}_{i\mu}\Phi_i)(\mathcal{D}_i^\mu\Phi_i)^*, \tag{12}$$

while the potential V is changed very essentially, and has to be formed by analogy with machinery of the $N = 1$ supersymmetric field theory [21] from a superpotential function $W(\Phi_i)$.[1] The scalar potential[2]

$$V(r) = \sum_i F_i F_i^* \tag{13}$$

is formed through derivatives of the function $W(\Phi_i)$,

$$F_i = \partial W / \partial \Phi_i \equiv \partial_i W, \tag{14}$$

where

$$W(\Phi_i, \bar{\Phi}_i) = Z(\Sigma\bar{\Sigma} - \eta^2) + (Z + \mu)H\bar{H}, \tag{15}$$

μ and η are real constants, and the special notations are introduced $(H, Z, \Sigma) \equiv (\Phi_1, \Phi_2, \Phi_3)$, to identify Φ_1 as the complex Higgs field

$$H = |H|e^{i\chi}, \tag{16}$$

which interacts with the KN vector field A_μ as $\mathcal{D}_{1\mu} = \nabla_{1\mu} + ieA_\mu$. The fields Φ_2 and Φ_3 are assumed uncharged, and $\mathcal{D}_{i\mu} = \nabla_{i\mu}$ for $i = 2, 3$.

The condition $F_i = \partial_i W = 0$ determines two vacuum states with $V = 0$:

(I) internal vacua: $r < R - \delta$, where the Higgs field $|H| = \eta$, and $Z = -\mu$, $\Sigma = 0$,

and

(E) external vacuum state: $r > R + \delta$, where the Higgs field $H = 0$, and $Z = 0$, $\Sigma = \eta$,

separated by spike of the potential $V > 0$ in zone

(R) – a domain wall, interpolating between zones **(I)** and **(E)**, in the full correspondence with the requirements **(I)**,**(E)**,**(R)**.

Reduction of the corresponding LG equations to Bogomolny form is performed by minimization of the energy density per unit area of the DW surface,

$$\mu = \frac{1}{2}\sum_{i=1}^{3} [\sum_{\mu=0}^{3} |\mathcal{D}_\mu^{(i)} \Phi_i|^2 + |\partial_i W|^2]. \tag{17}$$

The four dimensional domain wall solutions in supersymmetric Landau-Ginzburg model have paid attention in the works [28, 29, 30, 31, 32, 33, 34], where it was usually considered the static planar domain walls positioned in (x,y) plane, with the transverse to the wall z-direction. However, even in the simplest case of the one field $\Phi(z)$ and one coordinate z,

$$\mu = \frac{1}{2}(|\partial_z \Phi|^2 + |\partial_\Phi W|^2), \tag{18}$$

[1] It is really not only analogy, and as we shell see, only one step differs this model from the true supersymmetric Higgs model, which was obtained by Morris in [48] with the purpose to get supersymmetric generalization of the Witten superconducting string model [49]. This model was used in [22, 23] to describe superconducting core of the KN solution.

[2] The signs bar $^-$ and star * both are used for complex conjugation.

reduction of the LG equation to Bogomolny form turns out to be nontrivial, since it requires the introduction of an arbitrary phase factor α, so that (18) can be equivalently presented in the form

$$\mu = \frac{1}{2}|\partial_z \Phi - e^{i\alpha}\partial_{\bar{\Phi}}\bar{W}|^2 + Re\ e^{i\alpha}\partial_z W, \tag{19}$$

which is saturated by the Bogomolny equation

$$\partial_z \Phi = e^{i\alpha}\partial_{\bar{\Phi}}\bar{W}. \tag{20}$$

The domain wall forming the KN bag is much more complicated, since first of all it is not planar, but forms the spheroidal boundary profile of which is shown in Fig.2. Second, it is formed by three chiral fields Φ_i, and thirdly, the most important feature is that this domain wall is not static and has non-trivial dependence on the phases of the complex fields Φ_i. The corresponding BPS saturated solution was found in [15, 24], where it was shown that the phases α_i of the complex fields Φ_i should acquire nontrivial dependence from time and angular coordinate

$$\alpha_1 = 2\chi(t, \phi), \quad \alpha_2 = \alpha_3 = 0, \tag{21}$$

and the Higgs field becomes oscillating, showing that just in the KN bag model the transformation to Bogomolny form (19) begins to operate at full power.

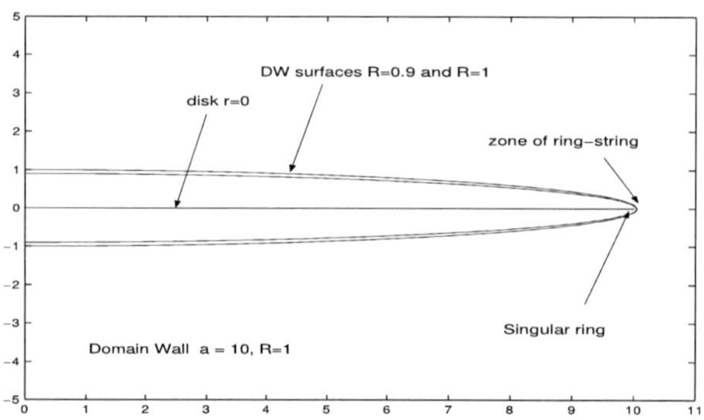

Figure 4: The domain wall profile (axial section) defined by the oblate spheroidal coordinate $r = R$.

3.2. Minimal Landau-Ginzburg model and quantization of the angular momentum

The non-trivial dependence (21) is fixed in zone **(I)**, where the generalized Landau-Ginzburg model is reduced to minimal LG model, and the Lagrangian (9) leads to equations

$$\Box A_\mu = J_\mu = e|H|^2(\chi,_\mu + eA_\mu). \tag{22}$$

One sees that vector potential A_μ acquires from the Higgs field the mass term $m_v = e|H|$, and the EM field becomes short-range, with the characteristic parameter $\lambda = 1/(e|H|)$ corresponding to the penetration depth of the EM field in superconductivity. As a consequence, the currents vanish inside the core, $J_\mu = 0$, leading to the equations

$$\Box A_\mu = 0, \quad \chi_{,\mu} + eA_\mu = 0, \tag{23}$$

showing that besides of the massive component $A_\mu^{m_v}$ which falls off receiving the mass m_v from the Higgs field, there are also the components of different behavior.

Vector-potential of the external KN solution (3) is

$$A_\mu dx^\mu = -\frac{er}{r^2 + a^2\cos^2\theta}(dr - dt - a\sin^2\theta d\phi). \tag{24}$$

It grows near the core and takes maximal value at the boundary of the disk, at $r = R = e^2/2m$, $\cos\theta = 0$,

$$A_\mu^{max} dx^\mu = -\frac{2m}{e}(dr - dt - ad\phi). \tag{25}$$

Note, that the component A_r is a perfect differential (as it is shown for example in [19]) and can be ignored. At the boundary, A_μ^{max} is dragged by the light-like direction of the Kerr singular ring (see Fig.3) and the component A_ϕ^{max} forms the closed Wilson loop, so that

$$e\oint A_\phi^{max} d\phi = 4\pi ma. \tag{26}$$

The right equation in (23) shows that penetrating inside the disk vector potential determines oscillating phase of the Higgs field as $\chi = 2mt + 2am\phi$. The condition of multiplicity of the periods χ and ϕ gives $2am = n,\ n = 1, 2, 3, ..$, which in view of $J = ma$, leads to quantization of angular momentum as

$$J = n/2, \ n = 1, 2, 3, ... \tag{27}$$

On the other hand (23) shows that phase of Higgs field

$$H = \Phi_1 = |H| e^{i(2mt + 2am\phi)} \tag{28}$$

oscillates with the frequency $\omega = 2m$ which supports extension of the components $A_t^{in} = \frac{2m}{e}$, $A_\phi^{in} = \frac{2ma}{e}$ inside the disk.[3] At the disk boundary (23) is broken, and according (22) there appear the surface currents J_μ.

[3]Note, that the left massless equation (23) is also satisfied, since $\Box A_t^{in} = 0$ is satisfied trivially. Also, $\Box A_\phi^{in} = 0$, because phase ϕ is analytic function of $(x + iy)$, leading to $\Box A_\phi^{in} = \partial\bar{\partial} A_\phi^{in} = 0$. These fields do not produce the field strength.

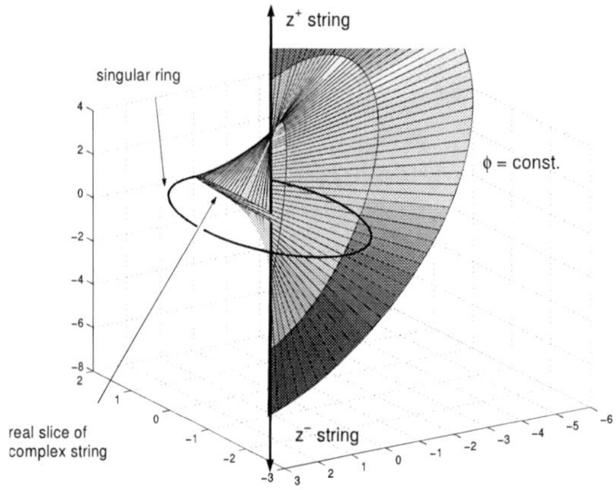

Figure 5: Kerr's coordinate $\phi = const.$ Kerr singular ring drags the vector potential, forming a closed Wilson loop along edge border of the DW.

4. SuperBag as nonperturbative solution of the SuperQED model

4.1. Bosonic sector of the supersymmetric LG model

As we noticed earlier, the generalized Lndau-Ginzburg model based on the superpotential (15) is not true supersymmetric model. The difference is that the true superpotential W is to be a chiral function of the chiral superfields Φ_i, while the scalar potential

$$V = F_i F_i^* \tag{29}$$

is formed from the chiral part

$$F_i^* = \partial W / \partial \Phi_i, \tag{30}$$

but also incudes the antichiral superpotential $W^+(\Phi_i^+)$ depending on the antichiral superfields Φ_i^+

$$F_i = \partial W^+ / \partial \Phi_i^+. \tag{31}$$

These relations are retained in the bosonic sector of the supersymmetric theory, where the fields Φ_i and Φ_i^+ turn into the complex conjugate scalar components of the superfields.

To get full correspondence with supersymmetric theory, the fields Φ_i and $\bar\Phi_i$ in (15), should be considered as independent chiral fields Φ_i and $\tilde\Phi_i$, and there should also be introduced an antichiral superpotential $W^+(\Phi_i^+, \tilde\Phi_i^+)$, which in the bosonic sector turns into complex conjugated superpotential, built of the complex conjugated fields $\bar W(\Phi_i^*, \tilde\Phi_i^*)$. From the complex point of view, the transition from (15) to supersymmeric Higgs model may be considered as *complexification* of the moduli space – analytical extension from the real section, fixed by condition $\bar{\Phi}_i = \Phi_i^*$, to its complex extension, the manifold with independent coordinates Φ_i and $\tilde{\Phi}_i$, supplemented with complex conjugate coordinates Φ_i^*, $\tilde{\Phi}_i^*$. Therefore, the transition to bosonic sector of the supersymmetric generalized LG model requires doubling of the chiral field to eliminate their degeneracy on the real slice.

Returning to the original work by Morris [48], where the potential (15) was suggested for super-generalization of the Witten's superconducting string model [49], we should double the charged chiral fields Σ and Φ, and consider five chiral superfields $\Sigma_\pm$, $\Phi_\pm$, and Z, which in Witten's interpretation of this model as the $U(I) \times U'(I)$ Higgs field model, acquire the charges $(\pm 1, 0)$ for $\Phi_\pm$, and charges for the $\Sigma_\pm$ fields as $(0, \pm 1)$. The chiral superpotential (15) takes the form

$$W(\Phi_i, \tilde{\Phi}_i) = Z(\Sigma_+\Sigma_- - \eta^2) + (Z + \mu)\Phi_+\Phi_-, \tag{32}$$

with identification

$$\Phi_i = (\Phi_+, \Phi_-, \Sigma_+, \Sigma_-, Z). \tag{33}$$

The auxiliary fields

$$F_i^* = \partial W / \partial \Phi_i = (F_+^*, F_-^*, F_{\Sigma+}^*, F_{\Sigma-}^*, F_Z^*) \tag{34}$$

take the form

$$\begin{aligned} F_\pm^* &= (Z + \mu)\Phi_\mp, & (35)\\ F_{\Sigma\pm}^* &= Z\Sigma_\mp, & (36)\\ F_Z^* &= \Sigma_+\Sigma_- + \Phi_+\Phi_- - \eta^2, & (37) \end{aligned}$$

Vacuum expectation values of fields Φ_i for which $F_i^* = 0$ give minima of the potential $V = 0$ corresponding to supersymmetric vacuum states. Just as in case (15), we obtain two isolated vacua

(I) $\Phi_-\Phi_+ = \eta^2, \quad Z = -\mu, \quad \Sigma_+ = \Sigma_- = 0;$
(E) $\Phi_- = \Phi_+ = 0, \; Z = 0, \; \Sigma_+\Sigma_- = \eta^2;$
separated by the zone
(R) of the positive potential

$$\begin{aligned} V = &|\Sigma_+\Sigma_- + \Phi_+\Phi_- - \eta^2|^2 + |(Z + \mu)\Phi_+|^2 \\ &+ |(Z + \mu)\Phi_-|^2 + |Z|^2(|\Sigma_+|^2 + |\Sigma_-|^2). \end{aligned} \tag{38}$$

4.2. Transition to SuperQED model

We note that two oppositely charged superfields Φ_+ and Φ_- give rise to correspondence of the supersymmetric Landau-Ginzburg model to kinetic part of the Wess-Zumino SuperQED model [21],

$$\mathcal{L}_{SQEDkin} = -\frac{1}{4} W^a W_a + \Phi_+^+ e^{eV} \Phi_+ |_{\theta\theta\bar{\theta}\bar{\theta}} + \Phi_-^+ e^{-eV} \Phi_- |_{\theta\theta\bar{\theta}\bar{\theta}}, \tag{39}$$

where V is vector superfield, and $W^a = -\frac{1}{4}\bar{D}\bar{D}D_\alpha V$. In the same time, the potential part (32) corresponds to the most general renormalizable supersymmetric Lagrangian and gives rise to nonperturbative generalization of the SuperQED model.

The chiral superfields $\Phi_\pm$, are expressed in the component form

$$\Phi_\pm(y) = H_\pm(y^\mu) + \sqrt{2}\theta\psi_\pm(y^\mu) + \theta\theta F_\pm(y^\mu), \tag{40}$$

as functions of the chiral coordinates $y^\mu = x^\mu + i\theta\sigma^\mu\bar{\theta}$ and θ, and the scalar components $H_\pm$ are *independent* Higgs fields, splitting of the complex conjugated Higgs field of the minimal Landau-Ginzburg model in (15) and (16). Interplay of the oppositely charged Higgs fields $H_\pm$ with vector potential in zone **(I)** is defined by (23) and yields

$$H_\pm = |H_\pm|e^{\pm i\chi}, \ \bar{H}_\pm = |H_\pm|e^{\mp i\chi}, \ \chi = 2mt + 2am\phi, \tag{41}$$

where the fields $\bar{H}_\pm$ are scalar components of the antichiral fields

$$\Phi^+_\pm(y^+) = \bar{H}_\pm(y^{+\mu}) + \sqrt{2}\bar{\theta}\bar{\psi}_\pm(y^{+\mu}) + \bar{\theta}\bar{\theta}\bar{F}_\pm(y^{+\mu}), \tag{42}$$

as functions of the antichiral coordinates $y^{+\mu} = x^\mu - i\theta\sigma^\mu\bar{\theta}$ and $\bar{\theta}$. The corresponding nonperturbative solution with doubled Higgs fields (41) can be obtained similar to [15].

In the Wess-Zumino SuperQED model, the two Weyl spinors $\psi^\pm$ in (40) combine into one massive Dirac spinor of the electron – superpartner of the Higgs doublet $H_\pm$, [21].

The nonperturbative super-bag solution generates in the core of spinning particle the flat Compton zone **(I)**, which is free from gravity and supersymmetric, representing the conditions for the work of the perturbative SuperQED model, while the remarkable perturbative properties of the WZ SuperQED model – "miracleous cancellations" of the component super-graphs [21] – form a link to perturbative QED. Note, that in the nonperturbative model of super-bag, the superpartners cannot be considered as separate particles, and are integrated as the superfield components of a single nonperturbative solution. The super-bag model reveals correspondence not only with gravity and electroweak sector of the SM, but also with a nonperturbative version of the SuperQED model.

5. Stringy structure

Usually, it is assumed that bags are deformed by rotations taking the shape of a string-like flux-tube joining the quark-antiquark pair [7].

In the KN Super-Bag, the spinning gravitational field controls disk-like shape of the bag, and string-like structure is formed for $a/R > 0$, at edge rim of the disk, as shown in Fig.4. In the equatorial plane, this string approaches very close to the Kerr singular ring, see Fig.4A, so, it is really just the singular ring regularized by the bag boundary.

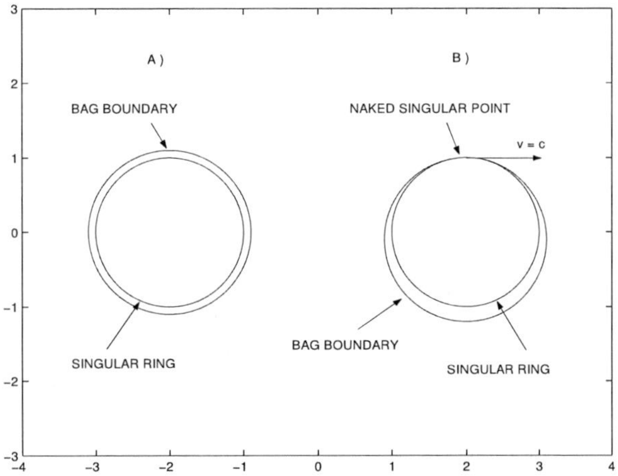

Figure 6: Regularization of the KN string. Boundary of bag fixes cut-off $R = r_e$ for the Kerr singular ring. A) The exact KN solution. B) The KN solution is excited by the lowest traveling mode: emergence of the singular pole.

Among diverse attempts to use nonperturbative models in the electroweak sector of the Standard Model (SM) [38, 39, 40, 41, 42], the central place takes the Nielsen-Olesen (NO) model [43, 44] of the string, which is created as a vortex line in a superconductor.

The assumption, that Kerr singular ring is similar to Nielsen-Olesen model of dual string was done very long ago in [46, 45], where it was noted that excitations of the KN solution create traveling waves along the Kerr ring. Later, it was obtained in [5, 6] close connection of the Kerr singular ring with the Sen fundamental string solution to low energy string theory.[4] In the KN bag model this string is formed at the sharp boundary of the superconducting disk, as a dual analog of the NO vortex line in superconductor.

In accordance with the condition (6), the KN gravity controls position of the bag boundary **(R)**, and also more thin effects, such as excitations of the KN gravity define dynamics of the bag and appearance of the traveling waves.

In particular, it has been shown [15], that the lowest EM excitation of the KN solution creates the traveling wave which has a circulating lightlike node. At this point, surface of the deformed bag touches the Kerr singular ring, as it is shown in Fig.4B, which breaks regularization at this point and creates the lightlike singular pole, which can be considered as emergence of the bare Dirac particle circulating inside the Compton zone of dressed

[4]Note also the complex N=2 critical string which was obtained in the complex Kerr geometry [47].

electron. On the other hand, this pole breaks homogeneity of the closed circular string, creating the frontal and rear ends turning this string in the open. As usual, the end points of an open string are associated with quarks, and the KN super-bag model turns into a single "bag-string-quark" system, 4D analog of D2-D1-D0-brane system of the string–M-theory.

6. Conclusions

We have considered principal features of the Kerr-Schild geometry which specify the supersymmetric bag model as a new way to particle physics consistent with gravity and electroweak sector of the SM. Two of these features are principally new, relative to the widespread belief:

– the spinning KN gravity is not weak, and becomes very strong at the Compton scale of the particle physics,

– compatibility between Quantum and Gravity can be achieved by means of supersymmetric generalization of the matter sector, without modification of the Einstein-Maxwell theory.

We considered interplay of the KN gravity with the matter sector based on the supersymmetric generalized LG field model, which is equivalent to supersymmetric Higgs mechanism of symmetry breaking, and give a nonperturbative solution to generalized Landau-Ginzburg field model in the form of a super-bag – nonperturbative version of the SuperQED model. By conception, the 4d super-bag model has to be soft and oscillating, similar to the conception of the superstring models [8, 25, 26].

Due to extreme high spin/mass ratio, impact of the gravitational KN field on the structure of space-time becomes very strong, and the consistent supersymmetric nonperturbative solutions become very sensible to the external Einstein-Maxwell field. As a result,

a) the super-bag model creates a free from gravity Compton core of spinning particle, where the supersymmetric vacuum state of the Higgs field provides the flat space, required for consistent work of quantum theory;

b) the super-bag takes the shape of a strongly oblate disk forming a circular string along its border;

c) gravitational and electromagnetic excitations of the KN solution create consistent stringy oscillations of the super-bag in the form of traveling waves.

Many problems remain to be solved. The closest is the so far unsolved problem of the exact nonstationary (oscillating or accelerating) generalization of the KN solution, the problem of the consistent solutions of the Dirac equation corresponding to confinement of quark inside the bag, and so on.

Nevertheless, the considered here features of the super-bag model are so intriguing that we risk to state that they really give the key to solution of the problem of unification of gravity with particle physics.

Finally, very important new aspect of this study is the direct link to non-perturbative Wess-Zumino SuperQED model, which provides remarkable cancellations between component diagrams, presenting a link between the nonperturbative bag-like solution and the conventional technics of the perturbative QED.

References

[1] J. Schwarz, *The Early History of String Theory and Supersymmetry*, CALT-68-2858 [arXiv:1201.0981]

[2] N.Arkani-Hamed1, S. Dimopoulos, G. Dvali and N. Kaloper, Infinitely Large New Dimensions, *Phys.Rev.Lett.* **84** 586 (2000) DOI: 10.1103/PhysRevLett.84.586 [hep-th/9907209].

[3] A. Dabholkar, J. P. Gauntlett, J. A. Harvey, D. Waldram, Strings as Solitons & Black Holes as Strings, *Nucl.Phys.* **B474**, 85 (1996) [arXiv:hep-th/9511053].

[4] A. Sen, Macroscopic Charged Heterotic String. *Nucl.Phys.* **B 388** 457 (1992), [arXiv:hep-th/9206016].

[5] Burinskii A., Some properties of the Kerr solution to low-energy string theory. *Phys. Rev. D* **52** 5826 (1995), [arXiv:hep-th/9504139].

[6] A. Sen, Rotating charged black hole soltion in heterotic string theory, *Phys. Rev. Lett.* **69**, 1006, (1992).

[7] Chodos A. et al. New extended model of hadrons. *Phys. Rev.* **D 9**, 3471 (1974).

[8] Bardeen W. A. at al., Heavy quarks and strong binding: A field theory of hadron structure. *Phys. Rev.* **D 11**, 1094 (1974).

[9] R. Dashen, B. Hasslacher, and A. Neveu, Nonperturbative methods and extended-hadron models in field theory: I,II, III. *Phys. Rev.* **D 10**, 4114 (1974), ibid. 4130 (1974), ibid 4138 (1974) .

[10] J.M. Maldacena, *Int.J.Mod.Phys.* **A15S1** , 840 (2000)

[11] A. Burinskii, Weakness of gravity as illusion which hides true path to unification of gravity with particle physics, *International Journal of Modern Physics D,* **26** (12) 1743022 (2017), DOI: 10.1142/S0218271817430222

[12] A. Burinskii, Illusion about weakness of gravity hides new way to unify gravity with particle physics, *Journal of Physics: Conf. Series* **942**, 012006 (2017), doi :10.1088/1742-6596/942/1/012006

[13] A. Burinskii, Spin shifts scale of gravitational interaction, opening new path to particle physics, *Journal of Physics: Conf. Series* **938** 012028 (2017), doi:10.1088/1742-6596/938/1/012028

[14] A. Burinskii, Gravitating lepton bag model, *JETP (Zh. Eksp. Teor. Fiz.)* **148**(8) (2015) 228, [arXiv:1505.03439].

[15] A. Burinskii, Stability of the lepton bag model based on the Kerr-Newman solution, *JETP (Zh. Eksp. Teor. Fiz.)* **148** (2015) 937.

[16] A. Burinskii, Source of the Kerr-Newman solution as a gravitating bag model: 50 years of the problem of the source of the Kerr solution, *Int. J. Mod. Phys.* **A31**, 1641002 (2016).

[17] A. Burinskii, Stringlike structures in the real and complex Kerr-Schild geometry, *Journal of Physics: Conference Series* **532** 012004 (2014), [arXiv:1410.2462].

[18] Carter B., Global structure of the Kerr family of gravitational fields, *Phys. Rev.* **174**, 1559 (1968).

[19] Debney G. C., Kerr R. P. and Schild A. Solutions of the Einstein and Einstein-Maxwell equations. *J. Math. Phys.* **10** 1842 (1969).

[20] Burinskii A., Emergence of the Dirac Equation in the Solitonic Source of the Kerr Spinning Particle. *Grav. Cosmol.* **21 (1)**, 28 (2015), arXiv:1404.5947, [arXiv:1404.5947].

[21] Wess J., Bagger J., *Supersymmetry and Supergravity*, Princeton Univ. Press, New Jersey, 1983.

[22] Burinskii A., Regularized Kerr-Newman Solution as a Gravitating Soliton. *J. Phys. A: Math. Theor.* **43**, 392001 (2010), [arXiv: 1003.2928].

[23] Burinskii A., Kerr-Newman electron as spinning soliton, *Int J. of Mod. Phys.* **A 29**, 1450133 (2014), [arXiv:1410.2888].

[24] A. Burinskii, Source of the Kerr-Newman solution as a supersymmetric domain-wall bubble: 50 years of the problem, *Phys Lett.* **B 754**, 99 (2016), [arXiv:1602.04215].

[25] Giles R.C., Semiclassical dynamics of the "SLAC bag", *Phys. Rev.* **D 70**, 1670 (1976).

[26] K. Johnson and C. B. Thorn, Stringlike solutions of the bag model, *Phys. Rev.* **D 13**, 1934 (1976).

[27] S.-H. H. Tye, Quark-binding string, *Phys. Rev.* **D 13**, 3416 (1976)

[28] P. Fendley, S. Mathur, C. Vafa and N.P. Warner, *Phys. Lett.* **B 243**, 257 (1990).

[29] Xinrui Hou, A. Losev, M. Shifman, *Phys. Rev.* **D 61**,085005 (2000), hep-th/9910071,

[30] E.R.C. Abraham and P.K. Townsend, *Nucl. Phys.* **B351**, 313 (1991).

[31] M. Cvetič, F. Quevedo and S. J. Rey, *Phys. Rev. Lett.* **67**, 1836, (1991).

[32] M. Cvetič, S. Griffies and S. J. Rey, *Nucl. Phys.* **B 381**, 301 (1992).

[33] G.W. Gibbons and P.K. Townsend, *Phys.Rev.Lett.* **83** 1727 (1999), [hep-th/9905196].

[34] B. Chibisov and M. Shifman, BPS-saturated walls in supersymmetric theories, *Phys. Rev.* **D 56**, 7990 (1997).

[35] López C.A. , An Extended Model Of The Electron In General Relativity, *Phys. Rev.* **D 30**, 313 (1984).

[36] M. Gürses and F. Gürsey, *J. Math. Phys.* **16**, 2385 (1975).

[37] P. A. M. Dirac, An Extensible Model of the Electron. *Proc. R. Soc. Lond. A* **268**, 57 (1962).

[38] Coleman S., Q-Balls, *Nuclear Physics* **B 262** (2) 263 (1985),
Rosen G., Particlelike Solutions to Nonlinear Complex Scalar Field Theories with Positive-Definite Energy Densities. *J. of Math. Phys.* **9** (7) 996 (1968), doi:10.1063/1.1664693 .

[39] A. Achucarro and T. Vachaspati, Semilocal and Electroweak Strings,*Phys.Rept.* **327** 347 (2000), [arXiv:hep-ph/9904229].

[40] Kusenko A., Solitons in the supersymmetric extensions of the standard model, *Phys.Lett.* **B405** 108 (1997).

[41] G. Dvali, A. Kusenko and M. Shaposhnikov, "New physics in a nutshell, or Q-ball as a power plant", *Phys.Lett.* **B417**, 99 (1998).

[42] Volkov M. and Wöhnert E., Spinning Q-balls, *Phys.Rev.* D **66**, 085003 (2002).

[43] H. B. Nielsen and P. Olesen, Vortex-line models for dual strings, *Nucl. Phys.* **B 61**, 45, (1973).

[44] J. Ambjorn and P. Olesen, On electroweak magnetism, *Nuclear Physics* **B315** 606 (1989).

[45] Burinskii A.Ya., Microgeons with spin. *Sov. Phys. JETP* **39** 193 (1974).

[46] Ivanenko D.D. and Burinskii A.Ya., Gravitational strings in the models of elementary particles, *Izv. Vuz. Fiz.*, **5**, 135 (1974).

[47] A. Burinskii, Twistorial analyticity and three stringy systems of the Kerr spinning particle, *Phys. Rev.* **D 70**, 086006 (2004) [arXiv:hep-th/0406063].
A. Burinskii, Orientifold D-String in the Source of the Kerr Spinning Particle, *Phys. Rev.* **D 68** (2003) 105004 [arXiv:hep-th/0308096].
A. Burinskii, Kerr spinning particle, strings, and superparticle models, *Phys. Rev.* **D 57,** 2392 (1998).

[48] J. R. Morris, *Phys. Rev.* **D 53** 2078 (1996) [arXiv:hep-ph/9511293].

[49] E. Witten, Superconducting strings. *Nucl.Phys.,* **B249**, 557, (1985).

Ricci-flat metrics on non-compact Calabi-Yau threefolds

Dmitri Bykov*
Arnold Sommerfeld Center for Theoretical Physics,
Department für Physik, Ludwig-Maximilians-Universität München,
Theresienstraße 37, 80333 München, Germany
Max-Planck-Institut für Gravitationsphysik, Albert-Einstein-Institut,
Am Mühlenberg 1, D-14476 Potsdam-Golm, Germany
Steklov Mathematical Institute of Russ. Acad. Sci.,
Gubkina str. 8, 119991 Moscow, Russia

ABSTRACT

We review the problem of constructing explicit Ricci-flat metrics on non-compact toric Calabi-Yau manifolds. The case when the manifold is the total space of the canonical bundle over the del Pezzo surface of rank one is analyzed in detail. We show that there exists a first-order deformation of the so called 'orthotoric metric', in line with the Calabi-Yau theorem. A deformation of the corresponding conformal Killing-Yano form does not exist.

1. Introduction

We will be considering Calabi-Yau threefolds $\mathcal{M}$:

- Complex manifolds of complex dimension three: $\dim_{\mathbb{C}} \mathcal{M} = 3$
- Zero first Chern class: $c_1(\mathcal{M}) = c_1(K) = 0$
  (K is the canonical bundle = bundle of 3-forms $\Omega \propto f(z)\, dz_1 \wedge dz_2 \wedge dz_3$), i.e. there exists a non-vanishing holomorphic 3-form Ω
- Such manifolds are used for supersymmetric compactifications in supergravity ($\mathbb{R}^{3,1} \times \mathcal{M}$), and serve as backgrounds for brane constructions ($AdS_5 \times Y^5$)

It is easy to show that compact Calabi-Yau's do not admit Killing vectors (apart from trivial cases), therefore explicit metrics are difficult to construct. For this reason we will be considering non-compact Calabi-Yau's, which **do** have symmetries. In this case the geometry of such manifolds may often be studied explicitly. These non-compact Calabi-Yau's may be thought of as describing singularities of compact Calabi-Yau's.

* e-mail address: dmitri.bykov@physik.uni-muenchen.de, bykov@mi.ras.ru

We will now describe a particular class of non-compact Calabi-Yau manifolds that we will study. Let X be a positively curved complex surface, $c_1(X) > 0$. Here one should recall that $c_1(X) = \left[\frac{i}{2\pi} R_{m\bar{n}}\, dz^m \wedge d\bar{z}^{\bar{n}}\right] \in H^2(X, \mathbb{R})$. We will be studying the case when

$$\mathcal{M} = \text{Total space of the canonical bundle of } X = \text{“Cone over } X\text{”}.$$

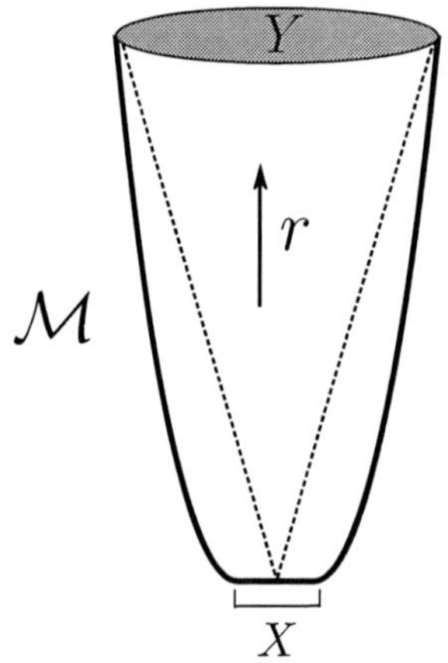

Figure 1: The corresponding singularity is pointlike and may be then resolved by gluing in a copy of X.

This is just like the prototypical $\mathbb{C}^2/\mathbb{Z}_2$-singularity (“$A_1$-singularity”, given by the equation $xy = z^2$) may be resolved by gluing in a copy of $\mathbb{CP}^1$ at the origin. The metric on the resolved space is then the Eguchi-Hanson metric. (However, this corresponds to $\mathcal{M}$ of complex dimension 2.)

1.1. First example. Calabi's ansatz.

If X admits a Kähler-Einstein metric, the metric on $\mathcal{M}$ may be found by means of an ansatz [1]

$$\mathcal{K} = \mathcal{K}(|u|^2\, e^K),$$

where $\mathcal{K}$ and K are the Kähler potentials of $\mathcal{M}$ and X respectively. The Ricci-flatness equation becomes in this case an ODE for the function $\mathcal{K}(x)$. For example, for $X = \mathbb{CP}^2$ one obtains in this way the (generalized) Eguchi-Hanson metric [2]. These metrics are asymptotically-conical, i.e. they have the form

$$ds^2 = dr^2 + r^2\, (\widetilde{ds^2})_Y \qquad \text{at} \qquad r \to \infty,$$

where $(\widetilde{ds^2})_Y$ is a Sasaki-Einstein metric on a 5D real manifold Y.

An important characteristic of a Kähler metric on $\mathcal{M}$ is the cohomology class $[\omega] \in H^2(\mathcal{M}, \mathbb{R})$ of the Kähler form. Since $\mathcal{M}$ is a total space of a line bundle, its cohomology is the same as that of the underlying surface X.

Therefore, for instance for $X = \mathbb{CP}^2$ we have $H^2(\mathcal{M}, \mathbb{R}) \simeq \mathbb{R}$, but for $X = \mathbb{CP}^1 \times \mathbb{CP}^1$ we have $H^2(\mathcal{M}, \mathbb{R}) \simeq \mathbb{R}^2$.

Calabi's ansatz gives a metric with a very particular and fixed $[\omega] \in H^2(\mathcal{M}, \mathbb{R})$. It turns out that in this case $[\omega] \in H^2_c(\mathcal{M}, \mathbb{R}) \subset H^2(\mathcal{M}, \mathbb{R})$, where H^2_c is the compactly supported cohomology. By Poincaré duality, the group $H^2_c(\mathcal{M}, \mathbb{R}) \simeq H_4(\mathcal{M}, \mathbb{R}) \simeq H_4(X, \mathbb{R}) \simeq \mathbb{R}$ is one-dimensional.

1.2. The Calabi-Yau theorem.

The Calabi-Yau theorem [3], [4] states, however, that, at least for compact $\mathcal{M}$, there is a unique Ricci-flat metric in **every** Kähler class $[\omega] \in H^2(\mathcal{M}, \mathbb{R})$.

For the case of interest $\mathcal{M}$ is not compact, but rather asymptotically-conical, and in this case there exists a proposal for a CY theorem due to [5], [6]. Moreover, one has the decay estimates

$$|g - g_0|_{g_0} = O\left(\frac{1}{r^6}\right) \qquad \text{for} \quad [\omega] \in H^2_c(\mathcal{M}, \mathbb{R})$$

$$|g - g_0|_{g_0} = O\left(\frac{1}{r^2}\right) \qquad \text{for} \quad [\omega] \in H^2(\mathcal{M}, \mathbb{R}) \setminus H^2_c(\mathcal{M}, \mathbb{R}),$$

where g_0 is the conical metric. Such estimates were introduced for the case of ALE-manifolds in [7].

1.3. Example. $X = \mathbb{CP}^1 \times \mathbb{CP}^1$.

The theory just described can be tested explicitly at the example of $X = \mathbb{CP}^1 \times \mathbb{CP}^1$. The ansatz for the Kähler potential on the cone over X is a generalized ansatz of Calabi constructed by [8], [9]:

$$\mathcal{K} = a \log(1 + |w^2|) + \mathcal{K}_0\left(|u^2|(1 + |w^2|)(1 + |x^2|)\right) .$$

The resulting metric, indeed, has two parameters that define the cohomology class of the Kähler form $[\omega] \in H^2(\mathcal{M}, \mathbb{R}) = \mathbb{R}^2$. These correspond to the sizes of the two spheres. The relevant Sasakian manifold Y at $r \to \infty$ is the conifold $T^{11} = \frac{SU(2)\times SU(2)}{U(1)}$, and the decay at infinity agrees with the predicted one.

2. The del Pezzo surface

We will be interested in the next-to-simplest example when X is the del Pezzo surface of rank one (also known as the Hirzebruch surface of rank one), i.e. the blow-up of $\mathbb{CP}^2$ at one point.

A blow-up means that we replace one point in $\mathbb{CP}^2$ by a sphere $\mathbb{CP}^1$. This $\mathbb{CP}^1$ 'remembers the direction', at which we approach the point. A 'good' metric on the new manifold should have two parameters, which describe the original size of the $\mathbb{CP}^2$ and the size of the glued in sphere $\mathbb{CP}^1$.

The del Pezzo surface is a toric manifold, and the best way to think of it is via its moment polygon.

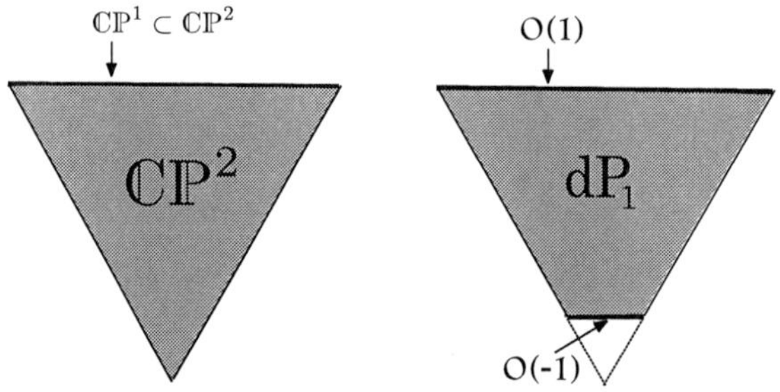

2.1. Metrics on the cone and toric geometry

A theorem of [10] says that there does **not** exist a Kähler-Einstein metric on $\mathbf{dP}_1$. How do we then construct a metric on the cone $\mathcal{M}$ over $\mathbf{dP}_1$? The only hope is to use its symmetries, which are those symmetries of $\mathbb{CP}^2$ that remain after the blow-up.

The relevant isometry group is $U(1) \times U(2)$, however for the moment let us focus on the toric $U(1)^3$ subgroup. Generally, the Kähler potential has the form

$$\mathcal{K} = \mathcal{K}\left(\underbrace{|z_1|^2}_{=e^{t_1}}, \underbrace{|z_2|^2}_{=e^{t_2}}, \underbrace{|z_3|^2}_{=e^{t_3}}\right) .$$

It is customary to introduce the symplectic potential $\mathcal{G}$ – the Legendre transform of the Kähler potential w.r.t. t_i:

$$\mathcal{G}(\mu_1, \mu_2, \mu_3) = \sum_{j=1}^{3} \mu_i \, t_i - \mathcal{K}$$

Here $\mu_i = \frac{\partial \mathcal{K}}{\partial t_i}$ are the moment maps for the $U(1)^3$ symmetries of the problem. The metric on $\mathcal{M}$ has the form

$$ds^2 = \frac{1}{4}\, \mathcal{G}_{ij} d\mu^i d\mu^j + (\mathcal{G}^{-1})^{ij} d\phi_i d\phi_j \, .$$

The Riemann tensor with all lower indices looks as follows:

$$R_{\bar{m}jk\bar{n}} = -\sum_{s,t} \mathcal{G}^{-1}_{ns} \, \frac{\partial^2 \mathcal{G}^{-1}_{jk}}{\partial \mu_s \partial \mu_t} \, \mathcal{G}^{-1}_{tm} \, .$$

The domain, on which $\mathcal{G}$ is defined, is the moment polytope. The potential $\mathcal{G}$ has singularities at the boundaries of the polytope. For instance, for flat space $\mathbb{C}^3$ the polytope is the octant, and $\mathcal{G}$ has the form

$$\mathcal{G}_{\text{flat}} = \sum_{k=1}^{3} \mu_k (\log \mu_k - 1) .$$

In general, at a boundary $\ell = 0$ the potential behaves as $\mathcal{G} = \ell (\log \ell - 1) + \ldots$

Quite generally, Kähler metrics on toric manifolds were constructed by [11]. They are built using Kähler quotients, and the corresponding symplectic potential exhibits the singularities just described.

In our problem we have more symmetry: $U(1) \times U(2)$ instead of $U(1)^3$. The Kähler potential is therefore

$$\mathcal{K} = \mathcal{K}\left(\underbrace{|w|^2}_{=e^t}, \underbrace{|z_1|^2 + |z_2|^2}_{=e^s} \right) ,$$

which means that the metric is of cohomogeneity-2. This implies the following form of $\mathcal{G}$:

$$\mathcal{G} = \left(\frac{\mu}{2} + \tau\right) \log\left(\frac{\mu}{2} + \tau\right) + \left(\frac{\mu}{2} - \tau\right) \log\left(\frac{\mu}{2} - \tau\right) - \mu \log \mu + G(\mu, \nu)$$

$$\mu = \mu_1 + \mu_2, \qquad \tau = \frac{\mu_1 - \mu_2}{2}, \qquad \nu = \mu_3 .$$

The Ricci-flatness equation is then a Monge-Ampère equation in two variables:

$$e^{G_\mu + G_\nu} \left(G_{\mu\mu} G_{\nu\nu} - G_{\mu\nu}^2 \right) = \mu$$

The domain of definition is the moment polytope of the cone $\mathcal{M}$:

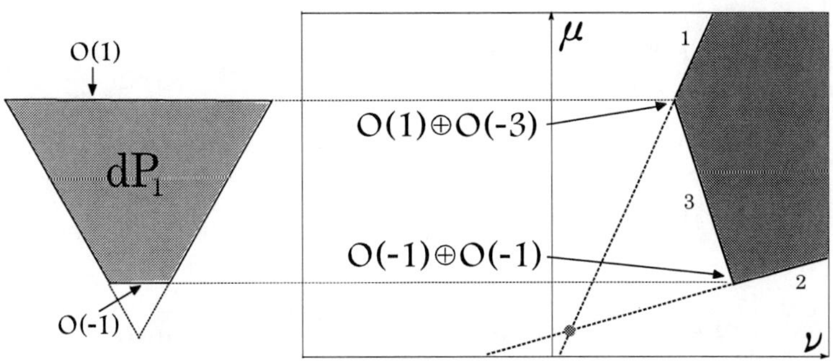

2.2. The asymptotic behavior of G

One can construct an **exact** solution of the above equation taking the conical ansatz for the metric $ds^2 = dr^2 + r^2\,\widetilde{ds^2}$. We make a change of variables $(\mu, \nu) \to (\nu, \xi = \frac{\mu}{\nu})$ and look for G in the form $(\nu \propto r^2)$

$$G = 3\,\nu\,(\log\nu - 1) + \nu\,P(\xi)$$

One obtains an ODE for $P(\xi)$ that can be solved exactly. As a result,

$$G = \sum_{i=0}^{2} \frac{\mu - \xi_i\,\nu}{1 - \xi_i}\,(\log(\mu - \xi_i\,\nu) - 1)\,,$$

where ξ_i are the roots of the cubic polynomial $Q(\xi) = \xi^3 - \frac{3}{2}\,\xi^2 + d$. Varying d, one arrives at the Sasakian manifolds called $Y^{p,q}$ that were discovered in [12]. The topology of the underlying del Pezzo surface forces us to pick the manifold $Y^{2,1}$.

2.3. Uniqueness

The conical metric constructed above is singular at $r = 0$. Constructing a smooth – resolved – metric is rather difficult. For the moment let us assume that, for a fixed moment polytope, we constructed one such metric with potential G_0. To check uniqueness, one can expand $G = G_0 + H$ to first order in H:

$$\triangle_{G_0} H = 0 \quad\Rightarrow\quad 0 = \int d\mu\,d\nu\;H\,\triangle_{G_0} H \stackrel{?}{=} -\int d\mu\,d\nu\;(\nabla\,H)^2$$

Whether we may integrate by parts depends on the behavior at infinity, where we have asymptotically

$$\triangle_{G_0} H = 0 \quad\to\quad -\frac{\partial}{\partial\xi}\left(Q(\xi)\,\frac{\partial H}{\partial\xi}\right) + \frac{\xi}{\nu}\,\frac{\partial}{\partial\nu}\left(\nu^3\,\frac{\partial H}{\partial\nu}\right) = 0$$

Substituting $H = \nu^m\,h(\xi)$, we get a Heun equation

$$-\frac{d}{d\xi}\left(Q(\xi)\,\frac{dh}{d\xi}\right) + m(m+2)\,\xi\,h(\xi) = 0$$

Therefore one needs to estimate the spectrum of the Laplacian on $Y^{2,1}$. We have the following result:
Proposition. [13]
For the smallest non-zero eigenvalue λ of the Laplacian $\triangle_\xi = -\frac{d}{d\xi}\left(Q(\xi)\,\frac{dh}{d\xi}\right)$, entering the equation $\triangle_\xi f + \lambda\,\xi\,f = 0$, one has the lower bound $\lambda \geq 3$.

As a result, we obtain uniqueness of the metric for a given moment polytope. Therefore all potential moduli of the metric have to be related to the moduli of the polytope, which in turn are the Kähler moduli.

3. Killing-Yano forms.

One approach to the explicit construction of a metric is to require that it admit a conformal Killing-Yano form (CKYF). First we recall some basic examples of conditions that lead to the reduction of the holonomy group:

$$\begin{array}{lcc} \nabla_i \xi_j = 0 & \Rightarrow & \text{Reduced holonomy} \\ \nabla_i \xi_j - \nabla_j \xi_i = 0 & \Rightarrow & \xi = d\chi \\ \nabla_i \xi_j + \nabla_j \xi_i = 0 & \Rightarrow & \text{Killing vector} \end{array}$$

The Killing-Yano form $\omega_{ij}\, dx^i \wedge dx^j$ is defined by the equation

$$\nabla_i \omega_{jk} + \nabla_j \omega_{ik} = 0$$

The conformal Killing-Yano form is defined analogously:

$$\nabla_i \omega_{jk} + \nabla_j \omega_{ik} - \text{trace parts} = 0$$

On a Kähler manifold we may expand $\omega = \omega^{(2,0)} \oplus \omega^{(1,1)} \oplus \omega^{(0,2)}$. The situation when ω is Hermitian, i.e. $\omega^{(2,0)} = 0$, is especially simple. Introducing the 'shifted' form $\Omega_{a\bar{b}} = \omega_{a\bar{b}} - h\, g_{a\bar{b}}$ ($h = g^{a\bar{b}} \omega_{a\bar{b}}$), one gets the equation [14]

$$\nabla_a \Omega_{b\bar{c}} = -2 g_{a\bar{c}}\, \partial_b h$$

The tensor Ω has various names, such as Hamiltonian two-form, twistor form, etc. One can show that its eigenvalue functions x_i have orthogonal gradients. They can be related to the 'moment map' variables μ_i corresponding to holomorphic isometries via the interesting formula:

$$\prod_{k=1}^{n} (\vartheta - x_k) = \sum_{k=0}^{n} \vartheta^k\, \mu_{k+1}.$$

3.1. The orthotoric metric.

At the end of the day the metric admitting a tensor Ω has the form (we set $x_1 = x, x_2 = y$, then $\mu = xy$, $\nu = x + y$)

$$ds^2 = x\, y\, g_{\mathbb{CP}^1} + (x - y) \left(\frac{dx^2}{P_1(x)} + \frac{dy^2}{P_2(y)} \right) + \text{angular part}$$

We call this metric the 'orthotoric metric'. We see that the variables separated. The requirement of Ricci-flatness fixes the functions P_1, P_2 to be cubic polynomials (one of which we encountered before):

$$P_1(x) = x^3 - \frac{3}{2}x^2 + c \qquad P_2(y) = y^3 - \frac{3}{2}y^2 + d\,.$$

The domain is $x \leq x_{min}$, $y \in [y_1, y_2]$. If we further require that the topology is that of the cone over $\mathbf{dP}_1$, the constants c and d are uniquely fixed. This metric was also obtained by [15], [16] and was extensively studied in [17].

The point is that the requirements of
(a) Ricci-flatness
(b) Cone over $\mathbf{dP}_1$ topology
(c) CKYF of type $(1, 1)$
completely fix the metric.

According to the Calabi-Yau theorem, however, the metric should contain additional parameters, corresponding to the deformation of the moment polytope. Altogether there are 2 parameters, since $H^2(\mathcal{M}, \mathbb{R}) = \mathbb{R}^2$.

3.2. Deformation of the metric and the CKYF.

One parameter is somewhat 'trivial', as it corresponds to a rescaling of the metric. We can still look for the other non-trivial parameter, which corresponds to the following deformation:

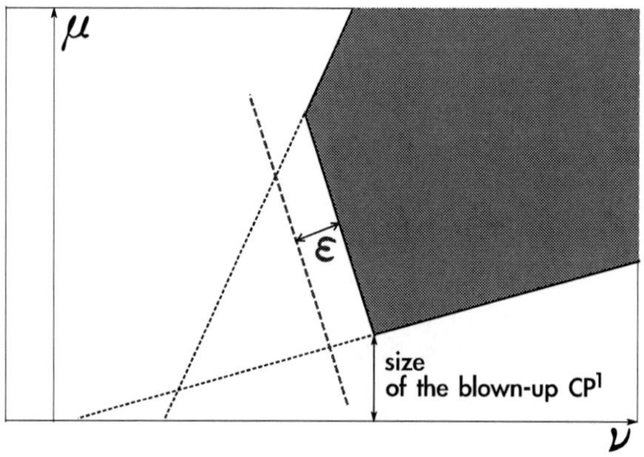

In the equation $\triangle_{G_0} H = 0$, if we substitute the orthotoric potential G_0, the variables separate:

$$\frac{1}{x}\frac{\partial}{\partial x}\left(P_1(x)\,\frac{\partial H}{\partial x}\right) - \frac{1}{y}\frac{\partial}{\partial y}\left(P_2(y)\,\frac{\partial H}{\partial y}\right) = 0$$

The unique solution compatible with the deformation of the moment polytope is

$$H(x, y) = \epsilon \int\limits_{x}^{\infty} \frac{d\hat{x}}{P_1(\hat{x})}\,.$$

For large x one has $H(x,y) = \frac{\epsilon}{2x^2} + \ldots$, and for the metric this implies the asymptotic behavior $|g - g_0|_{g_0} = O\left(\frac{1}{r^6}\right)$. This implies that the variation of the Kähler form has the property $[\delta\omega] \in H^2_c(\mathcal{M}, \mathbb{R})$.

The next question is: what happens to the Killing-Yano form? If it is deformed, it must acquire a non-zero $(2,0)$ part, i.e. $\omega^{2,0} = \omega_{mn}\, dz^m \wedge dz^n \neq 0$. On a Calabi-Yau manifold one has a nowhere vanishing three-form $\Omega_{mnp}\, dz^m \wedge dz^n \wedge dz^p$, and one can construct the 'inverse' 3-vector $\widetilde{\Omega}^{mnp}\, \partial_m \wedge \partial_n \wedge \partial_p$. We can then dualize $\omega^{2,0}$ to obtain a vector field $\omega^p := \widetilde{\Omega}^{mnp}\, \omega_{mn}$.

Using the fact that $\mathcal{M}$ is Ricci-flat, we can show that ω^p has to satisfy a rather stringent requirement

$$R^n_{\ mp\bar{k}}\, \omega^p = 0\,.$$

As we mentioned earlier, on a toric manifold the curvature tensor is $R_{\bar{m}jk\bar{n}} = -\sum_{s,t} \mathcal{G}^{-1}_{ns} \frac{\partial^2 \mathcal{G}^{-1}_{jk}}{\partial\mu_s \partial\mu_t} \mathcal{G}^{-1}_{tm}$. Using the explicit expression for the orthotoric potential $\mathcal{G}$, we can show that the only solution is $\omega^p = 0$. The final result may be summarized as follows:

Proposition. [13]
There exists a first-order deformation of the orthotoric metric that preserves Ricci-flatness and corresponds to a deformation of the moment polytope. Moreover, the deformation of the Kähler form has the property $[\delta\omega] \in H^2_c(\mathcal{M}, \mathbb{R})$. The deformed metric does not possess a conformal Killing-Yano tensor.

4. Summary.

- Metrics on non-compact Calabi-Yau manifold can be sometimes constructed explicitly
- Examples in $\dim_{\mathbb{C}}\mathcal{M} = 3$: Cones over $\mathbb{CP}^2$, $\mathbb{CP}^1 \times \mathbb{CP}^1$
- More complicated cases with conformal Killing-Yano tensors
- In the case of the cone over $\mathbf{dP}_1$ the corresponding metric is not the most general one, predicted by the Calabi-Yau theorem
- One can explicitly construct a first-order deformation of the metric
- Some of the questions, which are yet to be answered, are:
  What is the significance of the explicitly known (orthotoric) metric?
  Can one obtain a closed expression for the metric in the general case, or in other special cases?

Acknowledgements. I would like to cordially thank B. Dragovich, I. Salom and M. Vojinović for the invitation to the "9th Mathematical Physics Meeting" and for the warm hospitality in Belgrade. I am indebted to A.A.Slavnov and to my parents for support and encouragement.

References

[1] E. Calabi, "Métriques Kähleriennes et fibrés holomorphes," Annales Scientiques de l École Normale Supérieure, 12 (1979) pp. 268–294.

[2] T. Eguchi, A. J. Hanson, "Asymptotically Flat Self-dual Solutions to Euclidean Gravity", *Phys.Lett. B* **74** (1978) p. 249

[3] E. Calabi, "On Kähler manifolds with vanishing canonical class," in: Algebraic Geometry and Topology. A Symposium in Honor of S. Lefschetz, Princeton University Press (1957) pp. 78–89.

[4] S.-T. Yau, "On Ricci curvature of a compact Kähler manifold and complex Monge-Ampére equation I.", Comm. Pure and App. Math. 31, (1979) pp. 339–411.

[5] C. van Coevering, "Ricci-flat Kähler metrics on crepant resolutions of Kähler cones", *Journal Math. Ann.* **347:3** (2010), pp. 581–611

[6] R. Goto, "Calabi-Yau structures and Einstein-Sasakian structures on crepant resolutions of isolated singularities," J. Math. Soc. Japan 64:3 (2012) pp. 1005–1052

[7] D. Joyce, "Asymptotically locally Euclidean metrics with holonomy $SU(m)$," Ann. Global Anal. Geom. 19 (2001), no. 1, pp. 55–73.

[8] P. Candelas, X. C. de la Ossa, "Comments on Conifolds," Nucl. Phys. B **342** (1990), p. 246.

[9] L. A. Pando Zayas, A. A. Tseytlin, "3-branes on spaces with $R \times S^2 \times S^3$ topology," Phys. Rev. D **63** (2001) 086006

[10] G. Tian, S.-T. Yau, "Kahler-Einstein Metrics on Complex Surfaces with $c_1 > 0$", *Comm. Math. Phys.* **112:1** (1987), pp. 175–203

[11] V. Guillemin, "Kaehler structures on toric varieties," J. Differential Geom. 40 (1994), no. 2, pp. 285–309.

[12] J. P. Gauntlett, D. Martelli, J. Sparks, D. Waldram, "Sasaki-Einstein metrics on $S^2 \times S^3$," Adv. Theor. Math. Phys. **8** (2004) no.4, p. 711

[13] D. Bykov, "Ricci-flat metrics on the cone over $\mathbb{CP}^2\#\overline{\mathbb{CP}^2}$," arXiv:1712.07227 [hep-th].

[14] V. Apostolov, D. M. J. Calderbank, P. Gauduchon, "Hamiltonian 2-forms in Kähler geometry. I. General theory," J. Differential Geom. 73 (2006), no. 3, pp. 359–412.

[15] T. Oota, Y. Yasui, "Explicit toric metric on resolved Calabi-Yau cone," Phys. Lett. B **639** (2006) p. 54

[16] W. Chen, H. Lu, C. N. Pope, "Kerr-de Sitter black holes with NUT charges," Nucl. Phys. B **762** (2007) p. 38

[17] D. Martelli, J. Sparks, "Resolutions of non-regular Ricci-flat Kähler cones," J. Geom. Phys. **59** (2009) p. 1175

Algebraic structures in exceptional geometry

Martin Cederwall*
Dept. of Physics, Chalmers University of Technology,
SE 412 96, Gothenburg, Sweden

Abstract

Exceptional field theory (EFT) gives a geometric underpinning of the U-duality symmetries of M-theory. In this talk I give an overview of the surprisingly rich algebraic structures which naturally appear in the context of EFT. This includes Borcherds superalgebras, Cartan type superalgebras (tensor hierarchy algebras) and L_∞ algebras. This is the written version of a talk based mainly on refs. [52, 61, 55, 76, 62, 78], presented at ISQS25, Prague, June 2017, at QTS-10/LT-12, Varna, June 2017, at SQS 2017, Dubna, Aug. 2017, and at M∩Φ9, Belgrade, Sept. 2017.

Duality symmetries in string theory/M-theory mix gravitational and non-gravitational fields. Manifestation of such symmetries calls for a generalisation of the concept of geometry. It has been proposed that the compactifying space (torus) is enlarged to accommodate momenta (representing momenta and brane windings) in modules of a duality group. This leads to *double geometry* [1, 2, 4, 64, 6, 5, 8, 9, 10, 12, 11, 13, 14, 15, 45, 44, 28, 29, 60, 17, 18, 40, 41, 72, 73] in the context of T-duality, and *exceptional geometry* [7, 19, 20, 21, 22, 24, 25, 26, 27, 30, 31, 32, 33, 70, 34, 35, 56, 57, 49, 69, 71] in the context of U-duality. These classes of models are special cases of *extended geometries*, and can be treated in a unified manner [76]. The duality group is in a certain sense present already in the uncompactified theory. It becomes "geometrised".

In the present talk, I will

- Describe the basics of extended geometry, with focus on the gauge transformations;
- Describe the appearance of Borcherds superalgebras and Cartan-type superalgebras (tensor hierarchy superalgebras);
- Indicate why L_∞ algebras provide a good framework for describing the gauge symmetries.

* e-mail address: martin.cederwall@chalmers.se

- Point out some questions and directions.

The focus will thus be on algebraic aspects, and less on geometric ones. Consider compactification from 11 to $11 - n$ dimensions on T^n. As is well known, fields and charges fall into modules of $E_{n(n)}$.

n	$E_{n(n)}$	R_1
3	$SL(3) \times SL(2)$	$(\mathbf{3}, \mathbf{2})$
4	$SL(5)$	**10**
5	$Spin(5,5)$	**16**
6	$E_{6(6)}$	**27**
7	$E_{7(7)}$	**56**
8	$E_{8(8)}$	**248**
9	$E_{9(9)}$	fund

Table 1: A list of U-duality groups.

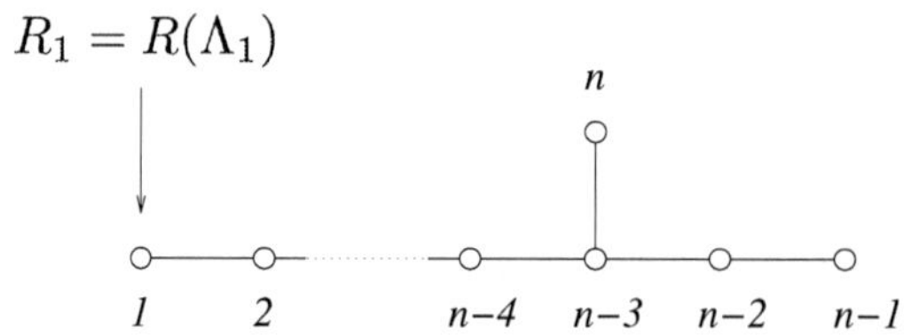

Figure 1: The module R_1.

To be explicit, take $n = 7$ as an example. The gauge parameters ξ^M in **56** of E_7 decompose as:

$$\begin{array}{ccccccccc} \xi^m & & \lambda_{mn} & & \tilde{\lambda}_{mnpqr} & & \tilde{\xi}_{m,n_1\ldots n_7} & \leftarrow & \xi^M \\ 7 & + & 21 & + & 21 & + & 7 & = & 56 \end{array} \tag{1}$$

We recognise the parameters for diffeomorphisms, gauge transformations of the 3-form and dual 6-form and a parameter for "dual diffeomorphisms". The scalar fields are in the coset $E_{7(7)}/K(E_{7(7)}) = E_{7(7)}/(SU(8)/\mathbb{Z}_2)$. The dimension of coset is: $133 - 63 = 70$, and it is parametrised by

$$\begin{array}{ccccccc} g_{mn} & & C_{mnp} & & \tilde{C}_{mnpqrs} & & \leftarrow G_{MN} \\ 28 & + & 35 & + & 7 & = & 70 \end{array} \tag{2}$$

From the point of view of $N = 8$ supergravity in $D = 4$, this is the scalar field coset. Now it becomes a generalised metric. There are also mixed fields (generalised graviphotons): 1-forms in R_1, etc.

The situation for T-duality is simpler. Compactification from 10 to $10-d$ dimensions gives the (continuous) T-duality group $O(d,d)$. The momenta are complemented with string windings to form the $2d$-dimensional module

Note that the continuous duality group is not to be seen as a global symmetry. Discrete duality transformations in $O(d,d;\mathbb{Z})$ or $E_{n(n)}(\mathbb{Z})$ arise as symmetries in certain backgrounds, roughly as the mapping class group $SL(n;\mathbb{Z})$ arises as discrete isometries of a torus. The rôle of the continuous versions of the duality groups is analogous to that of $GL(n)$ in ordinary geometry (gravity).

One has to decide how tensors transform. The generic recipe is to mimic the Lie derivative for ordinary diffeomorphisms:

$$L_U V^m = U^n{}_n V^m - {}_n U^m V^n \,. \tag{3}$$

The first term is a transport term, and the second one a $\mathfrak{gl}$ transformation, with parameter in red.

In the case of U-duality, the role of GL is assumed by $E_{n(n)} \times \mathbb{R}^+$, and

$$\begin{aligned} \mathcal{L}_U V^M &= L_U V^M + Y^{MN}{}_{PQN} U^P V^Q \\ &= U^N{}_N V^M + Z^{MN}{}_{PQN} U^P V^Q \,, \end{aligned} \tag{4}$$

where $Z^{MN}{}_{PQ} = -n P^M_{\text{adj}}{}_{Q,}{}^N{}_P + n {}_M{}_Q {}_N{}_P = Y^{MN}{}_{PQ} - {}_M{}_P {}_N{}_Q$ projects on the adjoint of $E_{n(n)} \times \mathbb{R}$, so that the red factor becomes a parameter for an $\mathfrak{e}_n \oplus \mathbb{R}$ transformation.

The transformations form an "algebra" for $n \leq 7$:

$$[\mathcal{L}_U, \mathcal{L}_V] W^M = \mathcal{L}_{[U,V]} W^M \,, \tag{5}$$

where the "Courant bracket" is $[U,V]^M = \frac{1}{2}(\mathcal{L}_U V^M - \mathcal{L}_V U^M)$, provided that the derivatives fulfil a "*section constraint*".

The section constraint ensures that fields locally depend only on an n-dimensional subspace of the coordinates, on which a $GL(n)$ subgroup acts. It reads $Y^{MN}{}_{PQ M} \cdots {}_N = 0$, or

$$(\otimes)|_{\overline{R}_2} = 0 \,. \tag{6}$$

For $n \geq 8$ more local transformations, so called "ancillary transformations" [76] emerge, which are constrained local transformations in $\mathfrak{g}$.

n	R_1	R_2
3	$(\mathbf{3},\mathbf{2})$	$(\overline{\mathbf{3}},\mathbf{1})$
4	$\mathbf{10}$	$\overline{\mathbf{5}}$
5	$\mathbf{16}$	$\mathbf{10}$
6	$\mathbf{27}$	$\overline{\mathbf{27}}$
7	$\mathbf{56}$	$\mathbf{133}$
8	$\mathbf{248}$	$\mathbf{1}\oplus\mathbf{3875}$

Table 2: A list of R_1 and R_2 for different E_n.

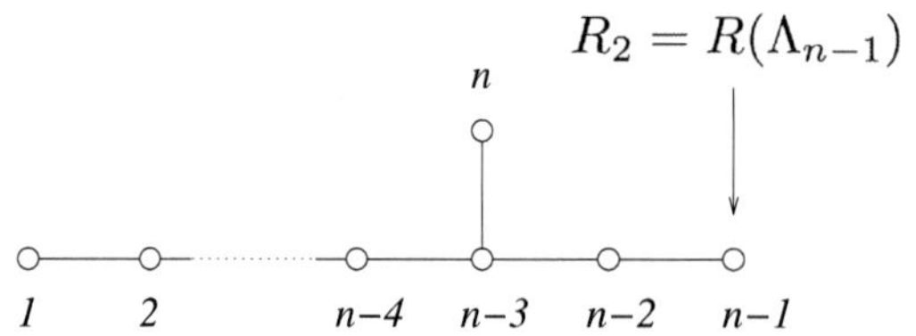

Figure 2: The module R_2.

The interpretation of the section condition is that the momenta locally are chosen so that they may span a linear subspace of cotangent space with maximal dimension, such that any pair of covectors p, p' in the subspace fulfil $(p \otimes p')|_{\overline{R}_2} = 0$.

The corresponding statement for double geometry is $\eta^{MN}{}_M \otimes {}_N = 0$, where η is the $O(d,d)$-invariant metric. The maximal linear subspace is a d-dimensional isotropic subspace, and it is determined by a pure spinor Λ. Once a Λ is chosen, the section condition can be written $\Gamma^M \Lambda_M = 0$. An analogous linear construction can be performed in the exceptional setting. The section condition in double geometry derives from the level matching condition in string theory. Locally, supergravity is recovered. Globally, non-geometric solutions are also obtained.

There is a universal form [52, 55, 76] of the generalised diffeomorphisms for any Kac–Moody algebra and choice of coordinate representation. Let the coordinate representation be $R(\lambda)$, for λ a fundamental weight dual to a simple root α (the construction can be made more general). Then

$$\sigma Y = -\eta_{AB} T^A \otimes T^B + (\lambda, \lambda) + \sigma - 1 \,, \tag{7}$$

where η is the Killing metric and σ the permutation operator, $\sigma(a \otimes b) = (b \otimes a)\sigma$.

This follows from the existence of a solution to the section constraint in the form of a linear space:

- Each momentum must be in the minimal orbit. Equivalently, $p \otimes p \in \overline{R(2\lambda)}$.
- Products of different momenta may contain $\overline{R(2\lambda)}$ and $\overline{R(2\lambda - \alpha)}$, where $R(2\lambda - \alpha)$ is the highest representation in the antisymmetric product. Expressing these conditions in terms of the quadratic Casimir gives the form of Y.

I will skip the detailed description of the generalised gravity. It effectively provides the local dynamics of gravity and 3-form, which are encoded by a vielbein $E_M{}^A$ in the coset $(E_{n(n)} \times \mathbb{R})/K(E_{n(n)})$.

n	$E_{n(n)}$	$K(E_{n(n)})$
3	$SL(3) \times SL(2)$	$SO(3) \times SO(2)$
4	$SL(5)$	$SO(5)$
5	$Spin(5,5)$	$(Spin(5) \times Spin(5))/\mathbb{Z}_2$
6	$E_{6(6)}$	$USp(8)/\mathbb{Z}_2$
7	$E_{7(7)}$	$SU(8)/\mathbb{Z}_2$
8	$E_{8(8)}$	$Spin(16)/\mathbb{Z}_2$
9	$E_{9(9)}$	$K(E_{9(9)})$

Table 3: A list of compact subgroups.

The T-duality case is described by a generalised metric in the coset $O(d,d)/(O(d) \times O(d))$, parametrised by the ordinary metric and B-field.

With some differences from ordinary geometry, one can go through the construction of connection, torsion, metric compatibility etc., and arrive at generalised Einstein's equations encoding the equations of motion for all fields. (This has been done for $n \leq 8$.)

For $n \geq 8$, the coset $E_{n(n)}/K(E_{n(n)})$ contains higher mixed tensors that do not carry independent physical degrees of freedom. They are removed by ancillary transformations that arise in the commutator between generalised diffeomorphisms [70, 56, 57, 55, 76].

One may introduce (local) supersymmetry. In the case of T-duality, the superspace is based on the fundamental representation of an orthosymplectic supergroup $OSp(d,d|2s)$. The exceptional cases are unexplored, but will be based on ∞-dimensional superalgebras [50].

The generalised diffeomorphisms do not satisfy a Jacobi identity. On general grounds, it can be shown that the "Jacobiator"

$$[[U,V],W] + \text{cycl} \neq 0 \,, \tag{8}$$

but is proportional to $([U,V],W) + \text{cycl}$, where $(U,V) = \frac{1}{2}(\mathcal{L}_U V + \mathcal{L}_V U)$.

It is important to show that the Jacobiator in some sense is trivial. It turns out that $\mathcal{L}_{(U,V)}W = 0$ (for $n \leq 7$), and the interpretation is that it is a gauge transformation with a parameter representing reducibility (for $n \leq 6$). (The limits on n in the statements here are due to non-covariance of the derivative arising at some point in the tensor hierarchy, see below. I will not go into details.)

In double geometry, this reducibility is just the scalar reducibility of a gauge transformation: $\delta B_2 = d\lambda_1$, with the reducibility $\delta\lambda_1 = d\lambda'_0$.

In exceptional geometry, the reducibility turns out to be more complicated, leading to an infinite (but well defined) reducibility, containing the modules of tensor hierarchies, and providing a natural generalisation of forms (having connection-free covariant derivatives).

The reducibility continues, and there are ghosts at all levels > 0. The representations are those of a "tensor hierarchy", the sequence of representations R_n of n-form gauge fields in the dimensionally reduces theory.

$$R_1 \xleftarrow{\partial} R_2 \xleftarrow{\partial} R_3 \xleftarrow{\partial} \ldots \tag{9}$$

Example, $n = 5$:

$$\mathbf{16} \xleftarrow{\partial} \mathbf{10} \xleftarrow{\partial} \overline{\mathbf{16}} \xleftarrow{\partial} \mathbf{45} \xleftarrow{\partial} \overline{\mathbf{144}} \xleftarrow{\partial} \ldots \tag{10}$$

$$16 - 10 + 16 - 45 + 144 - \ldots = 11 \tag{11}$$

(suitably regularised), which is the number of degrees of freedom of a pure spinor.

The representations $\{R_n\}_{n=1}^{\infty}$ agree with [51]

- The ghosts for a "pure spinor" constraint (a constraint implying an object lies in the minimal orbit);
- The positive levels of a Borcherds superalgebra $\mathcal{B}(E_n)$.

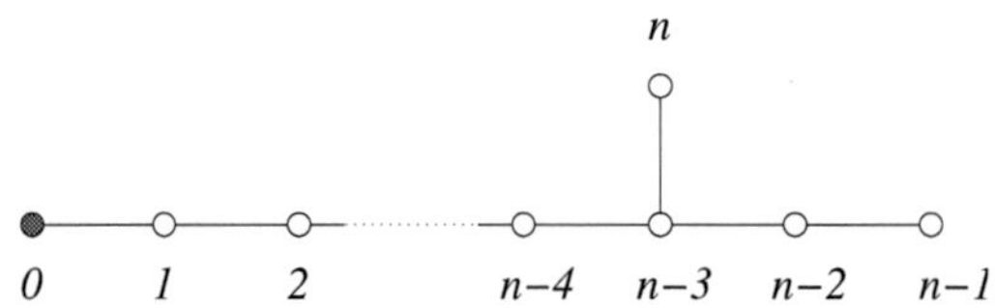

Figure 3: Dynkin diagram for $\mathcal{B}(E_n)$.

Indeed, the denominator appearing in the denominator formula for $\mathcal{B}(E_n)$ is identical to the partition function of a "pure spinor" [51].

$\mathcal{B}(D_n) \approx \mathfrak{osp}(n, n|2)$

$\mathcal{B}(A_n) \approx \mathfrak{sl}(n + 1|1)$

$$\ldots \xleftarrow{\partial} R_{-1} \xleftarrow{\partial} R_0 \xleftarrow{\partial} \underbrace{R_1 \xleftarrow{\partial} R_2 \xleftarrow{\partial} \ldots \xleftarrow{\partial} R_{8-n}}_{\text{covariant}} \xleftarrow{\partial} R_{9-n} \xleftarrow{\partial} R_{10-n} \xleftarrow{\partial} \ldots \tag{12}$$

The modules $R_1, \ldots, R_{8-n}$ behave like forms. The "exterior derivative" is connection-free (for a torsion-free connection), and there is a wedge product [32].

The modules show a symmetry: $R_{9-n} = \overline{R}_n$. There is another extension to negative levels that respects this symmetry, and seems more connected to geometry: tensor hierarchy algebras [61, 62]

In the classification of finite-dimensional superalgebras by Kac, there is a special class, "Cartan-type superalgebras". The Cartan-type superalgebra $W(n)$, which I prefer to call $W(A_{n-1})$, is asymmetric between positive and negative levels, and (therefore) not defined through generators corresponding to simple roots and Serre relations.

$W(A_{n-1})$ is the superalgebra of derivations on the superalgebra of (pointwise) forms in n dimensions.

Any operation $\omega \to \Omega \wedge \imath_V \omega$, where Ω is a form and V a vector, belongs to $W(A_{n-1})$. A basis is given by

level $= 1$	$\imath_a$
0	$e^b \imath_a$
-1	$e^{b_1} e^{b_2} \imath_a$
-2	$e^{b_1} e^{b_2} e^{b_3} \imath_a$
$\ldots$	$\ldots$

The level decomposition of $W(A_{n-1})$.

A subalgebra $S(A_{n-1})$ contains traceless tensors. The positive levels agree with $\mathcal{B}(A_{n-1}) \approx \mathfrak{sl}(n|1)$. Note that the representations of torsion and torsion Bianchi identity appear at levels -1 and -2.

In spite of the absence of a Cartan involution, there is a way to give a systematic Chevalley–Serre presentation of the superalgebra, based on the same Dynkin diagram as the Borcherds superalgebra [62].

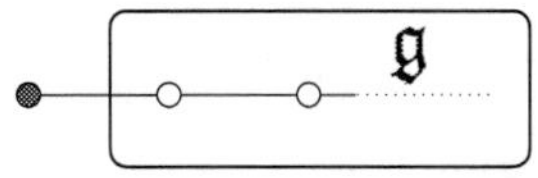

Figure 4: Dynkin diagram for $\mathcal{B}(\mathfrak{g})$ and $W(\mathfrak{g})$.

The construction can be extended to $W(D_n)$, and, most interestingly, $W(E_n)$ (and the corresponding $S(\mathfrak{g})$). The statements about torsion and Bianchi identities remain true (but we still lack a good geometric argument).

Back to the Jacobi identity. Expressed in terms of a fermionic ghost in R_1,

$$[[c, c], c] \neq 0 \,. \tag{13}$$

How is this remedied? The most general formalism for gauge symmetries is the Batalin–Vilkovisky formalism, where everything is encoded in the master equation $(S, S) = 0$.

If transformations are field-independent, one may consider the ghost action consistently. An L_∞ algebra is a (super)algebraic structure which provides a perturbative solution to the master equation.

Let C denote *all* ghosts. Then the master equation states the nilpotency of a transformation

$$\delta C = (S, C) = C + [C, C] + [C, C, C] + [C, C, C, C] + \ldots \qquad (14)$$

The identities that need to be fulfilled are:

$$\begin{array}{l} {}^2C = 0 \ , \\ [C, C] + 2[C, C] = 0 \ , \\ [C, C, C] + 2[[C, C], C] + 3[C, C, C] \ , \\ \ldots \end{array} \qquad (15)$$

Assuming $c = 0$, the non-vanishing of $[[c, c], c]$ can be compensated by the derivative of an element in R_2 (representing reducibility). One needs to introduce a 3-bracket

$$[c, c, c] \in R_2 \ . \qquad (16)$$

Then, there are more identities to check.

For double field theory, a 3-bracket is enough [77].

For exceptional field theory, there are signs, that one will in fact obtain arbitrarily high brackets [78]. There are also other issues concerning the non-covariance outside the "form window". I will not go into detail.

In conclusion, the area has rich connections to various areas of pure mathematics, some of which are under investigation:

- Group theory and representation theory
- Minimal orbits
- Superalgebras
- Cartan-type superalgebras
- Infinite-dimensional (affine, hyperbolic,...) Lie algebras
- Geometry and generalised geometry
- Automorphic forms
- L_∞ algebras
- ...

There are many open questions:

- Can the formalism be continued to $n > 9$? The transformations work for *e.g.* E_{10} [76], and there seems to be no reason (other than mathematical difficulties) that it stops there. Is there a connection to the "E_{10} proposal" [79] with emergent space?

- *Geometry from algebra?* What is the precise geometric relation between the tensor hierarchy algebra and the generalised diffeomorphisms?
- *Superspace/supergeometry?* And some formalism generalising that of pure spinor superfields, that manifests supersymmetry?
- *The section constraint:* Can it be lifted, or dynamically generated?
- What can be learnt about the full string theory/M-theory?
- ... ?

Thank you for your attention.

References

[1] A.A. Tseytlin, *"Duality symmetric closed string theory and interacting chiral scalars"*, Nucl. Phys. **B350** (1991) 395.

[2] W. Siegel, *"Two vierbein formalism for string inspired axionic gravity"*, Phys. Rev. **D47** (1993) 5453 [arXiv:hep-th/9302036].

[3] W. Siegel, *"Superspace duality in low-energy superstrings"*, Phys. Rev. **D48** (1993) 2826 [arXiv:hep-th/9305073].

[4] W. Siegel, *"Manifest duality in low-energy superstrings"*, in Berkeley 1993, Proceedings, Strings '93 353 [arXiv:hep-th/9308133].

[5] C.M. Hull, *"Doubled geometry and T-folds"*, J. High Energy Phys. **0707** (2007) 080 [arXiv:hep-th/0605149].

[6] C.M. Hull, *"A geometry for non-geometric string backgrounds"*, J. High Energy Phys. **0510** (2005) 065 [arXiv:hep-th/0406102].

[7] C.M. Hull, *"Generalised geometry for M-theory"*, J. High Energy Phys. **0707** (2007) 079 [arXiv:hep-th/0701203].

[8] C. Hull and B. Zwiebach, *"Double field theory"*, J. High Energy Phys. **0909** (2009) 99 [arXiv:0904.4664].

[9] O. Hohm, C.M. Hull and B. Zwiebach, *"Background independent action for double field theory"*, J. High Energy Phys. **1007** (2010) 016 [arXiv:1003.5027].

[10] O. Hohm, C.M. Hull and B. Zwiebach, *"Generalized metric formulation of double field theory"*, J. High Energy Phys. **1008** (2010) 008 [arXiv:1006.4823].

[11] O. Hohm and S.K. Kwak, *"$N = 1$ supersymmetric double field theory"*, J. High Energy Phys. **1203** (2012) 080 [arXiv:1111.7293].

[12] O. Hohm and S.K. Kwak, *"Frame-like geometry of double field theory"*, J. Phys. **A44** (2011) 085404 [arXiv:1011.4101].

[13] I. Jeon, K. Lee and J.-H. Park, *"Differential geometry with a projection: Application to double field theory"*, J. High Energy Phys. **1104** (2011) 014 [arXiv:1011.1324].

[14] I. Jeon, K. Lee and J.-H. Park, *"Stringy differential geometry, beyond Riemann"*, Phys. Rev. **D84** (2011) 044022 [arXiv:1105.6294].

[15] I. Jeon, K. Lee and J.-H. Park, *"Supersymmetric double field theory: stringy reformulation of supergravity"*, Phys. Rev. **D85** (2012) 081501 [arXiv:1112.0069].

[16] O. Hohm and B. Zwiebach, *"Large gauge transformations in double field theory"*, J. High Energy Phys. **1302** (2013) 075 [arXiv:1207.4198].

[17] J.-H. Park, *"Comments on double field theory and diffeomorphisms"*, J. High Energy Phys. **1306** (2013) 098 [arXiv:1304.5946].

[18] D.S. Berman, M. Cederwall and M.J. Perry, *"Global aspects of double geometry"*, J. High Energy Phys. **1409** (2014) 66 [arXiv:1401.1311].

[19] P.P. Pacheco and D. Waldram, *"M-theory, exceptional generalised geometry and superpotentials"*, J. High Energy Phys. **0809** (2008) 123 [arXiv:0804.1362].

[20] C. Hillmann, *"Generalized $E_{7(7)}$ coset dynamics and $D = 11$ supergravity"*, J. High Energy Phys. **0903** (2009) 135 [arXiv:0901.1581].

[21] D.S. Berman and M.J. Perry, *"Generalised geometry and M-theory"*, J. High Energy Phys. **1106** (2011) 074 [arXiv:1008.1763].

[22] D.S. Berman, H. Godazgar and M.J. Perry, *"SO(5,5) duality in M-theory and generalized geometry"*, Phys. Lett. **B700** (2011) 65 [arXiv:1103.5733].

[23] D.S. Berman, E.T. Musaev and M.J. Perry, *"Boundary terms in generalized geometry and doubled field theory"*, Phys. Lett. **B706** (2011) 228 [arXiv:1110.97].

[24] D.S. Berman, H. Godazgar, M. Godazgar and M.J. Perry, *"The local symmetries of M-theory and their formulation in generalised geometry"*, J. High Energy Phys. **1201** (2012) 012 [arXiv:1110.3930].

[25] D.S. Berman, H. Godazgar, M.J. Perry and P. West, *"Duality invariant actions and generalised geometry"*, J. High Energy Phys. **1202** (2012) 108 [arXiv:1111.0459].

[26] A. Coimbra, C. Strickland-Constable and D. Waldram, *"$E_{d(d)} \times \mathbb{R}^+$ generalised geometry, connections and M theory"* , J. High Energy Phys. **1402** (2014) 054 [arXiv:1112.3989].

[27] A. Coimbra, C. Strickland-Constable and D. Waldram, *"Supergravity as generalised geometry II: $E_{d(d)} \times \mathbb{R}^+$ and M theory"*, J. High Energy Phys. **1403** (2014) 019 [arXiv:1212.1586].

[28] I. Jeon, K. Lee, J.-H. Park and Y. Suh, *"Stringy unification of Type IIA and IIB supergravities under N=2 D=10 supersymmetric double field theory"*, Phys. Lett. **B723** (2013) 245 [arXiv:1210.5048].

[29] I. Jeon, K. Lee and J.-H. Park, *"Ramond–Ramond cohomology and O(D,D) T-duality"*, J. High Energy Phys. **1209** (2012) 079 [arXiv:1206.3478].

[30] D.S. Berman, M. Cederwall, A. Kleinschmidt and D.C. Thompson, *"The gauge structure of generalised diffeomorphisms"*, J. High Energy Phys. **1301** (2013) 64 [arXiv:1208.5884].

[31] J.-H. Park and Y. Suh, *"U-geometry: SL(5)"*, J. High Energy Phys. **1406** (2014) 102 [arXiv:1302.1652].

[32] M. Cederwall, J. Edlund and A. Karlsson, *"Exceptional geometry and tensor fields"*, J. High Energy Phys. **1307** (2013) 028 [arXiv:1302.6736].

[33] M. Cederwall, *"Non-gravitational exceptional supermultiplets"*, J. High Energy Phys. **1307** (2013) 025 [arXiv:1302.6737].

[34] O. Hohm and H. Samtleben, *"Exceptional field theory I: $E_{6(6)}$ covariant form of M-theory and type IIB"*, Phys. Rev. **D89** (2014) 066016 [arXiv:1312.0614].

[35] O. Hohm and H. Samtleben, *"Exceptional field theory II: $E_{7(7)}$ "*, Phys. Rev. **D89** (2014) 066016 [arXiv:1312.4542].

[36] O. Hohm and H. Samtleben, *"Exceptional field theory III: $E_{8(8)}$ "*, Phys. Rev. **D90** (2014) 066002 [arXiv:1406.3348].

[37] S. Kachru, M.B. Schulz, P.K. Tripathy and S.P. Trivedi, *"New supersymmetric string compactifications"*, J. High Energy Phys. **0303** (2003) 061 [arXiv:hep-th/0211182].

[38] C. Condeescu, I. Florakis, C. Kounnas and D. Lüst, *"Gauged supergravities and non-geometric Q/R-fluxes from asymmetric orbifold CFT's"*, J. High Energy Phys. **1310** (2013) 057 [arXiv:1307.0999].

[39] F. Hassler and D. Lüst, *"Consistent compactification of double field theory on non-geometric flux backgrounds"*, J. High Energy Phys. **1405** (2014) 085 [arXiv: 1401.5068].

[40] M. Cederwall, *"The geometry behind double geometry"*, J. High Energy Phys. **1409** (2014) 70 [arXiv:1402.2513].

[41] M. Cederwall, *"T-duality and non-geometric solutions from double geometry"*, Fortsch. Phys. **62** (2014) 942 [arXiv:1409.4463].

[42] M. Cederwall and J.A. Rosabal, "E_8 geometry", J. High Energy Phys. **1507** (2015) 007, [arXiv:1504.04843].

[43] O. Hohm, S.K. Kwak and B. Zwiebach, *"Unification of type II strings and T-duality"*, Phys. Rev. Lett. **107** (2011) 171603 [arXiv:1106.5452].

[44] O. Hohm, S.K. Kwak and B. Zwiebach, *"Double field theory of type II strings"*, J. High Energy Phys. **1109** (2011) 013 [arXiv:1107.0008].

[45] O. Hohm and B. Zwiebach, *"Towards an invariant geometry of double field theory"*, arXiv:1212.1736.

[46] E. Cartan, *"Leçons sur la théorie des spineurs"* (Hermann, Paris, 1937).

[47] V. G. Kac, *"Infinite-dimensional Lie algebras"*, Cambridge Univ. Press (1990).

[48] V.G. Kac, *"Classification of simple Lie superalgebras"*, Funktsional. Anal. i Prilozhen. **9** (1975) 91.

[49] M. Cederwall, *"Twistors and supertwistors for exceptional field theory"*, J. High Energy Phys. **1512** (2015) 123 [arXiv:1510.02298].

[50] M. Cederwall, *"Double supergeometry"*, J. High Energy Phys. **1606** (2016) 155 [arXiv:1603.04684].

[51] M. Cederwall and J. Palmkvist, *"Superalgebras, constraints and partition functions"*, J. High Energy Phys. **0815** (2015) 36 [arXiv:1503.06215].

[52] J. Palmkvist, *"Exceptional geometry and Borcherds superalgebras"*, J. High Energy Phys. **1511** (2015) 032 [arXiv:1507.08828].

[53] C. Strickland-Constable, *"Subsectors, Dynkin diagrams and new generalised geometries"*, J. High Energy Phys. **1708** (2017) 144 [arXiv:1310.4196].

[54] D. Baraglia. *"Leibniz algebroids, twistings and exceptional generalized geometry"*, J. Geom. Phys. **62** (2012) 903 [arXiv:1101.0856].

[55] G. Bossard, M. Cederwall, A. Kleinschmidt, J. Palmkvist and H. Samtleben, *"Generalised diffeomorphisms for E_9"*, arXiv:1708.08936.

[56] O. Hohm and H. Samtleben, *"Exceptional field theory III: $E_{8(8)}$"*, Phys. Rev. **D90** (2014) 066002 [arXiv:1406.3348].

[57] M. Cederwall and J.A. Rosabal, "E_8 geometry", J. High Energy Phys. **1507** (2015) 007, [arXiv:1504.04843].

[58] D.H. Peterson and V.G. Kac, *"Infinite flag varieties and conjugacy theorems"*, Proc. Natl. Acad. Sci. **80** (1983) 1778.

[59] J.-H. Park and Y. Suh, *"U-gravity: SL(N)"*, J. High Energy Phys. **1304** (2013) 102 [arXiv:1402.5027].]

[60] O. Hohm and B. Zwiebach, *"Large gauge transformations in double field theory"*, arXiv:1207.4198.

[61] J. Palmkvist, *"The tensor hierarchy algebra"*, J. Math. Phys. **55** (2014) 011701 [arXiv:1305.0018].

[62] L. Carbone, M. Cederwall and J. Palmkvist, *"Generators and relations for Lie superalgebras of Cartan type"*, in preparation.

[63] G. Bossard, A. Kleinschmidt, J. Palmkvist, C.N. Pope and E. Sezgin, *"Beyond E_{11}"*, J. High Energy Phys. **1705** (2017) 020 [arXiv:1703.01305].

[64] N. Hitchin, *"Lectures on generalized geometry"*, arXiv:1010.2526.

[65] B. de Wit, H. Nicolai and H. Samtleben, *"Gauged supergravities, tensor hierarchies, and M-theory*, J. High Energy Phys. **0802** (2008) 044 [arXiv:0801.1294].

[66] V. Kac, *"Lie superalgebras"*, Adv. Math. **26** (1977) 8.

[67] V. Kac, *"Simple irreducible graded Lie algebras of finite growth"*, Math. USSR Izv. **2** (1968) 1271.

[68] U. Ray, *"Automorphic forms and Lie superalgebras"*, Springer (2006).

[69] G. Bossard and A. Kleinschmidt, *"Loops in exceptional field theory"*, J. High Energy Phys. **1601** (2016) 164 [arXiv:1510.07859].

[70] O. Hohm and H. Samtleben, *"U-duality covariant gravity"*, J. High Energy Phys. **1309** (2013) 080 [arXiv:1307.0509].

[71] O. Hohm, E.T. Musaev and H. Samtleben, *"$O(d+1,d+1)$ enhanced double field theory"*, J. High Energy Phys. **1710** (2017) 086 [arXiv:1707.06693.].

[72] R. Blumenhagen, F. Hassler and D. Lüst, *"Double field theory on group manifolds"*, J. High Energy Phys. **1502** (2015) 001 [arXiv:1410.6374].

[73] R. Blumenhagen, P. du Bosque, F. Hassler and D. Lüst, *"Generalized metric formulation of double field theory on group manifolds"*, arXiv:1502.02428.

[74] F. Hassler, *"The topology of double field theory"*, arXiv:1611.07978.

[75] F. Hassler, *"Poisson-Lie T-duality in double field theory"*, arXiv:1707.08624.

[76] M. Cederwall and J. Palmkvist, *"Extended geometries"*, arXiv:1711.07694.

[77] O. Hohm and B. Zwiebach, *"L_∞ algebras and field theory"*, Fortsch. Phys. **65** (2017) 1700014 [arXiv:1701.08824].]

[78] D.S. Berman, M. Cederwall and C. Strickland-Constable, work in progress.

[79] T. Damour, M. Henneaux and H. Nicolai, *"E_{10} and a 'small tension expansion' of M theory"*, Phys. Rev. Lett. **89** (2002) 221601 [arXiv:hep-th/0207267].

Non-Riemannian generalizations of Born-Infeld models and trace free gravitational equations*

Diego Julio Cirilo-Lombardo†
CONICET- Universidad de Buenos Aires.
Instituto de Fisica del Plasma (INFIP).
Buenos Aires, Argentina, Universidad de Buenos Aires.
Facultad de Ciencias Exactas y Naturales.
Departamento de Fisica. Buenos Aires, Argentina
and
Bogoliubov Laboratory of Theoretical Physics,
Joint Institute for Nuclear Research,
141980 Dubna, Russian Federation

Abstract

Non-Riemannian generalization of the standard Born-Infeld (BI) Lagrangian is reviewded in this talk from a theory of gravitation with dynamical torsion field. The field equations derived from the proposed action lead to a trace free gravitational equation (non-riemannian analog to the trace free equation (TFE) from[1][2][3]) and the field equations for the torsion respectively. In this theoretical context, the fundamental constants arise all from the same geometry through geometrical invariant quantities (as from the curvature R). New results involving generation of primordial magnetic fields and the link with leptogenesis and baryogenesis are presented and possible explanations given. The physically admisible matter fields can be introduced in the model via the torsion vector h . Such fields include some dark matter candidates such as axion, right neutrinos and Majorana and moreover, physical observables as vorticity can be included in the same way. From a new wormhole soluton in a cosmological spacetime with torsion we also show that the primordial cosmic magnetic fields can originate from h with the axion field (that is con tained in h) the responsible to control the dynamics and stability of the cosmic magnetic field but not the magnetogenesis itself. The analisys of Grand Unified Theories (GUT) in the context of this model indicates that the group manifold candidates are based in SO(10), SU(5) or some exceptional groups as E(6),E (7), etc. Hints about astrophysical consequences of this formulation are given

* This work has been supported by *CONICET Argentina*
† e-mail address: diego777jcl@gmail.com

PACS: 04.50.-h, 02.20.Qs, 12.10.Dm

1. Introduction

In this talk we resume the results of previous works [1][2][3][4] and references therein. The idea to construct a complete geometrization of the physics is very old. The drawback of the Einstein GR (General Relativity) equations is the RHS: $R_{\alpha\beta} - \frac{g_{\alpha\beta}}{2}R = \kappa T_{\alpha\beta}$ with the symmetric tensor (non-geometrical) $\kappa T_{\alpha\beta}$ that introduces *heuristically* the energy-momentum distribution. Similar drawbacks are contained by the unimodular gravity. It is well known that the unimodular gravity is obtained from Einstein-Hilbert action in which the *unimodular condition:* $\sqrt{-\det g_{\mu\nu}} = 1$ is also imposed from the very beginning The resulting field equations correspond to the *traceless* Einstein equations and can be shown that they are equivalent to the full Einstein equations with the cosmological constant term Λ, where Λ enters as an integration constant and the equivalence between unimodular gravity and general relativity is given by the arbitrary value of lamda. On the other hand the idea that the cosmological term arises as an integration constant is one of the motivations for the study of the unimodular gravity and also in the context of supergravity. The fact that the determinant of the metric is fixed has clearly profound consequences on the structure of given theory. First of all, it reduces the full group of diffeomorphisms to invariance under the group of unimodular general coordinate transformations which are transformations that leave the determinant of the metric unchanged.

Similar thing happens in the non-Riemannian case, as pointed out in[?][?][?][?][?], where the corresponding affine geometrical structure induces naturally the following constraint: $\frac{K}{g} = constant$. This natural constraint impose a condition (ratio) between both basic tensors through their determinants: the metric determinant g and the fundamental one K (in the sense of a nonsymmetric theory that contains the antisymmetric structures), independently of the precise functional form of K or g. In this work our starting point will be precisely the last one, where a *metric affine structure* in the space-time manifold (as described in Section II) will be considered. We will also show that trace free gravitational equations can be naturally obtained when the Lagrangian function (geometrical action) is taken as a measure involving a particular combination of the fundamental tensors of the geometry:

$$\sqrt{|\det f\left(g_{\mu\nu}, f_{\mu\nu}, R_{\mu\nu}\right)|}$$

with the (0,2) tensors $g_{\mu\nu}, f_{\mu\nu}, R_{\mu\nu}$: the symmetric metric, the antisymmetric one(that acts as potential of the torsion field) and the generalized Ricci tensor (proper of the non Riemannian geometry). The three tensors are related with a Clifford structure of the tangent space (for details see[4]) where the explicit choice for $f\left(g_{\mu\nu}, f_{\mu\nu}, R_{\mu\nu}\right)$ is given in Section III. This type of Lagrangians, because are non-Riemannian generalizations of the well known Nambu-Goto and Born-Infeld (BI) ones, can be physically and geometrically analyzed. Due the pure geometrical structure of the theory, induced energy momentum tensors and fundamental constants (actually functions) *emerge* naturally. Consequently, this fact allows the physical realization of the Mach principle that is briefly treated in SectionVIII after the (trace free) dynamic equations in Section IV are obtained.

In Section V the trace free gravitational equations and the meaning of the cosmological term as integration constant are discussed from the physical point of view, meanwhile in Section VI the constancy of G (Newton constant) is similarly discussed. The important role played by the dual of the torsion field as geometrical energy-momentum tensor is given in Section VII. Some physical consequences of the model, as the geometrical origin of the $\alpha\Omega-$dynamo, is presented in Section IX that it is very important because establish the link between the mathematical structure of the model of the first part of the article and the physics of the early universe and the particle physics of the second half of this work. In Section X the direct relation between the torsion with axion electrodynamics and Chern-Simons (CS) theory is discussed considering the geometrical structure of the dual vector of the torsion field. In Section XI an explanation about the magnetogenesis in FRW scenario, the structure of the GUT where the SM is derived and the role of the axion in the dynamics of the cosmic magnetic field is presented. Finally some concluding remarks are given in Section XII.

2. Basis of the metrical-affine geometry

The starting point is a hypercomplex construction of the (metric compatible) spacetime manifold

$$M,\; g_{\mu\nu} \equiv e_{\mu} \cdot e_{\nu} \tag{1}$$

where for each point $p \in M$ there exists a local affine space A. The connection over A, $\widetilde{\Gamma}$, define a generalized affine connection Γ on M, specified by (∇, K), where K is an invertible $(1,1)$ tensor over M. We will demand for the connection to be compatible and rectilinear, that is

$$\nabla K = KT, \qquad \nabla g = 0 \tag{2}$$

where T is the torsion, and g the space-time metric (used to raise and lower the indices and determining the geodesics), that is preserved under parallel transport. This generalized compatibility condition ensures that the generalized affine connection Γ maps autoparallels of Γ on M into straight lines over the affine space A (locally). The first equation above is equal to the condition determining the connection in terms of the fundamental field in the UFT non-symmetric. Hence, K can be identified with the fundamental tensor in the non-symmetric fundamental theory. This fact gives us the possibility to restrict the connection to a (anti-)Hermitian theory.

The covariant derivative of a vector with respect to the generalized affine connection is given by

$$A^{\mu}{}_{;\nu} \equiv A^{\mu}{}_{,\nu} + \Gamma^{\mu}{}_{\alpha\nu} A^{\alpha} \tag{3}$$
$$A_{\mu;\nu} \equiv A_{\mu}{}_{,\nu} - \Gamma^{\alpha}{}_{\mu\nu} A_{\alpha} \tag{4}$$

The generalized compatibility condition (2) determines the 64 components of the connection by the 64 equations

$$K^{\mu}_{\nu;\alpha} = K^{\mu}_{\rho} T^{\rho}{}_{\nu\alpha} \qquad \text{where} \qquad T^{\rho}{}_{\nu\alpha} \equiv 2\Gamma^{\rho}{}_{[\alpha\nu]} \tag{5}$$

Notice that by contracting indices ν and α in the first equation above, an additional condition over this hypothetical fundamental (nonsymmetric) tensor K is obtained

$$K_{\mu\alpha;}{}^{\alpha} = 0$$

that, geometrically speaking, reads

$$d^{*}K = 0.$$

This is a current-free condition over the tensor K. Notice that the metric is used here to down the indices (metric compatible space-time) and consequently we can work also with $K_{\alpha\nu} = g_{\alpha\beta} K^{\beta}_{\nu}$

The metric is uniquely determined by the metricity condition, which puts 40 restrictions on the derivatives of the metric

$$g_{\mu\nu,\rho} = 2\Gamma_{(\mu\nu)\rho} \tag{6}$$

The space-time curvature tensor, that is defined in the usual way, has two possible contractions: the Ricci tensor $R^{\lambda}_{\mu\lambda\nu} = R_{\mu\nu}$, and the second contraction $R^{\lambda}_{\lambda\mu\nu} = 2\Gamma^{\lambda}{}_{\lambda[\nu,\mu]}$, which is identically zero due to the metricity condition (2).

In order to find a symmetry of the torsion tensor, let us denote the inverse of K by $\widehat{K}$. Therefore, $\widehat{K}$ is uniquely specified by condition $\widehat{K}^{\alpha\rho}$ $K_{\alpha\sigma} = K^{\alpha\rho}\widehat{K}_{\alpha\sigma} = \delta^{\rho}_{\sigma}$.

As it was pointed out in , inserting explicitly the torsion tensor as the antisymmetric part of the connection in (5), and multiplying by $\frac{1}{2}\widehat{K}^{\alpha\nu}$, results, after straighforward computations, in

$$\left(Ln\sqrt{-K}\right)_{,\mu} - \Gamma^{\nu}_{(\mu\nu)} = 0 \tag{7}$$

where $K = \det\,(K_{\mu\rho})$. Notice that from expression (7) we arrive at the relation between the determinants K and g:

$$\frac{K}{g} = \text{constant}$$

(strictly a constant scalar function of the coordinates). Now we can write

$$\Gamma^{\nu}{}_{\alpha\nu,\beta} - \Gamma^{\nu}{}_{\beta\nu,\alpha} = \Gamma^{\nu}{}_{\nu\beta,\alpha} - \Gamma^{\nu}{}_{\nu\alpha,\beta}\,, \tag{8}$$

as the first term of (7) is the derivative of a scalar. Then, the torsion tensor has the symmetry

$$T^{\nu}{}_{\nu[\beta,\alpha]} = T^{\nu}{}_{\nu[\alpha,\beta]} = 0 \tag{9}$$

This implies that the trace of the torsion tensor, defined as $T^{\nu}{}_{\nu\alpha}$, is the gradient of a scalar field

$$T_{\alpha} = \nabla_{\alpha}\phi \tag{10}$$

Expressions precisely as (1) and (2) ensure that the basic non-symmetric field structures (*i.e.* K) take on a definite geometrical meaning when interpreted in terms of affine geometry. Notice that the tensor K carries the 2-form (bivector) that will be associated with the fundamental antisymmetric form in the next Sections. Such antisymmetric form is introduced from the tangent space via the generalization of the Ambrose-Singer theorem by exponentiation.

3. Geometrical Lagrangians: the generalized Born-Infeld action

Let us start with the geometrical Lagrangian introduced in [[?][?][?][?][?]]

$$\mathcal{L}_g = \sqrt{\det\left[\lambda\left(g_{\alpha\beta} + F_{\alpha\beta}\right) + R_{\alpha\beta}\right]} \tag{11}$$

it can be rewritten as

$$\mathcal{L}_g = \sqrt{\det\left(G_{\alpha\beta} + \mathcal{F}_{\alpha\beta}\right)} \tag{12}$$

with the following redefinitions

$$G_{\alpha\beta} = \lambda g_{\alpha\beta} + R_{(\alpha\beta)} \qquad \text{and} \qquad \mathcal{F}_{\alpha\beta} = \lambda F_{\alpha\beta} + R_{[\alpha\beta]} \tag{13}$$

where a totally antisymmetric torsion tensor $T^{\alpha}_{\gamma\beta} = \varepsilon^{\alpha}_{\gamma\beta\delta}h^{\delta}$ is assumed(h^{δ} its dual vector field). Notice that the antisymmetric tensor $F_{\alpha\beta}$,that takes the role of the electromagnetic field, is proportional to the dual of the potential for the (totally antisymmetric) torsion field. A brief review on the origin of this type of Lagrangians in the context of unified theories in reductive geometries is in Appendix I of [1]. Consequently the generalized Ricci tensor splits into a symmetric and antisymmetric part, namely:

$$R_{\mu\nu} = \overbrace{\overset{\circ}{R}_{\mu\nu} - T_{\mu\rho}{}^{\alpha}T_{\alpha\nu}{}^{\rho}}^{R_{(\mu\nu)}} + \overbrace{\overset{\circ}{\nabla}_{\alpha}T_{\mu\nu}{}^{\alpha}}^{R_{[\mu\nu]}}$$

where $\overset{\circ}{R}_{\mu\nu}$ is the general relativistic Ricci tensor constructed with the Christoffel connection. The expansion of the determinant leads to the Born-Infeld generalization in the usual form :

$$\mathcal{L}_g = \sqrt{|G|}\sqrt{1+\frac{1}{2}\mathcal{F}_{\mu\nu}\mathcal{F}^{\mu\nu}-\frac{1}{16}\left(\mathcal{F}_{\mu\nu}\widetilde{\mathcal{F}}^{\mu\nu}\right)^2} \tag{14}$$

$$= \Lambda^2\sqrt{|g|}\sqrt{1+\frac{1}{2}\Lambda_1^2F_{\mu\nu}F^{\mu\nu}-\frac{1}{16b^4}\left(\Lambda_2^2F_{\mu\nu}\widetilde{F}^{\mu\nu}\right)^2} \tag{15}$$

where

$$\Lambda = \lambda + \frac{g_{\alpha\beta}R^{(\alpha\beta)}}{4} \tag{16}$$

$$\Lambda_1^2 = \lambda^2\left(1+\frac{2}{\lambda}\frac{F_{\mu\nu}R^{[\mu\nu]}}{F_{\mu\nu}F^{\mu\nu}}+\frac{1}{\lambda^2}\frac{R_{[\mu\nu]}R^{[\mu\nu]}}{F_{\mu\nu}F^{\mu\nu}}\right) \tag{17}$$

$$\Lambda_2^2 = \lambda^2\left(1+\frac{2}{\lambda}\frac{F_{\mu\nu}\widetilde{R}^{[\mu\nu]}}{F_{\mu\nu}\widetilde{F}^{\mu\nu}}+\frac{1}{\lambda^2}\frac{R_{[\mu\nu]}\widetilde{R}^{[\mu\nu]}}{F_{\mu\nu}\widetilde{F}^{\mu\nu}}\right) \tag{18}$$

Although the action is exact and have the correct limit, the analysis can be simplest and substantially improved using the following action

$$\mathcal{L}_{gs} = \sqrt{\det\left[\lambda g_{\alpha\beta}\left(1+\frac{R_s}{4\lambda}\right)+\lambda F_{\alpha\beta}\left(1+\frac{R_A}{\lambda}\right)\right]} \tag{19}$$

$$R_s \equiv g^{\alpha\beta}R_{(\alpha\beta)}; \qquad R_A \equiv f^{\alpha\beta}R_{[\alpha\beta]} \tag{20}$$

$\left(with\ f^{\alpha\beta} \equiv \frac{\partial \ln(\det F_{\mu\nu})}{\partial F_{\alpha\beta}}, \det F_{\mu\nu} = 2F_{\mu\nu}\widetilde{F}^{\mu\nu}\right)$ that contains all necessary information and is more suitable to manage. If the induced structure from the tangent space $T_p(M)$(via Ambrose-Singer theorem) is intrinsically related to a (super)manifold structure, we have seen that there exists a particular transformation where the details are given in [1][2]

4. Field equations

The geometry of the space-time Manifold is to be determined by the Noether symmetries

$$\frac{\delta L_G}{\delta g^{\mu\nu}} = 0, \qquad \frac{\delta L_G}{\delta f^{\mu\nu}} = 0 \tag{21}$$

where the functional (Hamiltonian) derivatives in the sense of Palatini (in this case with respect to the potentials), are understood. The choice "measure-like" form for the geometrical Lagrangian L_G (reminiscent of a nonlinear sigma model), as is evident, satisfy the following principles:

i) the principle of the natural extension of the Lagrangian density as square root of the fundamental line element containing also $F_{\mu\nu}$.

ii) the symmetry principle between $g_{\mu\nu}$ *and* $F_{\mu\nu}$*(e.g.* $g_{\mu\nu}$ *and* $F_{\mu\nu}$*should enter into* L_G *symmetrically)*

iii)the principle that the spinor symmetry, namely

$$\nabla_\mu g_{\lambda\nu} = 0, \qquad \nabla_\mu \sigma_{\lambda\nu} = 0 \tag{22,23}$$

with

$$g_{\lambda\nu} = \gamma_\lambda \cdot \gamma_\nu, \qquad \sigma_{\lambda\nu} = \gamma_\lambda \wedge \gamma_\nu \sim *F_{\lambda\nu} \tag{24,25}$$

should be derivable from L_G*(21)*

The last principle is key because it states that the spinor invariance of the fundamental space-time structure should be derivable from the dynamic symmetries given by (21). The fact that the L_G satisfies the 3 principles shows also that it has the simpler form.

Notice that the action density proposed by Einstein in in his nonsymmetric field theory satisfies i) and ii) but not iii).

Remark 1 *Due the totally antisymmetric character of the torsion field it is completely determined by the fundamental (structural 2-form) antisymmetric tensor, and consequently the variations must acquire the form given by expression (21):* ***metric and torsion have each one their respective potentials that are in coincidence with the fundamental structure of the geometry.***

4.1. $\delta_g L_G$

The starting point for the metrical variational procedure is in the same way as in the standard Born-Infeld theory: from the following factorization of the geometrical Lagrangian :

$$\mathcal{L} = \sqrt{|g|}\sqrt{\det(\alpha\lambda)}\sqrt{1+\frac{1}{2b^2}F_{\mu\nu}F^{\mu\nu}-\frac{1}{16b^4}\left(F_{\mu\nu}\widetilde{F}^{\mu\nu}\right)^2} \equiv \sqrt{|g|}\sqrt{\det(\alpha\lambda)}\mathbb{R} \tag{26}$$

where

$$b = \frac{\alpha}{\beta} = \frac{1+(R_S/4\lambda)}{1+(R_A/4\lambda)}, \qquad R_S = g^{\alpha\beta}R_{\alpha\beta}, \qquad R_A = f^{\alpha\beta}R_{\alpha\beta}, \tag{27,28,29}$$

and λ an arbitrary constant we perform the variational metric procedure with the following result (details see Appendix II)

$$\delta_g \mathcal{L} = 0 \Rightarrow R_{(\alpha\beta)} - \frac{g_{\alpha\beta}}{4} R_s = \frac{R_s}{2\mathbb{R}^2\alpha^2}\left[F_{\alpha\lambda}F_{\beta}{}^{\lambda} - F_{\mu\nu}F^{\mu\nu}\frac{R_{(\alpha\beta)}}{R_s}\right] + \qquad (30,\ 31)$$

$$+\frac{R_s}{4\mathbb{R}^2\alpha^2 b^2}\left[F_{\mu\nu}\widetilde{F}^{\mu\nu}\left(\frac{F_{\eta\rho}\widetilde{F}^{\eta\rho}}{8}g_{\alpha\beta} - F_{\alpha\lambda}\widetilde{F}_{\beta}{}^{\lambda}\right) + \frac{F_{\eta\rho}\widetilde{F}^{\eta\rho}}{2}\frac{R_{(\alpha\beta)}}{R_s}\right] +$$

$$+2\lambda\left[g_{\alpha\beta} + \frac{1}{\mathbb{R}^2\alpha^2}\left(F_{\alpha\lambda}F_{\beta}{}^{\lambda} + \frac{F_{\mu\nu}\widetilde{F}^{\mu\nu}}{2b^2}\left(\frac{F_{\eta\rho}\widetilde{F}^{\eta\rho}}{8}g_{\alpha\beta} - F_{\alpha\lambda}\widetilde{F}_{\beta}{}^{\lambda}\right)\right)\right],$$

Remark 2 *Notice that:*

1) The eq. (31) is trace-free *type, consequently the trace of the third term of the above equation (that is the cosmological one) is equal to zero. This happens trivially if* $\lambda = 0$ *or* $4\mathbb{R}^2\alpha^2 = -\left(F_{\alpha\lambda}F^{\ \alpha\lambda} - \frac{\left(F_{\mu\nu}\widetilde{F}^{\mu\nu}\right)^2}{4b^2}\right)$ *In terms of the Maxwell Lagrangian we have* $(\mathbb{R}\alpha)^2 = \left(L_{Maxwell} + \frac{\left(F_{\mu\nu}\widetilde{F}^{\mu\nu}\right)^2}{16b^2}\right) \equiv \mathcal{W}(I_S, I_P, b)$ *that allow us to simplify the eq. (31) once more as follows*

$$R_{(\alpha\beta)} - \frac{g_{\alpha\beta}}{4} R_s = \frac{R_s}{2\mathcal{W}}\left[F_{\alpha\lambda}F_{\beta}{}^{\lambda} - F_{\mu\nu}F^{\mu\nu}\frac{R_{(\alpha\beta)}}{R_s}\right] +$$

$$+\frac{R_s}{4\mathcal{W}b^2}\left[F_{\mu\nu}\widetilde{F}^{\mu\nu}\left(\frac{F_{\eta\rho}\widetilde{F}^{\eta\rho}}{8}g_{\alpha\beta} - F_{\alpha\lambda}\widetilde{F}_{\beta}{}^{\lambda}\right) + \frac{F_{\eta\rho}\widetilde{F}^{\eta\rho}}{2}\frac{R_{(\alpha\beta)}}{R_s}\right] +$$

$$+2\lambda\left[g_{\alpha\beta} + \frac{1}{\mathcal{W}}\left(F_{\alpha\lambda}F_{\beta}{}^{\lambda} + \frac{F_{\mu\nu}\widetilde{F}^{\mu\nu}}{2b^2}\left(\frac{F_{\eta\rho}\widetilde{F}^{\eta\rho}}{8}g_{\alpha\beta} - F_{\alpha\lambda}\widetilde{F}_{\beta}{}^{\lambda}\right)\right)\right],$$

2) b takes the place of limiting parameter (maximum value) for the electromagnetic field strength.

3)b is not a constant in general, in sharp contrast with the Born-Infeld or string theory cases.

4) Because b is the ratio $\frac{\alpha}{\beta} = \frac{1+(R_S/4\lambda)}{1+(R_A/\lambda 4)}$ *involving both curvature scalars from the contractions of the generalized Ricci tensor: it is preponderant when the symmetrical contraction of* $R_{\alpha\beta}$ *is greater than the skew one.*

5) The fact pointed out in ii), namely that the curvature scalar plays the role as some limiting parameter of the field strength, was conjectured by Mansouri in [?] in the context of gravity theory over group manifold (generally with symmetry breaking). In such a case, this limit was established after the explicit integration of the internal group-valuated variables that is not our case here.

6) In similar form that the Eddington conjecture: $R_{(\alpha\beta)} \propto g_{\alpha\beta}$, we have a condition over the ratios as follows:

$$\frac{R_{(\alpha\beta)}}{R_s} \propto \frac{g_{\alpha\beta}}{D} \tag{32}$$

that seems to be universal.

7)The equations are the simplest ones when $b^{-2} = 0$ $(\beta = 0)$,taking the exact "quasilinear" form

$$R_{(\alpha\beta)} - \frac{g_{\alpha\beta}}{4} R_s = \underbrace{\frac{R_s}{2\alpha^2}\left[F_{\alpha\lambda}F_{\beta}{}^{\lambda} - F_{\mu\nu}F^{\mu\nu}\frac{R_{(\alpha\beta)}}{R_s}\right]}_{Maxwell-like} + 2\lambda \underbrace{\left[g_{\alpha\beta} + \frac{1}{\mathcal{W}}F_{\alpha\lambda}F_{\beta}{}^{\lambda}\right]}_{\widetilde{g}_{eff}}, \tag{33}$$

this particular case (e.g. projective invariant) will be used through this work. Notice that when $b^{-2} = 0$ $(\beta = 0)$ all terms into the gravitational equation (31) involving the pseudoscalar invariant, namely $F_{\mu\nu}\widetilde{F}^{\mu\nu}$ or $F_{\alpha\lambda}\widetilde{F}_{\beta}{}^{\lambda}$,vanishes. Consequently we arrive to the simplest expression (33) that will be used in Section XI for example.

4.2. $\delta_f L_G$

Let us to take as starting point the geometrical Lagrangian (19)

$$\mathcal{L}_{gs} = \sqrt{\det\left[\lambda g_{\alpha\beta}\left(1 + \frac{R_s}{4\lambda}\right) + \lambda F_{\alpha\beta}\left(1 + \frac{R_A}{4\lambda}\right)\right]} \tag{34}$$

$$= \sqrt{|g|}\lambda^2\alpha^2\left(\sqrt{1 + \frac{1}{2}\mathcal{F}_{\mu\nu}\mathcal{F}^{\mu\nu} - \frac{1}{16}\left(\mathcal{F}_{\mu\nu}\widetilde{\mathcal{F}}^{\mu\nu}\right)^2}\right) \tag{35}$$

then, having into account that : $R_A = f^{\mu\nu}R_{\mu\nu}$ and $\frac{\partial \ln(\det F_{\mu\nu})}{\partial F_{\alpha\beta}} = f^{\alpha\beta}$ (due that b that contains R_A must be also included in the variation)we obtain

$$\frac{\delta L_G}{\delta F_{\sigma\omega}} = 0 \to \left(\frac{\sqrt{|g|}\lambda\beta}{2\mathbb{R}b}\right)\left[\mathbb{F}^{\sigma\omega}\beta - \frac{\mathbb{F}}{4\lambda}R_{[\mu\nu]}\chi^{\mu\nu\sigma\omega}\right] = 0 \tag{36}$$

where: $\mathbb{F} \equiv \left[F_{\mu\nu}F^{\mu\nu} - \frac{1}{4}b^{-2}\left(F_{\mu\nu}\widetilde{F}^{\mu\nu}\right)^2\right]$, $\mathbb{F}^{\sigma\alpha} \equiv \left[F^{\sigma\alpha} - \frac{1}{4}b^{-2}\left(F_{\mu\nu}\widetilde{F}^{\mu\nu}\right)\widetilde{F}^{\sigma\alpha}\right]$ and $\chi^{\mu\nu\sigma\omega} \equiv f^{\mu\omega}f^{\sigma\nu} - f^{\mu\sigma}f^{\omega\nu}$. Notice that the quantity $b = \alpha/\beta$ (concretely β) was also varied in the above expression given the second term in (36).

Contracting (36) with $F_{\alpha\beta}$, a condition over the curvature and the electromagnetic field invariants is obtained as

$$\left(\frac{\sqrt{|g|}\lambda\beta}{\mathbb{R}b}\right)\mathbb{F}\left[\beta - \frac{R_A}{2\lambda}\right] = 0$$

This condition is satisfied for $R_A = -4\lambda$ is the exact projective invariant case (that correspond with $\beta = 0$), and for $R_A = 2\lambda$.

Remark 3 *the variational equation (36) is a dynamic equation for the torsion field in complete analogy with the eqs. (31) for the curvature.*

5. Emergent trace free gravitational equations: the meaning of Λ

Starting from the trace free equation (31) *that is not assumed* but arises from the model, the task is to rewrite it as

$$\underbrace{\overset{\circ}{R}_{\alpha\beta} - \frac{g_{\alpha\beta}}{2}\overset{\circ}{R}}_{\equiv G_{\alpha\beta}} = \underbrace{6\left(-h_\alpha h_\beta + \frac{g_{\alpha\beta}}{2}h_\gamma h^\gamma\right)}_{\equiv T^h_{\alpha\beta}} + \frac{g_{\alpha\beta}}{2}R_s + T^F_{\alpha\beta} + 2\lambda\rho_{\alpha\beta} \tag{37}$$

where

$$\rho_{\alpha\beta} \equiv g_{\alpha\beta} + \frac{1}{\mathcal{W}}\left(F_{\alpha\lambda}F_\beta{}^\lambda + \frac{F_{\mu\nu}\widetilde{F}^{\mu\nu}}{2b^2}\left(\frac{F_{\eta\rho}\widetilde{F}^{\eta\rho}}{8}g_{\alpha\beta} - F_{\alpha\lambda}\widetilde{F}_\beta{}^\lambda\right)\right) \tag{38}$$

$$T^F_{\alpha\beta} \equiv \frac{R_s}{2\mathcal{W}}\left\{\left(F_{\alpha\lambda}F_\beta{}^\lambda - F_{\mu\nu}F^{\mu\nu}\frac{R_{(\alpha\beta)}}{R_s}\right) + \right. \tag{39}$$

$$\left. + \frac{1}{2b^2}\left[F_{\mu\nu}\widetilde{F}^{\mu\nu}\left(\frac{F_{\eta\rho}\widetilde{F}^{\eta\rho}}{8}g_{\alpha\beta} - F_{\alpha\lambda}\widetilde{F}_\beta{}^\lambda\right) + \frac{\left(F_{\eta\rho}\widetilde{F}^{\eta\rho}\right)^2}{2}\frac{R_{(\alpha\beta)}}{R_s}\right]\right\}$$

the LHS of (37) is the Einstein tensor. The "GR" divergence $\overset{\circ}{\nabla}{}^\alpha$ of $G_{\alpha\beta}$ is zero because is a geometrical geometrical identity and in an analog manner $\overset{\circ}{\nabla}{}^\alpha\left(T^h_{\alpha\beta} + T^F_{\alpha\beta}\right) = 0$ because both tensors have the same symmetry that the corresponding GR energy momentum tensors of a vector field and electromagnetic field respectively:

$$\overset{\circ}{\nabla}{}^\alpha G_{\alpha\beta} = \overset{\circ}{\nabla}{}^\alpha\left(T^h_{\alpha\beta} + T^F_{\alpha\beta}\right) = 0$$

consequently the remaining part must be a covariantly constant tensor that we *assume* proportional to $g_{\alpha\beta}$:

$$\nabla^\alpha\left(\frac{g_{\alpha\beta}}{2}R_s + 2\lambda\rho_{\alpha\beta}\right) = 0$$

$$\Rightarrow \left(\frac{g_{\alpha\beta}}{2}R_s + 2\lambda\rho_{\alpha\beta}\right) = \Lambda g_{\alpha\beta} \rightarrow R_s = 2\Lambda \tag{40}$$

Coming back to the original trace free expressions we have the expected formula

$$\underbrace{\overset{\circ}{R}_{\alpha\beta} - \frac{g_{\alpha\beta}}{2}\overset{\circ}{R}}_{\equiv G_{\alpha\beta}} = \underbrace{6\left(-h_{\alpha}h_{\beta} + \frac{g_{\alpha\beta}}{2}h_{\gamma}h^{\gamma}\right)}_{\equiv T^{h}_{\alpha\beta}} + T^{F}_{\alpha\beta} + \Lambda g_{\alpha\beta} \tag{41}$$

Remark 4 *Tracing the first expression in (40) we have* $R_s = 2\Lambda = \overset{\circ}{R} + 6h_{\mu}h^{\mu}$ *linking the value of the curvature and the norm of the torsion vector field. Consequently, if the dual of the torsion field have the role of the energy-matter carrier, the meaning of lambda as the vacuum energy is immediately established.*

Remark 5 *Notice that the LHS in expression (40) instead to be proportional to the metric tensor it can be proportional to the square of a Killing-Yano tensor.*

6. On the constancy of G

At this level, no assertion can state with respect to G or even with respect to c. The link with the general relativistic case is given by the identification of electromagnetic energy-momentum tensor with the term analogous $T^{F}_{\alpha\beta}$ in our metric variational equations:

$$\frac{8\pi G}{c^4}\left(F_{\alpha\lambda}F_{\beta}^{\ \lambda} - F_{\mu\nu}F^{\mu\nu}\frac{g_{\alpha\beta}}{4}\right) \rightarrow \frac{R_s}{2\mathcal{W}}\left(F_{\alpha\lambda}F_{\beta}^{\ \lambda} - F_{\mu\nu}F^{\mu\nu}\frac{R_{(\alpha\beta)}}{R_s}\right)$$

Consequently we have:

$$\kappa = \frac{8\pi G}{c^4} \rightarrow \frac{R_s}{2\mathbb{R}^2\alpha^2}$$

$$\text{and} \quad \frac{g_{\alpha\beta}}{4} = \frac{R_{(\alpha\beta)}}{R_s}$$

The above expression indicates that the ratio must remains constant due the Noether symmetries and conservation laws of the field equations. Notice that (as in the case of b) there exist a limit for all the physical fields coming from the *geometrical invariants quantities.*

7. The vector h_{μ} and the energy-matter interpretation

One of the characteristics that more attract the attention in unified field theoretical models is the possibility to introduce the energy and matter through its geometrical structure. In our case the torsion field takes the role of RHS of the standard GR gravity equation by mean its dual, namely h_μ.

Consequently, in order to explain the physical role of h_μ, we know (due the Hodge-de Rham decomposition [Appendix III]) that it can be decomposed as:

$$h_\alpha = \nabla_\alpha \Omega + \varepsilon_\alpha^{\beta\gamma\delta} \nabla_\beta A_{\gamma\delta} + \gamma_1 \overbrace{\varepsilon_\alpha^{\beta\gamma\delta} M_{\beta\gamma\delta}}^{axial\ vector} + \gamma_2 \overbrace{P_\alpha}^{polar\ vector} \tag{42}$$

where $\gamma_1$and γ_2 can be phenomenologically related to physical constants (e.g: $\gamma_1 = \frac{8\pi}{c}\sqrt{G}$ is a physical constant related to the Blackett formula). From the eq.motion for the torsion namely: $\nabla_\alpha T^{\alpha\beta\gamma} = -\lambda F^{\beta\gamma}$and coming back to (42) we obtain the following important equation

$$\overset{\circ}{\Box} A_{\gamma\delta} - \gamma \left[\nabla_\alpha M^\alpha{}_{\gamma\delta} + (\nabla_\gamma P_\delta - \nabla_\delta P_\gamma)\right] = -\lambda F_{\gamma\delta} \tag{43}$$

Let us consider, in particular, the case when $\lambda F_{\gamma\delta} \to 0$:

$$\overset{\circ}{\Box} A_{\gamma\delta} = \gamma \left[\nabla_\alpha M^\alpha{}_{\gamma\delta} + (\nabla_\gamma P_\delta - \nabla_\delta P_\gamma)\right] \tag{44}$$

We can immediately see that, if $M^\alpha{}_{\gamma\delta}$ is identified with the intrinsic spin angular momentum of the ponderable matter, P_δ is its lineal momentum vector and $A_{\gamma\delta}$ is the gravitational radiation tensor, then eq.(44) states that the sum of the intrinsic spin angular momentum and the orbital angular momentum of ponderable matter is conserved if the gravitational radiation is absent., if $M^\alpha{}_{\gamma\delta}$ is identified with the intrinsic spin angular momentum of the ponderable matter, P_δ is its lineal momentum vector and $A_{\gamma\delta}$ is the gravitational radiation tensor, then eq.(44) states that the sum of the intrinsic spin angular momentum and the orbital angular momentum of ponderable matter is conserved if the gravitational radiation is absent.

7.1. Killing-Yano systems and the vector h_μ

Without enter in many details (these will be treated somewhere) the antisymmetric tensor $A_{\gamma\delta}$ in the h_β composition is related with the Killing and Killing-Yano systems. Consequently we can introduce two types of couplings into the $A_{\gamma\delta}$ divergence : it correspond with the generalized current interpretation that also has h_μ.

i) Defining

$$A_{\gamma\delta} \equiv A_{[\gamma;\delta]} \tag{45}$$

such that

$$\overset{\circ}{\nabla}_\rho A_{[\gamma;\delta]} = \frac{4\pi}{3}\left(j_{[\gamma}\, g_{\delta]\rho}\right) \tag{46}$$

then , in this case we can identify $A_{\gamma\delta} = 2F_{\gamma\delta}$ because $F^{\delta}_{\gamma;\delta} = 4j_\gamma$ and $A_{[\gamma\delta;\rho]} = F_{[\gamma\delta;\rho]} = 0$

In this case the contribution of $A_{\gamma\delta}$ to h_β is null.
ii) Let us consider now a fully antisymmetric coupling as

$$A_{[\gamma;\delta];\rho} = \frac{4\pi}{3} j_{[\gamma} F_{\delta]\rho} \tag{47}$$

, having into account the vorticity vector also

$$\omega_\mu \equiv u^\lambda \varepsilon_{\lambda\mu\nu\rho} \nabla^\nu u^\rho \tag{48}$$

and considering a plasma with electrons, protons etc.

$$j^\gamma \sim A^\gamma + q_s n_s u_s^\gamma \tag{49}$$

where A_μ is the vector potential and q_s is the particle charge, $n_s$is the number density (in the rest frame) and the four-velocity of species s is u_s^γ. In this case $h_\alpha$takes the form

$$h_\alpha = \nabla_\alpha \Omega + \varepsilon_\alpha^{\beta\gamma\delta} \nabla_\beta A_{\gamma\delta} + \gamma_1 \varepsilon_\alpha^{\beta\gamma\delta} M_{\beta\gamma\delta} + \gamma_2 P_\alpha \rightarrow \tag{50}$$

$$h_\alpha = \nabla_\alpha \Omega + \varepsilon_\alpha^{\gamma\delta\rho} \frac{4\pi}{3} j_{[\gamma} F_{\delta]\rho} - \gamma_1 u^\lambda \varepsilon_{\lambda\alpha\nu\rho} \nabla^\nu u^\rho + \gamma_2 P_\alpha \tag{51}$$

$$h_\alpha = \nabla_\alpha \Omega + \varepsilon_\alpha^{\gamma\delta\rho} \frac{4\pi}{3} \left[A + q_s n_s u_s\right]_{[\gamma} F_{\delta]\rho} - \gamma_1 u^\lambda \varepsilon_{\lambda\alpha\nu\rho} \nabla^\nu u^\rho + \gamma_2 P_\alpha \tag{52}$$

Consequently in 3+1 decomposition we have (overbar correspond to spacial 3-dim. vectors)

$$h_0 = \nabla_0 \Omega + \frac{4\pi}{3} \overline{j} \cdot \overline{B} + \gamma_1 \overline{u} \cdot \left(\overline{\nabla} \times \overline{u}\right) + \gamma_2 P_0 \tag{53}$$

$$h_0 = \nabla_0 \Omega + \frac{4\pi}{3} \left[\overline{A} \cdot \left(\overline{\nabla} \times \overline{A}\right) + q_s n_s \overline{u}_s \cdot \overline{B}\right] + \gamma_1 \overline{u} \cdot \left(\overline{\nabla} \times \overline{u}\right) + \gamma_2 P_0 \tag{54}$$

and

$$h_i = \nabla_i \Omega + \frac{4\pi}{3} \left[-\left(\overline{j} \times \overline{E}\right)_i + j_0 \overline{B}_i\right] + \gamma_1 \left[u_0 \left(\overline{\nabla} \times \overline{u}\right) + \left(\overline{u} \times \overline{\nabla} u_0\right) + \left(\overline{u} \times \dot{\overline{u}}\right)\right]_i + \gamma_2 P_i \tag{55}$$

$$\begin{aligned} h_i = \nabla_i \Omega + \frac{4\pi}{3} \left[-\left(\left(\overline{A} + q_s n_s \overline{u}_s\right) \times \overline{E}\right)_i + \left(\Phi + q_s n_s u_{0s}\right) \overline{B}_i\right] + \\ + \gamma_1 \left[u_0 \left(\overline{\nabla} \times \overline{u}\right) + \left(\overline{u} \times \overline{\nabla} u_0\right) + \left(\overline{u} \times \dot{\overline{u}}\right)\right]_i + \gamma_2 P_i \end{aligned} \tag{56}$$

Notice that in h_0 we can recognize the magnetic and vortical helicities

$$h_0 = \nabla_0 \Omega + \frac{4\pi}{3} \left[h_M + q_s n_s \overline{u}_s \cdot \overline{B}\right] + \gamma_1 h_V + \gamma_2 P_0 \tag{57}$$

The above expression will be very important in the next sections, in particular to discuss magnetogenesis and particle generation. Notice the important fact that the symmetry of the vorticity can be associated to a 2-form bivector.

8. Physical consequences

In this Section we will make contact with the physical consequences of the model. Firstly we introduce the 3+1 splitting for axisymmetric space-times that is useful from the the physical viewpoint for the analysis of the electrodynamic equations with high degree of nonlinearity, as in our case. Secondly we take the 3+1 field equations in the in the linear limit where the induction equations (dynamo) are obtained, showing explicitly the important role of the torsion field as the generator of a purely geometric α-term. Thirdly, we derive the geometrical analog of the Lorentz force and the elimination of the electric field from the induction equations. Also, the origin of the seed magnetic field via the geometrical α-term generated by the torsion vector is worked out.

8.1. Electrodynamic structure in 3+1

The starting point will be the line element in $3+1$ splitting(Appendix IV of [1]): the 4-dimensional space-time is split into 3-dimensional space and 1-dimensional time to form a foliation of 3-dimensional spacelike hypersurfaces. The metric of the space-time is consequently, given by $ds^2 = -\alpha^2 dt^2 + \gamma_{ij}\left(dx^i + \beta^i dt\right)\left(dx^j + \beta^j dt\right)$ where γ_{ij} is the metric of the 3-dimensional hypersurface,α is the lapse function, and β^i is the shift function (see Appendix IV of[1] for details) . For any nonlinear Lagrangian, in sharp contrast with the Einstein-Maxwell case, the field equations $d{*}\mathbb{F} = {*}\mathbb{J}$ and the Bianchi-geometrical condition $dF = 0$ (where we have defined the Hodge dual $*$ and $\mathbb{F} = \frac{\partial \mathcal{L}}{\partial F}$) are expressed by the vector fields

$$E, B, \mathbb{E} = \frac{\partial \mathcal{L}}{\partial E}, \mathbb{B} = \frac{\partial \mathcal{L}}{\partial B} \tag{58}$$

that live into the slice. In our case given by the geometrical Lagrangian $\mathcal{L}_g$ (not be confused with the Lie derivative$\mathcal{L}_\beta$!)

$$\nabla \cdot \mathbb{E} = -\overline{h} \cdot \mathbb{B} + 4\pi\rho_e \tag{59}$$

$$\nabla \cdot B = 0 \tag{60}$$

$$\begin{aligned} \nabla \times (\alpha E) &= -(\partial_t - \mathcal{L}_\beta)B \\ &= -\partial_0 B + (\beta \cdot \nabla) B - (B \cdot \nabla)\beta \end{aligned} \tag{61}$$

$$\begin{aligned} \nabla \times (\alpha \mathbb{B}) + h_0 \mathbb{B} - \overline{h} \times \mathbb{E} &= -(\partial_t - \mathcal{L}_\beta)\mathbb{E} + 4\pi\alpha j \\ &= \partial_0 \mathbb{E} - (\beta \cdot \nabla)\mathbb{E} + (\mathbb{E} \cdot \nabla)\beta + 4\pi\alpha j \end{aligned} \tag{62}$$

where h^μ is the torsion vector. Notice that, here and the subsequent Sections, the overbar indicates 3-dimensional space vectors.

8.2. Dynamo effect and geometrical origin of $\alpha\Omega$ term[3]

In the case of weak field approximation and $\left(F^{01} \to E^{i},\ F^{jk} \to B^{i}\right)$the electromagnetic Maxwell-type equations in 3+1 take the form

$$\nabla_{\nu} F^{\nu\mu} = T^{\mu\nu\rho} F_{\nu\rho} = \varepsilon^{\mu\nu\rho}{}_{\delta} h^{\delta} F_{\nu\rho} \qquad (d^{*}F =^{*} J) \tag{63}$$

$$\overline{\nabla} \cdot \overline{E} = -\overline{h} \cdot \overline{B} \tag{64}$$

$$\partial_t \overline{E} - \overline{\nabla} \times \overline{B} = h^{0} \overline{B} - \overline{h} \times \overline{E} \tag{65}$$

and

$$\nabla_{\nu}{}^{*}F^{\nu\mu} = 0 \qquad (dF = 0) \tag{66}$$

$$\overline{\nabla} \cdot \overline{B} = 0 \tag{67}$$

$$\partial_t \overline{B} = -\overline{\nabla} \times \overline{E} \tag{68}$$

Putting all together, the set of equations is

$$\overline{\nabla} \cdot \overline{E} + \overline{h} \cdot \overline{B} = \rho_{ext} \tag{69}$$

$$\partial_t \overline{E} - \overline{\nabla} \times \overline{B} = h^{0} \overline{B} - \overline{h} \times \overline{E} - \sigma_{ext} \left[\overline{E} + \overline{v} \times \overline{B}\right] \tag{70}$$

$$\overline{\nabla} \cdot \overline{B} = 0 \tag{71}$$

$$\partial_t \overline{B} = -\overline{\nabla} \times \overline{E} \tag{72}$$

where we have introduced external charge density and current. Following the standard procedure we take the rotational to the second equation above obtaining straightforwardly the modified dynamo equation

$$\overline{\nabla} \times \partial_t \overline{E} + \overline{\nabla}^{2} \overline{B} = \overline{\nabla} \times \left(h^{0} \overline{B}\right) + \left(\overline{h} \cdot \overline{B} - \rho_{ext}\right) \overline{h} + \left(\overline{\nabla} \cdot \overline{h}\right) \overline{E} - \sigma_{ext} \left[\partial_t \overline{B} + \left(\overline{\nabla} \cdot \overline{v}\right) \overline{B}\right] \tag{73}$$

where the standard identities of the vector calculus plus the first, the third and the fourth equations above have been introduced. Notice that in the case of the standard approximation and (in the spirit of this research) without any external or additional ingredients, we have

$$\overline{\nabla}^{2} \overline{B} = h^{0} \left(\overline{\nabla} \times \overline{B}\right) + \left(\overline{h} \cdot \overline{B}\right) \overline{h} + \left(\overline{\nabla} \cdot \overline{h}\right) \overline{E} \tag{74}$$

Here we can see that there exist and α- term with a pure geometrical origin (and not only a turbulent one) that is given by h^{0} (the zero component of the dual of the torsion tensor).

8.3. The generalized Lorentz force

An important point in any theory beyond relativity is the concept of force. As is known, general relativity has deficiencies at this point. Now we are going to show that it is possible to derive from our proposal the Lorentz force as follows. From expression (32) the geometrical induced current is recognized

$$\partial_t \overline{E} - \overline{\nabla} \times \overline{B} = h^0 \overline{B} - \overline{h} \times \overline{E} \equiv \overline{J} \tag{77}$$

$$\overline{J} \times \overline{B} = \left(h^0 \overline{B} - \overline{h} \times \overline{E} - \overline{j}_{ext}\right) \times \overline{B} \tag{78}$$

$$= -\left[\left(\overline{h} \cdot \overline{B}\right) \overline{E} - \left(\overline{E} \cdot \overline{B}\right) \overline{h}\right] - \overline{j}_{ext} \times \overline{B} \tag{79}$$

we assume j_{ext} proportional to the velocity and other contributions. Consequently, reordering terms from above, a geometrically induced Lorentz-like force arises

$$\left(\overline{J} + \overline{j}_{ext}\right) \times \overline{B} = -\left[\underbrace{\left(\overline{h} \cdot \overline{B}\right)}_{\rho_{geom}} \overline{E} - \left(\overline{E} \cdot \overline{B}\right) \overline{h}\right] \rightarrow \tag{80}$$

$$\underbrace{\left(\overline{h} \cdot \overline{B}\right)}_{\rho_{geom}} \overline{E} + \underbrace{\left(\overline{J} + \overline{j}_{ext}\right)}_{j_{gen}} \times \overline{B} = \left(\overline{E} \cdot \overline{B}\right) \overline{h} \rightarrow \text{Lorentz induced force} \tag{81}$$

being the responsible of the induced force, the torsion vector itself. Notice, from the above equation, the following issues:

1)The external currents are identified with $\overline{J}$

2) We can eliminate the electric field in standard form

$$\overline{E} = \frac{\left(\overline{E} \cdot \overline{B}\right) \overline{h} - \overline{j}_{gen} \times \overline{B}}{\left(\overline{h} \cdot \overline{B}\right)} \tag{82}$$

being the above expression very important in order to replace the electric field into the dynamo equation, introducing naturally the external current in the model.

8.4. Generalized current and α-term

In previous paragraph we have derived a geometrical induced Lorentz force where the link between the physical world and the proposed geometrical model is through a generalized current j_{gen}. An important fact of that expression is that it is possible to eliminate the electric field (and insert it into the equation of induction) as follows.

From the formula of the induction, namely

$$\underbrace{\overline{\nabla}^2\overline{B}+\overline{\nabla}\times\left(-h^0\overline{B}+\overline{h}\times\overline{E}\right)}_{\mathcal{E}_{Geom}}=0 \tag{83}$$

and using the eq. (49) to eliminate the electric field as function of the torsion, the generalized current and the magnetic field respectively:

$$\overline{h}\times\overline{E}=\frac{-\overline{h}\times\left(\overline{j}_{gen}\times\overline{B}\right)}{\left(\overline{h}\cdot\overline{B}\right)}=-\frac{\left(\overline{h}\cdot\overline{B}\right)\overline{j}_{gen}-\left(\overline{h}\cdot\overline{j}_{gen}\right)\overline{B}}{\left(\overline{h}\cdot\overline{B}\right)} \tag{84}$$

$$\overline{h}\times\overline{E}=-\overline{j}_{gen}+\frac{\left(\overline{h}\cdot\overline{j}_{gen}\right)}{\left(\overline{h}\cdot\overline{B}\right)}\overline{B}=\left|\overline{j}_{gen}\right|\left(-n_{\overline{j}_{gen}}+\frac{\cos\alpha}{\cos\beta}n_B\right) \tag{85}$$

being α the angle between the vector torsion $\overline{h}$ and the generalized current $\overline{j}_{gen}$and β the angle between $\overline{h}$ and the magnetic field $\overline{B}$.Above, n_B and $n_{\overline{j}_{gen}}$ are unitary vectors in the direction of $\overline{B}$ and $\overline{j}_{gen}$ respectively. Notice the important fact that the RHS of (68) is independent of the torsion and the magnetic field. Consequently we obtain

$$\underbrace{\overline{\nabla}^2\overline{B}+\overline{\nabla}\times\left[-\overline{j}_{gen}+\left(\frac{\left(\overline{h}\cdot\overline{j}_{gen}\right)}{\left(\overline{h}\cdot\overline{B}\right)}-h^0\right)\overline{B}\right]}_{\mathcal{E}_{Geom}}=0 \tag{86}$$

We introduce the explicitly the physical scenario via the generalized current $\overline{j}_{gen}$

$$-\overline{j}_{gen}\sim\sigma_{ext}\left[\overline{E}+\overline{v}\times\overline{B}\right]+\left(\frac{c}{e}\frac{\overline{\nabla}p}{n_e}\right) \tag{87}$$

then

$$\underbrace{\overline{\nabla}^2\overline{B}+\overline{\nabla}\times\left[\sigma_{ext}\left[\overline{E}+\overline{v}\times\overline{B}\right]+\left(\frac{c}{e}\frac{\overline{\nabla}p}{n_e}\right)+\left(\frac{\left(\overline{h}\cdot\overline{j}_{gen}\right)}{\left(\overline{h}\cdot\overline{B}\right)}-h^0\right)\overline{B}\right]}_{\mathcal{E}_{Geom}}=0 \tag{88}$$

$$\underbrace{\overline{\nabla}^2\overline{B}+\sigma_{ext}\left[\left(-\partial_t\overline{B}\right)+\overline{\nabla}\times\left(\overline{v}\times\overline{B}\right)\right]+\overline{\nabla}\times\left(\frac{c}{e}\frac{\overline{\nabla}p}{n_e}\right)+\overline{\nabla}\times\left(\frac{\left(\overline{h}\cdot\overline{j}_{gen}\right)}{\left(\overline{h}\cdot\overline{B}\right)}-h^0\right)\overline{B}}_{\mathcal{E}_{Geom}}=0 \tag{89}$$

finally the expected geometrically induced expression is obtained:

$$\partial_t\overline{B}=\eta\overline{\nabla}^2\overline{B}+\overline{\nabla}\times\left(\overline{v}\times\overline{B}\right)+\eta\overline{\nabla}\times\left[\left(\frac{c}{e}\frac{\overline{\nabla}p}{n_e}\right)+\alpha\overline{B}\right]=\partial_t\overline{B} \tag{90}$$

$$\rightarrow \eta \underbrace{\overline{\nabla}^2 \overline{B}}_{diffusive} + \underbrace{\overline{\nabla} \times (\overline{v} \times \overline{B})}_{advective} + \underbrace{\eta \overline{\nabla} \times (\alpha \overline{B})}_{\alpha - term} + \underbrace{\frac{c}{e} \frac{\overline{\nabla} p \times \overline{\nabla} n_e}{n_e^2}}_{Biermann\ battery} = -\partial_t \overline{B} \quad (91)$$

where $\eta \equiv \frac{1}{\sigma_{ext}}$ as usual and the geometric α:

$$\begin{aligned} \alpha &\equiv \left(\frac{(\overline{h} \cdot \overline{j}_{gen})}{(\overline{h} \cdot \overline{B})} - h^0 \right) \\ &= \left(\frac{\cos \alpha \left| \overline{j}_{gen} \right|}{\cos \beta \left| \overline{B} \right|} - h^0 \right) \end{aligned} \quad (92)$$

8.5. Seed magnetic field

Notice from the last expression that $\alpha \overline{B}$ is explicitly

$$\alpha \overline{B} = \frac{\cos \alpha \left| \overline{j}_{gen} \right|}{\cos \beta} n_B - h^0 \overline{B} \quad (93)$$

or (via elimination of the unitary vector)

$$\alpha \left| \overline{B} \right| = \frac{\cos \alpha \left| \overline{j}_{gen} \right|}{\cos \beta} - h^0 \left| \overline{B} \right| \quad (94)$$

we see *clearly* the first term in RHS independent of the intensity of the magnetic field. Considering only the terms of interest without the diffusive and advective term in the induction equation(only time-dependence for the magnetic field is preserved) namely

$$\underbrace{\eta \overline{\nabla} \times (\alpha \overline{B})}_{\alpha - term} = -\partial_t \overline{B} \quad (96)$$

$$\eta \overline{\nabla} \left(\frac{\cos \alpha \left| \overline{j}_{gen} \right|}{\cos \beta} \right) = -\partial_t \left| \overline{B} \right| \quad (97)$$

we see that the currents given by the fields (related to the geometry via h_α) originate the magnetic field.

If we consider all the currents of the fields of theory (fermions, bosons, etc.) the seed would be precisely these field currents. The other missing point is to derive the fluid (hydrodynamic) equations (which as is known does not have a definite Lagrangian formulation) from the same unified formulation. Notice that there are, under special conditions, analogous formulas for vorticity ω than for the magnetic field B. This would mean that the 2-form of vorticity must also be included in the fundamental antisymmetric tensor, together with the electromagnetic field.

8.6. Comparison with the mean field formalism

Now we compare the obtained equations with respect to the mean field formalism. Starting from expressions (69-72) as before, we have:

$$\eta\overline{\nabla}^2\overline{B}+\overline{\nabla}\times\left(\overline{v}\times\overline{B}\right)-\partial_t\overline{B}+\underbrace{\eta\overline{\nabla}\times\left(-h^0\overline{B}+\overline{h}\times\overline{E}\right)}_{\mathcal{E}_{Geom}}=0 \tag{98}$$

$\mathcal{E}_{Geom}$ takes the place of electromotive force due the torsion field with full analogy as $\mathcal{E}=\langle u\times b\rangle$ is the mean electromotive force due to fluctuations. Also as in the mean field case that there are the splitting

$$\mathcal{E}=\mathcal{E}^{\langle 0\rangle}+\mathcal{E}^{\langle\overline{B}\rangle} \tag{99}$$

with $\mathcal{E}^{\langle 0\rangle}$ independent of $\langle\overline{B}\rangle$ and $\mathcal{E}^{\langle\overline{B}\rangle}$linear and homogeneous in $\overline{B}$, we have in the torsion case the following correspondence

$$-h^0\overline{B}\longleftrightarrow\mathcal{E}^{\langle\overline{B}\rangle}$$
$$\overline{h}\times\overline{E}\longleftrightarrow\mathcal{E}^{\langle 0\rangle}$$
$$geometrical\longleftrightarrow turbulent$$

Consequently, the problems of mean-field dynamo theory that are concerned with the generation of a mean EMF by turbulence, have in this model a pure geometric counterpart. In the past years, attention has shifted from kinematic calculations, akin to those familiar from quasilinear theory for plasmas, to self-consistent theories which account for the effects of small scale magnetic fields (including their back-reaction on the dynamics) and for the constraints imposed by the topological conservation laws, such as that for magnetic helicity. Here the torsion vector generalize (as we can see from above set of equations) the concept of helicity. The consequence of this role of the dual torsion field is that the traditionally invoked mean-field dynamo mechanism (i.e. the so-called alpha effect) may be severely quenched or increased at modest fields and magnetic Reynolds numbers, and that spatial transport of this generalized magnetic helicity is crucial to mitigating this quench. Thus, the dynamo problem is seen in our model as one of generalized helicity transport, and so may be tackled like other problems in turbulent transport. A key element in this approach is to understand the evolution of the torsion vector field besides of thc turbulence energy and the generalized helicity profiles in space-time. This forces us to confront the problem of spreading of strong MHD turbulence, and a spatial variant or analogue of the selective decay problem with the dynamics of the torsion field.

9. Torsion, axion electrodynamics vs. Chern Simons theory

Let us review briefly the electromagnetic sector of the theory QCD based in a gauge symmetry $SU(3) \times U(1)$

$$L_{QCD/QED} = +\sum \overline{\psi}_f \left[\gamma^\mu \left(\partial_\mu - ig_f t^\alpha A^\alpha_\mu - iq_f A_\mu\right) - m_f\right] \psi_f - \\ -\frac{G^\alpha_{\mu\nu} G^{\alpha\mu\nu}}{4} - \frac{F_{\mu\nu} F^{\mu\nu}}{4} - \frac{g^2 \theta G^\alpha_{\mu\nu} \widetilde{G}^{\alpha\mu\nu}}{32\pi^2} - \frac{g^2 \theta F_{\mu\nu} \widetilde{F}^{\mu\nu}}{32\pi^2}, \tag{100}$$

As is well know, electromagnetic fields will couple to the electromagnetic currents, namely:$J_\mu = \sum\limits_f q_f \overline{\psi}_f \gamma_\mu \psi_f$ consequently , there appear term will induce through the quark loop the coupling of $F_{\mu\nu}\widetilde{F}^{\mu\nu}$ (the anomaly) to the QCD topological charge . The effective Lagrangian can be written as

$$L_{MCS} = -\frac{F_{\mu\nu} F^{\mu\nu}}{4} - A_\mu J^\mu - \frac{c}{4} \theta F_{\mu\nu} \widetilde{F}^{\mu\nu} \tag{101}$$

where a pseudo-scalar field $\theta = \theta(\overline{x}, t)$ (playing the role of the axion field) is introduced and $c = \sum\limits_f \frac{(q_f e)^2}{2\pi^2}$. This is the Chern-Simons Lagrangian where, if $\theta$is constant, the last term is a total divergence: $F_{\mu\nu}\widetilde{F}^{\mu\nu} = \partial_\mu J^\mu_{CS}$.The question appear if θ is not a constant $\theta F_{\mu\nu}\widetilde{F}^{\mu\nu} = \theta \partial_\mu J^\mu_{CS} = \partial_\mu \left(\theta J^\mu_{CS}\right) - J^\mu_{CS} \partial_\mu \theta$

Now we can see from the previous section that if, from the general decomposition of the four dimensional dual of the torsion field via the Hodge de Rham theorem we retain b_α as gradient of a pseudoscalar (e.g: axion) these equations coincide in form with the respective equation for MCS theory. Precisely because under this condition $h_\alpha = \nabla_\alpha \theta$, in flat space (curvature=0 but torsion$\neq 0$) the equations become the same as in namely

$$\overline{\nabla} \cdot \overline{E} - c\overline{P} \cdot \overline{B} = \rho_{ext} \tag{102}$$

$$\partial_t \overline{E} - \overline{\nabla} \times \overline{B} = -c\dot{\theta}\overline{B} + c\overline{P} \times \overline{E} - \sigma_{ext}\left[\overline{E} + \overline{v} \times \overline{B}\right] \tag{103}$$

$$\overline{\nabla} \cdot \overline{B} = 0 \tag{104}$$

$$\partial_t \overline{B} = -\overline{\nabla} \times \overline{E} \tag{105}$$

provided:

$$h^0 \to -c\dot{\theta} \tag{106}$$

$$\overline{h} \to -c\overline{P} \tag{107}$$

where from QCD the constant c is determined as $c = \frac{e^2}{2\pi}$ and the $\partial_\mu \theta = \left(\dot{\theta}, \overline{P}\right)$. The main difference is that while in the case of photons in axion ED was given by the Lagrangian where that above equations are derived is

$$L_{MCS} = -\frac{F_{\mu\nu}F^{\mu\nu}}{4} - A_\mu J^\mu + \frac{c}{4}P_\mu J^\mu_{CS}, \qquad J^\mu_{CS} \equiv \varepsilon^{\mu\sigma\rho\nu} A_\sigma F_{\rho\nu} \quad (108)$$

in our case is the dual of the torsion field (that we take as the gradient of a pseudoscalar) responsible of the particular structure of the set of equations.

10. Magnetic helicity generation and cosmic torsion field

Here we consider the projective invariant case: $\beta = 0$ ($R_A = -4\lambda$)where the gravitational and field equations are considerably simplified because $\mathbb{R} = 1$ and $b^{-1} = 0$. Scalar curvature R and the torsion 2-form field $T^a_{\mu\nu}$ with a $SU(2)$ −Yang-Mills structure are defined in terms of the affine connection $\Gamma^\lambda_{\mu\nu}$ and the $SU(2)$ valuated (structural torsion potential) $f^a{}_\mu$ by

$$R = g^{\mu\nu} R_{\mu\nu} \qquad R_{\mu\nu} = R^\lambda_{\mu\lambda\nu} \quad (109)$$

$$R^\lambda_{\mu\lambda\nu} = \partial_\nu \Gamma^\lambda_{\mu\rho} - \partial_\rho \Gamma^\lambda_{\mu\nu} + ...$$

$$T^a_{\mu\nu} = \partial_\mu f^a{}_\nu - \partial_\nu f^a{}_\mu + \varepsilon^a_{bc} f^b{}_\mu f^c{}_\nu$$

G and Λ are the geometrically induced Newton gravitational constant (as we have been discussed before) and the integration cosmological constant, respectively. From the last equation for the totally antisymmetric Torsion 2-form, the potential $f^a{}_\mu$ define the affine connection $\Gamma^\lambda_{\mu\nu}$ .Similarly to the case of Einstein-Yang-Mills systems, for our new UFT model it can be interpreted as a prototype of gauge theories interacting with gravity (*e.g.* QCD, GUTs, etc.). We stress here the important fact that all the fundamental constants are really geometrically induced as required by the Mach principle. After varying the action, we obtain the gravitational equation (41), namely

$$\overset{\circ}{R}_{\alpha\beta} - \frac{g_{\alpha\beta}}{2}\overset{\circ}{R} = 6\left(-h_\alpha h_\beta + \frac{g_{\alpha\beta}}{2} h_\gamma h^\gamma\right) + \kappa_{geom}\left[F_{\alpha\lambda}F_\beta{}^\lambda - F_{\mu\nu}F^{\mu\nu}\frac{g_{\alpha\beta}}{4}\right] + \Lambda g_{\alpha\beta} \quad (110)$$

with the "gravitational constant" geometrically induced as

$$\kappa_g \equiv \frac{R_s}{2\mathcal{W}} = \left.\frac{8\pi G}{c^4}\right|_{today} \quad (111)$$

and the field equation for the torsion 2-form in differential form

$$d^*T^a + \tfrac{1}{2}\varepsilon^{abc}\left(f_b \wedge^* T_c -^* T_b \wedge f_c\right) = -\lambda\, {}^*f^a \quad (112)$$

Notice that κ_g and Λ are not independent, but related by $R_s = 2\Lambda$. In this case $\beta = 0$ we have the simplest expression:

$$\kappa_g \equiv \frac{R_s}{2\left(1+\frac{R_s}{4\lambda}\right)^2} = \frac{\Lambda}{\left(1+\frac{2\Lambda}{4\lambda}\right)^2}$$

in consequence, generalizing the conjecture of Markov **if Λ is proportional to the energy,** κ goes as Λ if $|\Lambda| \leq 1$, and as Λ^{-1} in other case.

We are going to seek for a classical solution with the following ansatz for the metric and gauge connection

$$ds^2 = d\tau^2 + a^2(\tau)\,\sigma^i \otimes \sigma^i \equiv d\tau^2 + e^i \otimes e^i. \tag{113}$$

Here τ is the euclidean time and the dreibein is defined by $e^i \equiv a(\tau)\,\sigma^i$. The gauge connection is

$$f^a \equiv f^a_\mu dx^\mu = f\sigma^a, \tag{114}$$

for $a, b, c = 1, 2, 3$, and for $a, b, c = 0$ we have

$$f^0 \equiv f^0_\mu dx^\mu = s\sigma^0. \tag{115}$$

This choice for the potential torsion is accordingly to the symmetries involved in the problem.

The σ^i 1-form satisfies the $SU(2)$ Maurer-Cartan structure equation

$$d\sigma^a + \varepsilon^a{}_{bc}\sigma^b \wedge \sigma^c = 0 \tag{116}$$

Notice that in the ansatz the frame and $SU(2)$ (isospin-like) indices are identified (as for the case with the non-abelian-Born-Infeld (NBI) Lagrangian) The torsion 2-form

$$T^\gamma = \frac{1}{2}T^\gamma{}_{\mu\nu}dx^\mu \wedge dx^\nu \tag{117}$$

becomes

$$\begin{aligned} T^a &= df^a + \frac{1}{2}\varepsilon^a{}_{bc}f^b \wedge f^c \\ &= \left(-f + \frac{1}{2}f^2\right)\varepsilon^a{}_{bc}\sigma^b \wedge \sigma^c \end{aligned} \tag{118}$$

$$\begin{aligned} d^*T^a + \tfrac{1}{2}\varepsilon^{abc}\left(f_b \wedge^* T_c -^* T_b \wedge f_c\right) &= -2\lambda\, {}^*f^a \\ (-2f + f^2)(1-f)d\tau \wedge e^b \wedge e^c &= -2\lambda d\tau \wedge e^b \wedge e^c \end{aligned} \tag{119}$$

$${}^*T^a \equiv h(-2f + f^2)d\tau \wedge \frac{e^a}{a^2} \tag{120}$$

$$^{*}f^{a} = -f\frac{d\tau \wedge e^{b} \wedge e^{c}}{a^{3}} \tag{121}$$

Note that to be complete in our description of the possible physical scenarios, we include f^0 as an $U(1)$ component of the torsion potential (although does not belong to the space $SU(2)/U(1)$). Having all the above issues into account, the expression for the torsion is analogous to the non-abelian 2-form strength field.

Inserting T^a from (118) into the dynamic equation (112) we obtain

$$(-2f + f^2)(1 - f)d\tau \wedge e^b \wedge e^c = -\lambda d\tau \wedge e^b \wedge e^c, \tag{122}$$

and from expression (122) we have an algebraic cubic equation for f

$$(-2f + f^2)(1 - f) + \lambda = 0 \tag{123}$$

We can see that, in contrast with our previous work with a dualistic theory where the NBI energy-momentum tensor of Born-Infeld was considered, there exist three non trivial solutions for f, depending on the cosmological constant λ. In this preliminary analysis of the problem, only the values of f that make the quantity $\left(-f + \frac{1}{2}f^2\right) \in \mathbb{R}$. Consequently for $\lambda = 2$ we find $f = 2.35$ then

$$T^a_{bc} = \frac{2}{5}\frac{\varepsilon^a_{bc}}{a^2}; \qquad T^a_{0c} = 0 \tag{124}$$

That is, only spatial torsion field is non vanishing while cosmic time torsion field vanishes. Substituting the expression for the torsion 2-form [1] into the symmetric part of the variational equation we reduce the gravitational equations to an ordinary differential equation for the scale factor a,

$$3\left[\left(\frac{\dot{a}}{a}\right)^2 - \frac{1}{a^2}\right] - \Lambda = \frac{3\kappa_g}{4a^2}\left(f^2 + s^2\right) + \frac{3}{2a^4}f^2\left(f - 2\right)^2 \tag{125}$$

that in the case for the computed value for $f \sim 2.35$ with $s = 10$ and $\Lambda \lesssim 1$ the scale factor is described in Figure 1 and the scale factor goes as:

$$a(\tau) = \Lambda^{-1/2}\sqrt{\left(1 - \frac{12\kappa_g^2\Lambda}{\alpha}\right)^{1/2} \sinh\left(\sqrt{\Lambda/3}\,(\tau - \tau_0)\right) - 1 + \kappa_g\left(f^2 + s^2\right)/4} \tag{126}$$

where we define the geometrically induced fine structure function $\alpha \equiv \kappa_g\left(f^2 + s^2\right)/4$

[1] in the tetrad: $\overset{\circ}{R}_{00} = -3\frac{\ddot{a}}{a}$, $\overset{\circ}{R}_{ab} = -\left[\frac{\ddot{a}}{a} + 2\left(\frac{\dot{a}}{a}\right)^2 - \frac{2}{a^2}\right]$

10.1. Primordial symmetries of standard model and torsion field

Recently (see ref. in [1][2][4]the cross section for neutrino helicity spin flip obtained from this type of $f(R;T)$ model of gravitation with dynamic torsion field introduced by us was phenomenologically analyzed using the relation with the axion decay constant f_a (Peccei-Quinn parameter) due the energy dependence of the cross section, Consequently, the link with the phenomenological energy/mass window was found from the astrophysical and high energy viewpoints. The important point is that, in relation with the torsion vector interaction Lagrangian, the f_a parameter gives an estimate of the torsion field strength that can variate with time within cosmological scenarios as the described above, potentially capable of modifying the overall leptogenesis picture, the magnetogenesis, the bariogenesis and also to obtain some indication about the primordial (super) symmetry of the early universe.

In FRW scenario given here we saw that the torsion through its dual vector, namely:

$$h^0 = \frac{2}{5}\frac{\delta_a^0 C_\tau}{a^2} d\tau \wedge e^a \tag{127}$$

goes as $\sim a^{-2}$ with C_τ a covariantly constant vector field$\left(\text{e.g.:}\overset{\circ}{\nabla} C_\tau = 0\right)$ that we take of the form $C_\tau \sim \left(\dot{\theta} + q_\tau\right)$ (due the Hodge-de Rham decomposition of h_μ, expression(42))where $\theta$is a pseudoscalar field playing the obvious role of axion and q_τ :vector field linking h^0 with the magnetic field via the equation of motion for the torsion. Consequently, the torsion dual vector h^ihas the maximum value when the radius of the universe is $a_{\min}$, e.g. $a_{\min} = a(\tau_0)$ increases to the maximum value the spin-flip neutrino cross section and, for instance, the quantity of right neutrinos compensating consequently the actual (e.g. $a_{today} = a(\tau)$)assymetry of the electroweak sector of the SM (see the behaviour of a in Fig.1). This fact indicates that the original symmetry group contains naturally $SU_R(2) \times SU_L(2) \times U(1)$ tipically inside GUT's structurally based generally in $SO(10)$, $SU(5)$ or some exceptional groups as $E(6)$,$E(7)$, *etc.*

Also it is interesting to note that from the FRW line element written in terms of the cosmic time the Hubble flow electromagnetic fields $E_\mu \equiv$

$(0, E_i) = a^{-2}(0, \partial_\tau A_i)$ and $B_\mu \equiv (0, B_i) = a^{-2}(0, \varepsilon_{ijk}\partial_j A_k)$

$$\nabla \cdot \overline{E} + \left(\frac{\alpha}{f}\overline{\nabla}\theta + \overline{\Pi}\right) \cdot \left(a^2\overline{B}\right) = 0 \tag{111}$$

$$\partial_\tau\left(a^2\overline{E}\right) - \overline{\nabla} \times \left(a^2\overline{B}\right) = \left(\frac{\alpha}{f}\partial_\tau\theta + \Pi_0\right)\left(a^2\overline{B}\right) - \left(\frac{\alpha}{f}\overline{\nabla}\theta + \overline{\Pi}\right) \times \overline{E} \tag{112}$$

$$\overline{\nabla} \cdot \overline{B} = 0 \tag{113}$$

$$\partial_t \overline{B} = -\overline{\nabla} \times \overline{E} \tag{114}$$

where $\Pi_\mu \equiv f_\mu\left(u_\mu, \gamma^5 b_\mu, eA_\mu.....\right)$ is a vector function of physical entities as potential vector, vorticity, angular velocity, axial vector etc etc. as described by expression (42). In principle we can suppose that it is zero (low back reaction) then

$$\overline{h} = \frac{\alpha}{f}\overline{\nabla}\theta, \qquad h^0 = \frac{\alpha}{f}\partial_\tau\theta \tag{115}$$

being $\left[\partial_\tau^2 - \overline{\nabla}^2 - \frac{\alpha}{f}\partial_\tau\theta\overline{\nabla}\times\right]\left(a^2\overline{B}\right) = 0$ the second order equation for the magnetic field that shows the chiral character of the plasma particles.

10.2. Magnetogenesis and cosmic helicity

Now we pass to see which role plays the torsion field in the magnetic field generation in a FRW cosmology. Taking as the starting point the (hyper) electrodynamic equations and introducing a Fourier mode decomposition $\overline{B}(\overline{x}) = \int d^3\overline{k}\overline{B}\left(\overline{k}\right)e^{-i\overline{k}\cdot\overline{x}}$ with $\overline{B}\left(\overline{k}\right) = h_i\overrightarrow{e}_i$ where $i = 1, 2$, $\overrightarrow{e}_i^2 = 1$, $\overrightarrow{e}_i \cdot \overrightarrow{k} = \overrightarrow{e}_1 \cdot \overrightarrow{e}_2 = 0$ the torsion-modified dynamical equations for the expanding FRW become

$$\dot{\overline{z}} + \left[\left(2\dot{a} + \frac{k^2}{\sigma}\right) + \frac{ah^0\left|k\right|}{\sigma}\right]\overline{z} = 0 \tag{116}$$

$$\dot{z} + \left[\left(2\dot{a} + \frac{k^2}{\sigma}\right) - \frac{ah^0\left|k\right|}{\sigma}\right]z = 0 \tag{117}$$

where the magnetic field is written in terms of complex variable $z\left(\overline{z}\right)$ as

$$z = h_1 + ih_2 \tag{118}$$

$$\overline{z} = h_1 - ih_2 \tag{119}$$

from equation (117) we see that the solution for z namely:

$$z = z_0 e^{-\left(2a + \frac{k^2}{\sigma}\tau\right) + \int \frac{ah^0\left|k\right|}{\sigma}d\tau} \tag{120}$$

contains the instable mode in the sense of $\frac{k}{\sigma}\tau < \int \frac{ah^0}{\sigma} d\tau$. Consequently a defined polarization of the magnetic field appear and from the dynamical equation for the torsion field: $\nabla_{[\mu} h_{\nu]} = -\lambda \widetilde{F}_{\mu\nu}$ that in this case we have

$$\nabla_{[i} h_{\tau]} = \nabla_i \left(a^{-2} q_\tau\right) = -\lambda B_i \tag{121}$$

that implies a relation between the vector part of the h^0 (namely q_τ) with the vector potential A^k of the magnetic field as follows:

$$\nabla_i q_\tau \approx -\lambda \varepsilon_{ijk} \nabla^j A^k \tag{122}$$

Consequently, the primordial magnetic field (or seed) would be connected in a self-consistent way with the torsion field by means of the dual vector h_0.It (h_μ) in turn, would be connected phenomenologically with the physical fields (matter) of theory through Hodge-de Rham decomposition expression (42). We note from expression (120) that the pseudo-scalar (axion) controls the stability, growth and dynamo effect but not the generation of the magnetic field (primordial or seed) as is clear from expression (122) where the (pseudo) -vector part of h_0 contributes directly to the generation of the magnetic field as clearly given by eq. (121)

10.3. Magnetogenesis and cosmic helicity II

In the case to include the *complete alpha term* given by equations (92) and in the same analytical conditions (e.g.: Fourier decomposition) from the previous paragraph, the torsion-modified dynamic equations for the expanding FRW become

$$\dot{\overline{z}} + \left[\left(2\dot{a} + \frac{k^2}{\sigma}\right) + \frac{a\,|k|}{\sigma}\left(h^0 - \frac{\cos\alpha\,\left|\overline{\jmath}_{gen}\right|}{\cos\beta\,|\overline{z}|}\right)\right]\overline{z} = 0 \tag{123}$$

$$\dot{z} + \left[\left(2\dot{a} + \frac{k^2}{\sigma}\right) - \frac{a\,|k|}{\sigma}\left(h^0 - \frac{\cos\alpha\,\left|\overline{\jmath}_{gen}\right|}{\cos\beta\,|z|}\right)\right] z = 0 \tag{124}$$

where in this case the magnetic field is written (by convenience) in terms of complex variable $z\,(\overline{z})$ as

$$z = |z|\,e^{i\rho} \rightarrow \dot{z} = \left(\dot{|z|} + i\dot{\rho}\,|z|\right) e^{i\rho} \tag{125}$$

$$\overline{z} = |\overline{z}|\,e^{-i\rho} \rightarrow \dot{\overline{z}} = \left(\dot{|\overline{z}|} - i\dot{\rho}\,|\overline{z}|\right) e^{-i\rho} \tag{126}$$

From equation (124) we see that the solution for z namely:

$$z = z_0 \exp\left[-\left(2a + \frac{k^2}{\sigma}\tau\right) + \int \frac{a\,|k|}{\sigma}\left(h^0 - \frac{\cos\alpha\,\left|\overline{j}_{gen}\right|}{\cos\beta\,|z_0|}\right) d\tau\right] \tag{127}$$

with $z_0 = |z_0|\, e^{i\rho_0}$ $\quad (|z_0| = const)$

contains the instable mode in the sense of Joice and Shaposchnikov, for example (117) $\frac{k}{\sigma}\tau < \int \frac{a}{\sigma}\left(h^0 - \frac{\cos\alpha\left|\overline{j}_{gen}\right|}{\cos\beta|z|}\right) d\tau$. But now *there are not a definite polarization for the magnetic field*, but now all depends on the difference:

$$\int \frac{a}{\sigma}\left(h^0 - \frac{\cos\alpha\,\left|\overline{j}_{gen}\right|}{\cos\beta\,|z_0|}\right) d\tau$$

Replacing explicitly h_α from the decomposition (42) we can see in a clear form, the interplay between the physical entities, as the vortical and magnetic helicities for example:

$$\left(\nabla_0\Omega + \frac{4\pi}{3}\left[h_M + q_s n_s \overline{u}_s \cdot \overline{B}\right] + \gamma_1 h_V + \gamma_2 P_0\right) - \frac{\cos\alpha\,\left|\overline{j}_{gen}\right|}{\cos\beta\,|z_0|}$$

Now considering in $\left|\overline{j}_{gen}\right|$ the fermionic current $\sum_f q_f \overline{\psi}_f \gamma_\mu \psi_f$, Ω as the axion a , $|z_0| = \frac{\cos\beta}{\cos\alpha}$ and putting $\gamma_2 = 0$ we have an interesting expression:

$$\nabla_0 a + \frac{4\pi}{3}\left[h_M + q_s n_s \overline{u}_s \cdot \overline{B}\right] + \gamma_1 h_V = \left|\sum_f q_f \overline{\psi}_f \gamma_\mu \psi_f\right|$$

The above expression it is very important because establishes the desired connection between helicities, magnetic field and fermionic fields and axion. We can order it as

$$\nabla_0 a - \left|\sum_f q_f \overline{\psi}_f \gamma_\mu \psi_f\right| = -\left[\frac{4\pi}{3}\left(h_M + q_s n_s \overline{u}_s \cdot \overline{B}\right) + \gamma_1 h_V\right]$$

We now clearly see the link between the axion and the fermionic fields (the dynamics of the interacting fields and the involved currents) in the LHS and the macroscopic physical observables in the RHS giving an indication of the origin of leptogenesis and bariogenesis in the context of this non Riemannian gravitational model.

11. Discussion and perspectives

In this paper we have introduced a simple geometric Lagrangian in the context of a unified theory based on affine geometry. From the functional action proposed, that is as square root or measure, the dynamic equations were derived: an equation analogous to trace free Einstein equations TFE and a dynamic equation for the torsion (which was taken totally antisymmetric). Although the aim of this paper was to introduce and to analize the model from the viewpoint of previous research, we bring some new results and possible explanations about the generation of primordial magnetic fields and the link with the leptogenesis and baryogenesis. The physically admisible analysis of the torsion vector h_{μ}, from the point of view of the symmetries, has allowed us to see how matter fields can be introduced in the model. These fields include some dark matter candidates such as axion, right neutrinos and Majorana. Also the vorticity can be included in the same way and, as the torsion vector is connected to the magnetic field, both vorticity and magnetic field can be treated with equal footing. The other point is that from the wormhole soluton in a cosmological spacetime with torsion we show that primordial cosmic magnetic fields can be originated by the dual torsion field h_{μ} being the axion field contained in h_{μ}, the responsible to control the dynamics and stability of the cosmic magnetic field, but is not responsible of the magnetogenesis itself. Also the energy conditions in the wormhole solution are fulfilled. The last important point to highlight is that the dynamic torsion field h_{μ} acts as mechanism of the reduction of an original (early, primordial) GUT (Grand Unified Theory) symmetry of the universe containing $\sim$SU(3) $\times$SU$(2)_R$ $\times$SU$(2)_L$ $\times$U(1) to SU(3) $\times$SU$(2)_L$ $\times$U(1) today. Consequently, the GUT candidates are $SO(10)$, $SU(5)$ or some exceptional groups as $E(6)$,$E(7)$ for example. Also we can give some hints on the astrophysical consequences of the affine structure underlying the gravitational theory presented here, as the new dynamo effects manifestation by mean a modified Grad Shafranov equation, acrretion mechanisms and magnetosphere structucture driven by axion fields, etc

12. Acknowlegments

I am very grateful to the organizers of this Meeting, particularly Professors Igor Salom, Marko Vojinovic and Djorde Sijacki for fruitful discussions.and CONICET-ARGENTINA for financial support.

References

[1] Diego Julio Cirilo-Lombardo, JHEAp 16 (2017) 1-14

[2] Diego Julio Cirilo-Lombardo, Int.J.Geom.Meth.Mod.Phys. 14 (2017) no.07, 1750108

[3] Diego Julio Cirilo-Lombardo, Phys.Part.Nucl.Lett. 14 (2017) no.6, 799-810

[4] David Alvarez-Castillo, Diego Julio Cirilo-Lombardo, Jilberto Zamora-Saa .JHEAp 13-14 (2017) 10-16,, Andrej B. Arbuzov and Diego Julio Cirilo-Lombardo, Int.J.Geom.Meth.Mod.Phys. 15 (2017) no.01, 1850005

Power law method for finding soliton solutions of the $2D$ Ricci flow model

Radu Constantinescu
Aurelia Florian
Carmen Ionescu
Alina-Maria Pauna
University of Craiova, 13 A. I. Cuza, 200585 Craiova, Romania

Abstract

The paper reviews few general methods which are used for investigating the integrable models and for finding their analytic solutions.Two specific methods, the symmetry method and the auxiliary equation method, will be especially considered. They will be applied to the Ricci flow model expressed as a nonlinear PDE in $2+1$ dimensions. Both methods have a similar philosophy: replacing the model by an ODE obtained through similarity reduction (in the approach based on symmetry), respectively by passing to the wave variable. The focus will be on the auxiliary equation method and a new approach, called power law method will be propsed.

1. Introduction

An important question in the study of the nonlinear differential equations is related to their integrability, that is establishing if solutions of the considered equations exist. There is not a general theory/procedure allowing to completely solve nonlinear ODEs or PDEs. Sometimes it is quite enough to decide if the system is integrable or not. There are many methods which have been proposed for deciding on the integrability of the nonlinear equations such as: the inverse scattering method [1, 2], the Hirota bilinear transformation [3], the generalized Riccati equation method [4], the $\left(\frac{G'}{G}\right)$-expansion method [5, 6], the Lie symmetry method [7, 8, 9, 10, 11], the soliton ansatz method [12, 13], the generalized conditional symmetry approach [14, 15] and other techniques. In this paper we will focus on two such methods: the symmetry approach and the expansion method applied to a specific nonlinear PDE, namely the 2D Ricci flow equation [16], [17]. It has been used by mathematicians in order to understand special geometries which admit 3-manifolds. These mathematical developments have been useful in physics too. Many papers dealing with physical applications of the Ricci flow, among which [18], [19], [20], [21], have been published. In [22] the Lie point symmetries are calculated in terms of two arbitrary functions. Conservation laws for the $2D$ Ricci flow model via the direct construction method [23], have been established in [24] and some new invariant solutions have been derived.

The paper is organized as follow: after this introductory section, in Section 2 the symmetry method will be presented. Section 3 will review the auxiliary equation method, while in Section 4 the $2D$ Ricci flow model will be analyzed. As a novelty, a most general polynomial expansion which includes almost all the approaches usually applied in literature will be proposed.

2. Symmetries and their applications in nonlinear dynamics

Many natural phenomena are described by a system of nonlinear partial differential equations (PDEs) which is often difficult to be solved analytically, since there is no general theory for completely solving nonlinear pdes. One of the most useful techniques for finding exact solutions for the dynamical systems described by nonlinear pdes is *the symmetry method.* On the one hand, we can consider the symmetry reduction of differential equations and thus obtain classes of exact solutions. On the other hand, by definition, a symmetry transforms solutions into solutions, and thus symmetries can be used to generate new solutions from known ones. The classical Lie method [25], the nonclassical method [26], the direct method [27], the differential constraint approach [28], the new nonclassical algorithm, the generalized symmetry method [29] are important methods for finding special classes of solutions for nonlinear PDEs.

The **Lie symmetry method** has been proven to be a powerful tool for studying a remarkable number of PDEs arising in mathematical physics [30], [31].

Let us consider a dynamical system described by the n-th order partial differential equation:

$$\Delta_\nu(x, u^{(n)}[x]) = 0, \tag{1}$$

where $x \equiv \{x^i,\ i = \overline{1,p}\} \subset R^p$ represent independent variables, while $u \equiv \{u^\alpha, \alpha = \overline{1,q}\} \subset R^q$ dependent ones. The notation $u^{(n)}$designates the set of variables which includes u and the partial derivatives of u up to the n-th order.

The general infinitesimal symmetry operator has the form:

$$U = \sum_{i=1}^{p} \xi^i(x, u)\frac{\partial}{\partial x^i} + \sum_{\alpha=1}^{q} \phi_\alpha(x, u)\frac{\partial}{\partial u^\alpha}. \tag{2}$$

The n-th extension of (2) is given by:

$$U^{(n)} = U + \sum_{\alpha=1}^{q} \sum_{J} \phi_{\alpha}^{J}(x, u^{(n)}) \frac{\partial}{\partial u_J^{\alpha}}, \tag{3}$$

where

$$u_J^{\alpha} = \frac{\partial^m u^{\alpha}}{\partial x^{j_1} \partial x^{j_2} .. \partial x^{j_m}}. \tag{4}$$

Also, in (4) the second summation refers to all the multi-indices $J = (j_1, ...j_m)$, with $1 \leq j_m \leq p, 1 \leq m \leq n$ .The coefficient functions ϕ_{α}^{J} are given by the following formula:

$$\phi_{\alpha}^{J}(x^i, u^{(n)}) = \mathcal{D}_J[\phi_{\alpha} - \sum_{i=1}^{p} \xi^i u_i^{\alpha}] + \sum_{i=1}^{p} \xi^i u_{J,i}^{\alpha}, \ \alpha = \overline{1, q}, \tag{5}$$

in which

$$u_i^{\alpha} = \frac{\partial u^{\alpha}}{\partial x^i}, \ i = \overline{1, p}, \tag{6}$$

$$u_{J,i}^{\alpha} = \frac{\partial u_J^{\alpha}}{\partial x^i} = \frac{\partial^{m+1} u^{\alpha}}{\partial x^i \partial x^{j_1} \partial x^{j_2} .. \partial x^{j_m}}, \tag{7}$$

$$\mathcal{D}_J = \mathcal{D}_{j_1} \mathcal{D}_{j_2} ... \mathcal{D}_{j_m} = \frac{d^{\ m}}{dx^{j_1} dx^{j_2} .. dx^{j_m}}. \tag{8}$$

The Lie symmetries represent the set of all the infinitesimal transformations which keep invariant the differential system. The invariance condition is:

$$U^{(n)}[\Delta_{\nu}] \mid_{\Delta_{\nu}=0} = 0. \tag{9}$$

The characteristic equations associated to the general symmetry generator (2) have the form:

$$\frac{dx^1}{\xi^1} = ... = \frac{dx^p}{\xi^p} = \frac{du^1}{\phi_1} = ... = \frac{du^q}{\phi_q}. \tag{10}$$

By integrating the characteristic system of ordinary differential equations (10), the invariants I_r, $r = \overline{1, (p+q-1)}$ of the analyzed system can be found. They are identified the constants of integration. Following this way, the set of similarity variables is found in terms through which the original evolutionary equation with p independent variables and q dependent ones can be reduced to a set of differential equations with $(p+q-1)$ variables. These are the similarity reduced equations which generate the similarity solution of the analyzed model.

The **nonclassical symmetry method** (NMS) may be used to derive nonclassical symmetries which are the one-parameter groups of transformations acting on the space of the independent and dependent variables of the systems that leave only a subset of the set of all analytical solutions invariant.

The basic idea of the nonclassical method is to augment (1) with the invariance surface condition:

$$\Omega_\alpha \equiv \phi_\alpha(x,u) - \sum_{i=1}^{p} \xi^i(x,u)\frac{\partial u^\alpha}{\partial x^i} = 0, \ \alpha = \overline{1,q}. \tag{11}$$

The $q-$tuple $Q = (Q^1, Q^2, ...Q^q)$ is known as the characteristic of the symmetry operator (2).

The infinitesimal invariance of the equation (1) and the invariant surface condition (11) can be written:

$$U^{(n)}(\Delta_\nu)_{\Delta_\nu=0,\Omega_\alpha=0} = 0, \ U^{(n)}(\Omega_\alpha)_{\Delta_\nu=0,\Omega_\alpha=0} = 0. \tag{12}$$

By extending the conditions (12), we see that the number of determining equations for the infinitesimals $\xi^i(x,u)$, $\phi_\alpha(x,u)$ appearing in the nonclassical method is smaller than the one for the classical method. The main difficulty of this approach is that the determining equations are no longer linear. On the other hand, the NSM may produce more solutions than the CSM, since any classical symmetry is a nonclassical one, but conversely this does not apply [32].

For $\xi^p \neq 0$ in [33] is proposed a new algorithm for finding nonclassical symmetries. This algorithm is based on the following remark: since U is a symmetry operator so is αU, for any function $\alpha = \alpha(x,u)$. Thereby, if $\xi^p \neq 0$ we could multiplicate U by $(1/\xi^p)$ and write down the invariant surface condition under the equivalent expression:

$$\frac{\partial u}{\partial x^p} = \eta(x,u) - \sum_{i=1}^{p-1} \xi^i(x,u)\frac{\partial u}{\partial x^i}. \tag{13}$$

By substituting (13) and its derivatives with respect to x in (1) a new partial differential equation results:

$$\Delta' \equiv \Delta'(F_\beta(x,u), u^{[l]}, ..., u^{[n]}), \tag{14}$$

for the unknown function $u = u(x^1, ..., x^{p-1}; x^p)$ of $x^1, ..., x^{p-1}$ (here x^p is considered as a parameter); the functions $F_\beta(x,u)$ are the coefficients of $u^{[l]}$, where $u^{[l]}$ denotes the whole set of partial derivatives of u with respect to $x = (x^1, ..., x^{p-1})$ up to order N.

The next step consists in applying the classical Lie method to (14). Thus, let us consider a Lie group of point transformations associated with the following infinitesimal symmetry generator:

$$V = \sum_{i=1}^{p} s^i(x,u)\frac{\partial}{\partial x^i} + r(x,u)\frac{\partial}{\partial u}. \tag{15}$$

By imposing the invariance of (14) under the action of operator (15), the determining differential equations for the infinitesimals $s^i(x,u), r(x,u)$.will be obtained. By resetting the coefficient functions $s^p = 1$, $s^i = \xi^i$, $i = 1,...,p-2$, $r = \eta$ and substituting the functions F_β into the determining equations, we could find determining equations for nonclassical symmetries of the original dynamical system described by PDE of type (1).

3. The expansion method

The second method we are focusing on is the so called "expansion method". Its idea consists in reducing the governing PDE to an ODE. To do that, the first step to be followed is the introduction of the wave variable $\xi = \xi(t, x^1, ..., x^p)$. The second step is to look for solutions of the ODE in terms of the solutions of another ODE, called "auxiliary equation", with already known solutions.

Let us consider a governing equation of the form:

$$F(u, u_t, u_x, u_{xx}, u_{tt}, ...) = 0 \tag{16}$$

We define the wave coordinate:

$$\xi = x - Vt \tag{17}$$

By that, the equation (16) becomes the following ODE:

$$Q(u, u', u'', u''', ...) = 0 \tag{18}$$

where the derivatives are considered in respect with ξ.

There are many versions related to the expansion method, depending on the auxiliary equation which is chosen. Two of the most used versions are:

i)The tanh method - solutions of (18) in terms of $\tanh(\xi), \cosh(\xi), \sinh(\xi)$, etc. which are solutions $\phi(\xi)$ of equations, as Riccati, so:

$$u(\xi) = \sum_{i=0}^{N} a_i \varphi^i \tag{19}$$

ii) The G'/G method, where $G(\xi)$ solution of an auxiliary equation. In this case:

$$u(\xi) = \sum_{i=0}^{N} a_i \left(\frac{G'}{G}\right)^i \tag{20}$$

The approach we are proposing is an unifying one. More precisely, the solution of the master equation will be asked to be a polynomial expansion in terms of the solutions $G(\xi)$ of an "auxiliary equation":

$$u(\xi) = \sum_{i=0}^{N} P_i(G)(G')^i \tag{21}$$

where $P_i(G)$ are polynomials in G to be determined. Computing the derivatives of $u(\xi)$ higher order derivatives $G', G'', G\prime''$, ... could appear. So we might look to a more general solution depending on higher derivatives of $G(\xi)$:

$$u(\xi) = P_0(G) + P_1(G)G' + P_2(G, G')G'' + ... \tag{22}$$

Although, the higher derivatives $G''.G'''$,... can be expressed in terms of G, G' by using an adequate auxiliary equation. The form and the order of the auxiliary equations are important. The most used auxiliary equations which appear in the papers published in the last years are:

- Riccati Equation (first order nonlinear equation):

$$G' = \alpha + \beta G^2 \tag{23}$$

- Second order linear ODE:

$$G'' + AG' + BG = 0 \tag{24}$$

-Second order nonlinear ODE:

$$AGG'' - B(G')^2 - CGG' - EG^2 = 0 \tag{25}$$

-Third order nonlinear ODE:

$$AG^2G''' - B(G')^3 - CG(G')^2 - DG^2G' - FG^3 = 0 \tag{26}$$

The next step after choosing the auxiliary equation is to determine the limit N of the expansion (21) by a standard "balancing" procedure: replace (21) in (18) and take into account the higheast nonlinearity and the term with the maximal order of derivation.

In our case a new requirement is imposed: polynomial expantions for the functions $P_0(G), P_1(G), P_2(G)$,... To have a true balance and compatibility, we have to consider expansions of the form:

$$P_2(G) = \sum_{i=-2}^{0} a_i G^i \tag{27}$$

$$P_1(G) = \sum_{j=-1}^{0} b_j G^j \tag{28}$$

$$P_0(G) = c_0 \tag{29}$$

From the algebraic system generated by these choices, we can determine the coefficients a_i, b_i and c_i. After that, we can write down the form of the solutions $u(\xi)$. These solutions have to be discussed for various possible values of the coefficients A, B, C,...appearing in the master equation.

4. Application to the 2D Ricci flow equation

The $2D$ Ricci flow equation in has the form:

$$u_t = \frac{u_{xy}}{u} - \frac{u_x u_y}{u^2} \tag{30}$$

It was analyzed through symmetry method in [34].Here we will focus on the auxiliary equation method and its use in the direct finding of soliton type solutions. We will apply the wave transformation which transforms the equation (30) in:

$$U'U^2 + \frac{\alpha\beta}{\lambda}(UU'' - U'^2) = 0 \tag{31}$$

We will solve equation (31) by four different methods, in order to compare the solutions themselves and the efficiency of the methods. A general approach, unifying methods as tanh or G'/G, will be now proposed. It will be denominated as the *power law method.*

4.1. Double integration method

The equation (31) can be solved directly by double integration and the form of the solution is:

$$\begin{aligned} U &= \frac{e^A}{-1 + \lambda c_1 e^A} \\ A(\xi) &= \frac{\xi + c_2}{c_1 \alpha\beta} \end{aligned} \tag{32}$$

The direct integration leads to singular solution which are not of Physical interest.

4.2. Solution of tanh type

The simplest way of finding soliton solutions for (26) is to use Riccati as auxiliary equation and to look for solutions of (31) . More precisely, we will consider:

$$U(\xi) = \sum_{i=0}^{N} a_i G^i \tag{33}$$

The Riccati equation has the form:

$$G' = k + G^2 \tag{34}$$

with k a real constant.

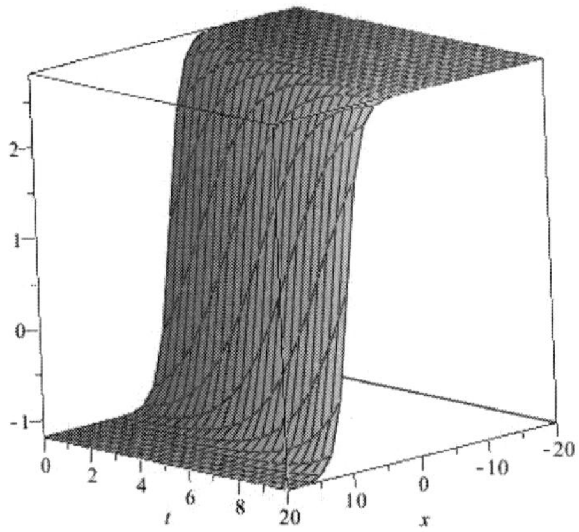

Figure 1: Tanh method $y = 0$

Considering integrating constant as zero,

$$\begin{aligned} \xi &= \{ {}^{\frac{1}{\sqrt{k}} \tan^{-1}(\frac{G}{\sqrt{k}})}_{-\frac{1}{\sqrt{k}} \cot^{-1}(\frac{G}{\sqrt{k}})}, k > 0 \\ \xi &= -\frac{1}{G}, k = 0 \\ \xi &= \{ {}^{-\frac{1}{\sqrt{-k}} \tanh^{-1}(\frac{G}{\sqrt{-k}})}_{-\frac{1}{\sqrt{-k}} \coth^{-1}(\frac{G}{\sqrt{-k}})}, k < 0 \end{aligned} \tag{35}$$

The balancing procedure leads to the maximal value $N = 1$, that is the solution we are looking for will have the form:

$$U(\xi) = a_0 + a_1 G \tag{36}$$

$$U(\xi) = \frac{\alpha\beta}{\lambda} \left(\sqrt{-k} - \sqrt{-k} \tanh \sqrt{-k}\xi \right) \tag{37}$$

4.3. Solution of G'/G type

We solve now the equation by using the G'/G method. It imposes to look for solutions of the form:

$$U(\xi) = \sum_{i=0}^{N} d_i \left(\frac{G'}{G} \right)^i \tag{38}$$

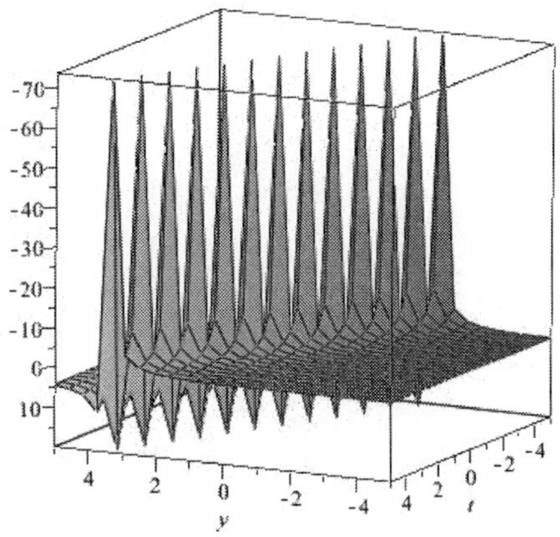

Figure 2: Polynomial expansion for $\alpha = 1, \beta = 2, \lambda = 1, x = 0$

We will consider that d_i are constant coefficients, while this time,$G(\xi)$ is a solution of the auxiliary equation of the form:

$$G'' + mG' + nG = 0 \tag{39}$$

Again, the balancing procedure leads to the same limit $N = 1$.

By introducing (38) in (39) we get a polynomial equation in G' containing monomyals until G'^7. Equating with zero the coefficients for all this monomials we get a system of 8 ODE with the unknown quantities $a_0(G), a_1(G)$ which satisfy the following equations:

$$a_0 = \frac{\alpha\beta m}{a_1} + Ce^{-\frac{a_1\lambda}{\alpha\beta}G} \tag{40}$$

$$a_1' a_1 + \frac{\alpha\beta}{\lambda}(a_1'' a_1 - a_1') = 0 \tag{41}$$

4.4. Power law expansion

Another quite similar approach for solving the "master" equation (31) consists in looking for polynomial solutions of the type:

$$U' = V(U) \tag{42}$$

Then:

$$\alpha\beta U V(U)\frac{\mathrm{d}V}{\mathrm{d}U} + \lambda U^2 V(U) - \alpha\beta V(U)^2 = 0. \tag{43}$$

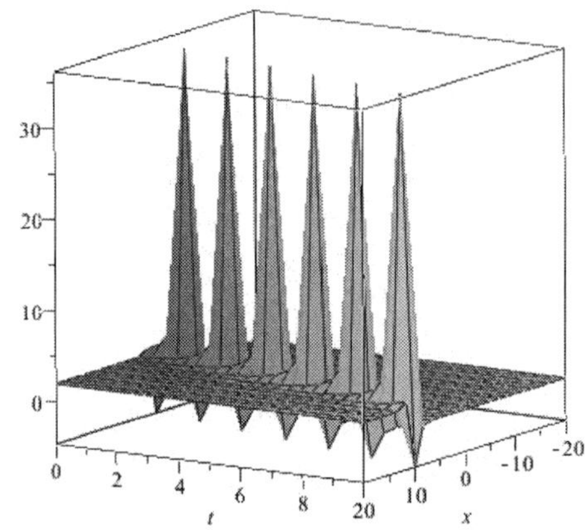

Figure 3: Power law method for $\alpha = 1, \beta = 1, \lambda = 1, y = 0, k = 1/2$

The solution of (43) is

$$V(U) = -\frac{\lambda U(2U + k)}{2\alpha\beta}, \tag{44}$$

where k is the integration constant. From the above relation we can find that

$$U = \frac{k}{\sinh(\frac{\lambda k\zeta}{2\alpha\beta}) + \cosh(\frac{\lambda k\zeta}{2\alpha\beta}) - 2} \tag{45}$$

The final solution is quite similar with the ones got through previous approaches and it is presented in the figure below.

5. Conclusions

We proposed a general algorithm for finding solutions of nonlinear PDEs by using polynomial expansions in terms of auxiliary equations solutions. It includes all the methods proposed in literature, known as tanh, cosh, sinh, G'/G, etc. The main idea is quite similar with what symmetry method offers: to reduce a complicated equation to a simpler one, to solve this last equation, and to transfer its solutions to the master (complicated) equation. We pointed out the importance of three main factors: - the choice of the auxiliary equation; - the choice of the form of solution; - the balancing procedure. For the specific model we tackled, we get that the polynomials from (21) have the form:

$$P_2(G) = a_2 G^{-2} \tag{46}$$

$$P_1(G) = b_1 G^{-1} \tag{47}$$

$$P_0(G) = 0 \tag{48}$$

It appears in a natural way that really the largest class of solutions can be expressed as (G'/G) expansions. Why this expansion was choosen in previous approaches was not at all clear. The method we proposed is purelly analytic and it open the doors for finding other solutions which do not belong to the class of (G'/G) class.

References

[1] M. J. Ablowitz, P. A. Clarkson, Solitons, (Cambridge University Press, Cambridge, 1990).

[2] V. O. Vakhnenko, E. J. Parkes, A. J. Morrison, A Bäcklund transformation and the inverse scattering transform method for the generalised Vakhnenko equation, Chaos Soliton Fract. 17, 683–692 (2003).

[3] R. Hirota, Backlund Transformations, (Springer, Berlin, 1980).

[4] D. Lu, B. Hong, L. Tian, New explicit exact solutions for the generalized coupled Hirota–Satsuma KdV system, Comput. Math. Appl. 53, 1181–1190 (2007).

[5] C. Bian, J. Pang, L. Jin, X. Ying, Solving two fifth order strong nonlinear evolution equations by using the G'/G-expansion method, Commun. Nonlinear Sci. Numer. Simul. 15, 2337–2343 (2010).

[6] N. A. Kudryashov, A note on the G'/G-expansion method, Appl. Math. Comput. 217(4), 1755–1758 (2010).

[7] P. J. Olver, Applications of Symmetry Methods to Partial Differential Equations, (Springer-Verlag, New-York, 1986).

[8] R. Cimpoiasu, R. Constantinescu, V. M. Cimpoiasu, Integrability of dynamical systems with polynomial Hamiltonians, Rom. J. Phys. 50(3-4), 317–324 (2005).

[9] R. Cimpoiasu, R. Constantinescu, Lie Symmetries for Hamiltonian Systems Methodological Approach, Int. J.Theor. Phys. 45(9), 1769–11782 (2006).

[10] Z. Y. Ma, H. WU, Q. Y. Zhu, Lie symmetry, full symmetry group, and exact solutions to the (2+1) dissipative AKNS equation, Rom. J. Phys. 62, 114–127 (2017).

[11] F. Jian-hui, C. Pei-sheng, Z. Jun, Form invariance and Lie symmetry of variable mass nonholonomic mechanical system, Appl. Math. Mech. 26(2), 204–209 (2005).

[12] A. H. Bhrawy, M. A. Abdelkawy, S. Kumar, A. Biswas, Solitons and other solutions to Kadomtsev-Petviashvili equation of B-type, Rom. J. Phys. 58, 729–748 (2013).

[13] A. Biswas, Solitary waves for power-law regularized long-wave equation and R (m, n) equation, Nonlinear Dyn. 59, 423–426 (2010).

[14] R. Cimpoiasu, Generalized conditional symmetries and related solutions of the Grad-Shafranov equation, Phys. Plasmas 21(4), 042118-1–042118-7 (2014).

[15] K. Su, J. Cao, Third-Order Conditional Lie–Bäcklund Symmetries of Nonlinear Reaction-Diffusion , Adv. Math. Phys. 2017 2825416-1–2825416-9 (2017).

[16] Hamilton, R.S.: Three-manifolds with positive Ricci curvature. J. Diff. Geom. 17, 255–306 (1982)

[17] Perelman G.: The entropy formula for the Ricci flow and its geometric applications. arXiv preprint math.DG/0211159 (2002)

[18] Woolgar, E.: Some applications of Ricci flow in physics. Can. J. Phys. 86(4), 645–651 (2008)

[19] Oliynyk, T., Suneeta V., Woolgar E.: A metric for gradient RG flow of the worldsheet sigma model. Phys. Rev. D 76, 045001-1–045001-8 (2008).

[20] Oliynyk, T., Suneeta V., Woolgar E.: A gradient flow for world sheet nonlinear sigma models. Nucl. Phys. B 739, 441–458 (2006)

[21] Bakas I.: Ricci fows and infnite dimensional algebras. Fortschritte der Physik 52(6-7), 464–471 (2004)

[22] Cimpoiasu, R., Constantinescu R.: Symmetries and invariants for the 2D-Ricci flow model. J. Nonlinear Math. Phys. 13, 285–292 (2006)

[23] Bluman G.W., Cheviakov A.F., Anco S.C.: Applications of Symetry Methods to Partial Differential Equations. Springer, New York, (2010)

[24] Cimpoiasu, R.: Conservation Laws and Associated Lie Symmetries for the 2D Ricci Flow Mode. Rom. J. Phys. 58, 519–528, (2013)

[25] S. Lie, In: Gesammelte Abhandlungen, Band 4, B. G. Teubner, Leipzig, Germany, 1929, 320.

[26] G. W. Bluman and J. D. Cole, J. Math. Mech. 18 (1969) 1025.

[27] P. A. Clarkson and M. D. Kruskal, J. Math. Phys. 30 (1989) 2201.

[28] P. J. Olver and P. Rosenau, Phys. Lett. A 114 (1986) 172.

[29] C. Z. Qu, Stud. Appl. Math. 99 (1997) 107.

[30] N. Ibragimov, Handbook of Lie Group Analysis of Differential Equations, third ed., CRC Press, Boca Raton, FL, USA, 1996.

[31] L. V. Ovsiannikov, Group Analysis of Differential Equations, Academic Press, New York, USA, 1982.

[32] P. A. Clarkson and E. L. Mansfield, Physica D70 (1994) 250.

[33] N. Bîlă and J. Niesen, J. Symbol Comp. 38 (2004) 1523.

[34] R.Cimpoiasu, R.Constantinescu, J.Nonlin.Math.Phys., vol 13, no. 2, (2006), 285-292.

Symmetries of a bosonic string*

Ljubica Davidović[†]
Institute of Physics, University of Belgrade, Serbia
Branislav Sazdović[‡]
Institute of Physics, University of Belgrade, Serbia

Abstract

We consider the string theory of a closed and an open string and search for the transformations of the space-time fields which do not change the physical content of the theory, i.e. the symmetry transformations. In the open string theory we start with a modified action which has an additional surface term which enables the invariance of the complete action to the general coordinate transformations and the gauge transformations. The string theory is conformally invariant world sheet field theory. Therefore, the physics is preserved if the conformal field theories corresponding to the initial and the transformed field configurations are isomorphic. We show that the general coordinate transformations are T-dual to the gauge transformations.

1. Bosonic string actions

The dynamics of the bosonic string, moving in a curved background associated with the massless bosonic fields [1], a metric field $G_{\mu\nu} = G_{\nu\mu}$, a Kalb-Ramond field $B_{\mu\nu} = -B_{\nu\mu}$ and a dilaton field Φ is described by the string sigma model action. The background fields in which the string moves have to satisfy the space-time equations of motion

$$
\begin{aligned}
&R_{\mu\nu} - \frac{1}{4}B_{\mu\rho\sigma}B_{\nu}{}^{\rho\sigma} + 2D_\mu\partial_\nu\Phi = 0,\\
&D_\rho B^\rho{}_{\mu\nu} - 2\partial_\rho\Phi B^\rho{}_{\mu\nu} = 0,\\
&4(\partial\Phi)^2 - 4D_\mu\partial^\mu\Phi + \frac{1}{12}B_{\mu\nu\rho}B^{\mu\nu\rho} + 4\pi\kappa(D-26)/3 - R = 0, \quad (1)
\end{aligned}
$$

in order for the conformal invariance of the quantum theory to be preserved. Here $B_{\mu\nu\rho} = \partial_\mu B_{\nu\rho} + \partial_\nu B_{\rho\mu} + \partial_\rho B_{\mu\nu}$ is the field strength of the field $B_{\mu\nu}$,

* This work has been supported in part by The Serbian Ministry of Education, Science and Technological Development, under contract No. 171031.
† e-mail address: ljubica@ipb.ac.rs
‡ e-mail address: sazdovic@ipb.ac.rs

$R_{\mu\nu}$ is a Ricci tensor, and D_μ is a covariant derivative with respect to the space-time metric. We broadly investigated the string sigma model for the solution of the space-time equations of motion called *a weakly curved background* [2] (of the first and second order which differ by the term in brackets), given by

$$G_{\mu\nu}(x) = g_{\mu\nu} + \left\{3h^2_{\mu\nu}(x)\right\},$$
$$B_{\mu\nu}(x) = b_{\mu\nu} + h_{\mu\nu}(x), \quad h_{\mu\nu}(x) = \frac{1}{3}B_{\mu\nu\rho}x^\rho. \tag{2}$$

For the open bosonic string we searched for the solution of the boundary conditions, for which we obtained the effective theory describing a closed effective string and investigated the commutativity of the effective string coordinates. We obtained that noncommutativity exists along the entire string when the Kalb-Ramond field is coordinate dependent.

For the closed string moving in a weakly curved background we developed the T-dualization procedure, based on the standard procedure, giving a prescription how to find theories T-dual to a given theory. The generalization of our procedure was made for the weakly curved background of the second order, which does not posses the global shift symmetry. Our procedure enabled T-dualization of an arbitrary space-time coordinate. For the first order weakly curved background we performed the T-dualization of the arbitrary set of the initial coordinates [3]. If we choose d coordinates and mark the T-dualizations performed along them, along the rest of the coordinates and along all coordinates by

$$\mathcal{T}^a = \circ_{n=1}^{d} T^{\mu_n}, \quad \mathcal{T}^i = \circ_{n=d+1}^{D} T^{\mu_n}, \quad \mathcal{T} = \circ_{n=1}^{D} T^{\mu_n}, \tag{3}$$

and analogously mark the T-dualizations in the dual space

$$\mathcal{T}_a = \circ_{n=1}^{d} T_{\mu_n}, \quad \mathcal{T}_i = \circ_{n=d+1}^{D} T_{\mu_n}, \quad \tilde{\mathcal{T}} = \circ_{n=1}^{D} T_{\mu_n}, \tag{4}$$

then we obtain a set of string sigma models connected by the diagram

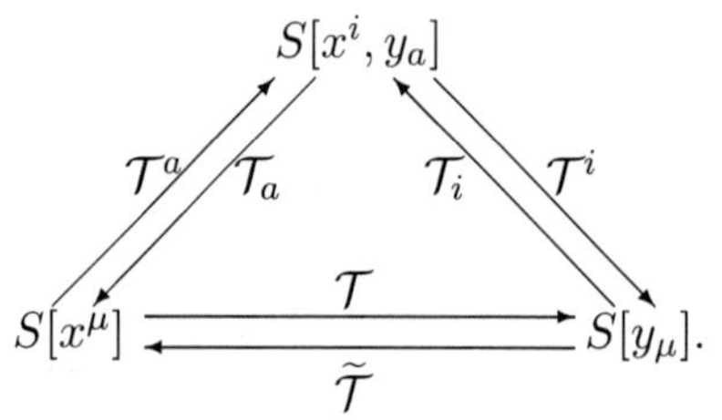

In a left vertex is the initial action, the bosonic string action in a conformal gauge $g_{\alpha\beta} = e^{2F}\eta_{\alpha\beta}$,

$$S[x] = \kappa \int_\Sigma d^2\xi \; \partial_+ x^\mu \Pi_{+\mu\nu}(x) \partial_- x^\nu, \quad \partial_\pm = \partial_\tau \pm \partial_\sigma, \tag{5}$$

with the background field composition

$$\Pi_{\pm\mu\nu}(x) = B_{\mu\nu}(x) \pm \frac{1}{2}G_{\mu\nu}(x).$$

The totally T-dualized theory [4], in the right vertex of the T-duality diagram is

$$^{\star}S[y] = \frac{\kappa^2}{2}\int d^2\xi\ \partial_+ y_\mu \Theta_-^{\mu\nu}[\Delta V^{(0)}(y)]\partial_- y_\nu, \tag{6}$$

with the background

$$\Theta_\pm^{\mu\nu}[\Delta V] = -\frac{2}{\kappa}(G_E^{-1}\Pi_\pm G^{-1})^{\mu\nu} = \theta^{\mu\nu}[\Delta V] \mp \frac{1}{\kappa}(G_E^{-1})^{\mu\nu}[\Delta V], \tag{7}$$

which is the inverse of the initial background field composition

$$\Theta_\pm^{\mu\nu}\Pi_{\mp\nu\rho} = \frac{1}{2\kappa}\delta_\rho^\mu, \tag{8}$$

and its argument is equal to

$$\Delta V^{(0)\mu}(y) = -\kappa\theta_0^{\mu\nu}\Delta y_\nu^{(0)} + (g^{-1})^{\mu\nu}\Delta\tilde{y}_\nu^{(0)}, \tag{9}$$

where the tilde coordinates represent the flat space-time T-duals of the corresponding T-dual coordinates.

The central action is a partially T-dualized action, obtained applying the T-dualization procedure to coordinates x^a, given by

$$\begin{aligned} S[x^i, y_a] = \kappa \int d^2\xi \Big[& \partial_+ x^i \overline{\Pi}_{+ij}(x^i, \Delta V^a(x^i, y_a))\partial_- x^j \\ & -\kappa\, \partial_+ x^i \Pi_{+ia}(x^i, \Delta V^a(x^i, y_a))\widetilde{\Theta}_-^{ab}(x^i, \Delta V^a(x^i, y_a))\partial_- y_b \\ & +\kappa\, \partial_+ y_a \widetilde{\Theta}_-^{ab}(x^i, \Delta V^a(x^i, y_a))\Pi_{+bi}(x^i, \Delta V^a(x^i, y_a))\partial_- x^i \\ & +\frac{\kappa}{2}\, \partial_+ y_a \widetilde{\Theta}_-^{ab}(x^i, \Delta V^a(x^i, y_a))\partial_- y_b \Big]. \end{aligned} \tag{10}$$

The new background field compositions are defined in terms of the initial field composition restricted to corresponding subspaces and their inverses and the inverses of the totally T-dualized field compositions restricted to corresponding subspaces

$$\widetilde{\Theta}_\pm^{ab}\Pi_{\mp bc} = \Pi_{\mp cb}\widetilde{\Theta}_\pm^{ba} = \frac{1}{2\kappa}\delta_c^a, \tag{11}$$

and

$$\overline{\Pi}_{\pm ij}\Theta_\mp^{jk} = \Theta_\mp^{kj}\overline{\Pi}_{\pm ji} = \frac{1}{2\kappa}\delta_i^k. \tag{12}$$

The argument of the background fields is defined by

$$\Delta V^{(0)a}(x^i, y_a) = -\kappa\Big[\tilde{\Theta}^{ab}_{0+}\Pi_{0-bi} + \tilde{\Theta}^{ab}_{0-}\Pi_{0+bi}\Big]\Delta x^{(0)i}$$
$$-\kappa\Big[\tilde{\Theta}^{ab}_{0+}\Pi_{0-bi} - \tilde{\Theta}^{ab}_{0-}\Pi_{0+bi}\Big]\Delta\tilde{x}^{(0)i}$$
$$-\frac{\kappa}{2}\Big[\tilde{\Theta}^{ab}_{0+} + \tilde{\Theta}^{ab}_{0-}\Big]\Delta y^{(0)}_b - \frac{\kappa}{2}\Big[\tilde{\Theta}^{ab}_{0+} - \tilde{\Theta}^{ab}_{0-}\Big]\Delta\tilde{y}^{(0)}_b,$$

where the tilde coordinates represent the flat space-time T-duals of the appropriate initial or T-dual coordinate. Despite the fact the T-dualization procedure was applied to x^a, there appear the T-duals of the undualized directions.

If the same procedure is applied to the effective theory of the open bosonic string, one can investigate the open string T-duality [5]. There are four relevant theories in this context. The initial open string theory

$$S = \kappa\int_\Sigma d^2\xi\,\partial_+ x^\mu \Pi_{+\mu\nu}(x)\partial_- x^\nu, \tag{13}$$

its effective closed string theory, obtained for the solution of the boundary conditions

$$S^{eff} = \kappa\int_{\Sigma^\star} d^2\xi\,\partial_+ q^\mu\,\Pi^{eff}_{+\mu\nu}(q, 2b\tilde{q})\,\partial_- q^\nu, \tag{14}$$

the T-dual of the effective theory

$${}^\star S^{eff} = \frac{\kappa^2}{2}\int_{\Sigma^\star} d^2\xi\,\partial_+\varrho_\mu(\Theta^{eff}_-)^{\mu\nu}(g_E^{-1}\tilde{\varrho}, 2bg_E^{-1}\varrho)\partial_-\varrho_\nu, \tag{15}$$

and the open string theory having its effective theory equal to this T-dual theory

$$\tilde{S} = \kappa\int_\Sigma d^2\xi\,\partial_+ y_\mu\tilde{\Pi}^{\mu\nu}_+(y)\partial_- y_\nu.$$

The effective theory background $\Pi^{eff}_{\pm\mu\nu}$ is composed of the effective metric $G^{eff}_{\mu\nu}(q) = G^E_{\mu\nu}(q) = G_{\mu\nu} - 4B_{\mu\rho}(q)(G^{-1})^{\rho\sigma}B_{\sigma\nu}(q)$ and the effective Kalb-Ramond field $B^{eff}_{\mu\nu}(2b\tilde{q}) = -\frac{\kappa}{2}[g\Delta\theta(2b\tilde{q})g]_{\mu\nu}$, where $\Delta\theta$ is an infinitesimal part of the noncommutativity parameter for the initial coordinates. Its T-dual background $(\Theta^{eff}_\mp)^{\mu\nu}$ is just its inverse. They are both defined in the doubled spaces, given in terms of the appropriate coordinates and their doubles. The relation between these background fields resembles the relation of the closed string initial field composition and its T-dual. However, in that case T-dualization transformed a geometrical space into the double space. In the open string T-duality, this change is not present.

The T-dual space remains the geometrical space as the initial space and the T-dual background fields keep the same form as the initial open string background. In the open string case, the important role in the relation between the T-dual backgrounds plays a matrix C, which is introduced to define the connection between the variables of the open string theory T-dual and the effective theory T-dual.

2. The symmetries

So, we found a number of physically equivalent string sigma models. Enough to rise again the old question of what is the symmetry transformation of fields. In a classical theory it is the change in fields which does not change the classical action. If the string theory is defined by the σ-model action, the space-time fields appear in this world-sheet action as the coupling constants. So, the standard technique for obtaining the symmetries is not applicable for finding the symmetry transformation of these fields. One should instead consider the conformal field theories corresponding to different field configurations [7]. In order for the transformation of field to be a symmetry (which is in fact the symmetry of the space-time action) the change in fields should correspond to change in energy-momentum tensors of two isomorphic conformal field theories.

Let us therefore consider the energy-momentum tensor

$$T_{\pm} = \mp \frac{1}{4\kappa}(G^{-1})^{\mu\nu} j_{\pm\mu} j_{\pm\nu}, \tag{16}$$

given in terms of currents $j_{\pm\mu} = \pi_\mu + 2\kappa \Pi_{\pm\mu\nu}(x) x'^\nu$. The hamiltonian corresponding to the initial string sigma model lagrangian is

$$\mathcal{H}_C = T_- - T_+. \tag{17}$$

The space-time equations of motion for the background fields (1), come from the condition that the energy-momentum tensor on a quantum level should satisfy the Virasoro algebras

$$\begin{aligned}
\Big[\hat{T}_\pm(\varphi(\sigma)), \hat{T}_\pm(\varphi(\bar{\sigma}))\Big] &= i\hbar \Big[\hat{T}_\pm(\varphi(\sigma)) + \hat{T}_\pm(\varphi(\bar{\sigma}))\Big]\delta'(\sigma - \bar{\sigma}), \\
\Big[\hat{T}_\pm(\varphi(\sigma)), \hat{T}_\mp(\varphi(\bar{\sigma}))\Big] &= 0.
\end{aligned} \tag{18}$$

It is known that the similarity transformation $\hat{T}_\pm \to e^{-i\hat{\Gamma}} \hat{T}_\pm e^{i\hat{\Gamma}}$, which causes the transformation of energy-momentum tensor

$$\delta \hat{T}_\pm(\varphi) = -i\Big[\hat{\Gamma}, \hat{T}(\varphi)\Big], \tag{19}$$

preserves the Virasoro algebra.

Let us consider the classical theory and suppose the coordinates and momenta satisfy the standard Poisson brackets

$$\{x^\mu(\sigma), \pi_\nu(\bar\sigma)\} = \delta^\mu_\nu \delta(\sigma - \bar\sigma), \tag{20}$$

consequently the Poisson brackets of currents are

$$\begin{aligned}
\{j_{\pm\mu}(\sigma), j_{\pm\nu}(\bar\sigma)\} &= \pm 2\kappa \Gamma_{\mp\mu,\nu\rho}\, x'^\rho(\sigma)\delta(\sigma-\bar\sigma) \pm 2\kappa G_{\mu\nu}(x(\sigma))\delta'(\sigma-\bar\sigma), \\
\{j_{\pm\mu}(\sigma), j_{\mp\nu}(\bar\sigma)\} &= \pm 2\kappa \Gamma_{\mp\rho,\mu\nu}\, x'^\rho(\sigma)\delta(\sigma-\bar\sigma),
\end{aligned} \tag{21}$$

where the generalized connection is given in terms of the Christoffel symbol $\Gamma_{\mu,\nu\rho} = \frac{1}{2}(\partial_\nu G_{\rho\mu} + \partial_\rho G_{\mu\nu} - \partial_\mu G_{\nu\rho})$ and the field strength of the field $B_{\mu\nu}$, $B_{\mu\nu\rho} = \partial_\mu B_{\nu\rho} + \partial_\nu B_{\rho\mu} + \partial_\rho B_{\mu\nu}$ by

$$\Gamma_{\pm\mu,\nu\rho} = \Gamma_{\mu,\nu\rho} \pm B_{\mu\nu\rho}. \tag{22}$$

One can further calculate the Poisson brackets between the energy-momentum tensor and currents

$$\begin{aligned}
\{T_\pm(\sigma), j_{\pm\mu}(\bar\sigma)\} &= \pm\frac{1}{2\kappa}\Gamma_{\mp\rho,\mu\nu} j^\nu_\mp j^\rho_\pm \delta(\sigma-\bar\sigma) - j_{\pm\mu}(\sigma)\delta'(\sigma-\bar\sigma), \\
\{T_\pm(\sigma), j_{\mp\mu}(\bar\sigma)\} &= \pm\frac{1}{2\kappa}\Gamma_{\mp\rho,\nu\mu} j^\nu_\pm j^\rho_\mp \delta(\sigma-\bar\sigma),
\end{aligned} \tag{23}$$

to finally obtain that $T_\pm$ satisfy the Virasoro algebra

$$\begin{aligned}
&\{T_\pm(\sigma), T_\pm(\bar\sigma)\} = -(T_\pm(\sigma) + T_\pm(\bar\sigma))\delta'(\sigma-\bar\sigma), \\
&\{T_\pm(\sigma), T_\mp(\bar\sigma)\} = 0.
\end{aligned} \tag{24}$$

The classical analogue of transformation (19) is just

$$\delta T_\pm(\varphi) = \Big\{\Gamma, T(\varphi)\Big\}. \tag{25}$$

Let us demonstrate that the known symmetry of the string theory, a gauge transformation of the Kalb-Ramond field

$$\begin{aligned}
&\delta_\Lambda B_{\mu\nu} = \partial_\mu \Lambda_\nu - \partial_\nu \Lambda_\mu, \\
&\delta_\Lambda G_{\mu\nu} = 0,
\end{aligned} \tag{26}$$

given in terms of the vector gauge parameters Λ_μ, causes the transformation of the energy-momentum tensor which can be expressed in terms of the generator of the symmetry Γ as in (25). The change in $G_{\mu\nu}$ and $B_{\mu\nu}$ causes the following transformation of $T_\pm$

$$\delta T_\pm = \frac{1}{2\kappa}\Big(\frac{1}{\kappa}\delta_\Lambda B_{\mu\nu} \pm \frac{1}{2}\delta_\Lambda G_{\mu\nu}\Big) j^\mu_\pm j^\nu_\mp. \tag{27}$$

We consider a following expression

$$\Gamma_\Lambda = 2\int_{-\pi}^{\pi} d\sigma \Lambda_\mu x'^\mu = \frac{1}{\kappa}\int_{-\pi}^{\pi} d\sigma \Lambda_\mu (j_+^\mu - j_-^\mu), \tag{28}$$

and calculate its bracket with the energy-momentum tensor

$$\{\Gamma_\Lambda, T_\pm(\sigma)\} = \frac{1}{2\kappa^2}(\partial_\mu\Lambda_\nu - \partial_\nu\Lambda_\mu) j_\pm^\mu j_\mp^\nu. \tag{29}$$

Comparing (29) and (27), keeping in mind (26), we conclude that Γ_Λ is the generator of this symmetry.

2.1. The closed string symmetry and generators

Now, let us start the other way around. Let us suppose the background fields undergo a small change in value

$$\Pi_{\pm\mu\nu} \to \Pi_{\pm\mu\nu} + \delta\Pi_{\pm\mu\nu}. \tag{30}$$

This causes the following change of currents

$$\delta j_{\pm\mu} = 2\kappa\delta\Pi_{\pm\mu\nu}(x) x'^\nu, \tag{31}$$

and consequently the energy-momentum tensor $T_\pm = \mp\frac{1}{4\kappa}(G^{-1})^{\mu\nu} j_{\pm\mu} j_{\pm\nu}$ changes by

$$\delta T_\pm = \frac{1}{2\kappa}\delta\Pi_{\pm\mu\nu} j_\pm^\mu j_\mp^\nu. \tag{32}$$

If we demand this change to be equal to the change in the energy momentum tensor (25), we will obtained the corresponding symmetry transformation laws and their generators.

We assume the form of the generator of the general transformation (32), is the same as in (28), and therefore consider the expression $\mathcal{G} = \mathcal{G}_+ + \mathcal{G}_-$ with

$$\mathcal{G}_\pm = \int d\sigma\, \Lambda_\pm^\mu(x(\sigma)) j_{\pm\mu}(\sigma). \tag{33}$$

The Poisson brackets between energy-momentum tensor and these quantities are

$$\begin{aligned} \{T_\pm(\sigma), \mathcal{G}_\pm(\bar\sigma)\} &= \pm\frac{1}{2\kappa}\Big(D_{\mp\nu}\Lambda_\pm^\mu\Big) j_\mp^\nu j_{\pm\mu}, \\ \{T_\pm(\sigma), \mathcal{G}_\mp(\bar\sigma)\} &= \pm\frac{1}{2\kappa}\Big(D_{\pm\nu}\Lambda_\mp^\mu\Big) j_\pm^\nu j_{\mp\mu}, \end{aligned} \tag{34}$$

where the covariant derivatives are given by $D_{\pm\mu}\Lambda^\nu = \partial_\mu\Lambda^\nu + \Gamma^\nu_{\pm\rho\mu}\Lambda^\rho = D_\mu\Lambda^\nu \pm B^\nu{}_{\rho\mu}\Lambda^\rho$. One demands the energy-momentum tensor transforms only by a similarity transformation, and therefore the change (32) is in fact

$$\delta T_\pm = \{\mathcal{G}, T_\pm\} = \frac{1}{2\kappa}\delta\Pi_{\pm\mu\nu} j_\pm^\mu j_\mp^\nu.$$

This law determines the closed string symmetry transformations, allowed by a similarity transformation. Separating the parameters in the generators (33) into $\Lambda_{\pm\mu} = \xi_\mu \pm \Lambda_\mu$, one obtains that the symmetry transformation of fields are

$$\begin{aligned} \delta G_{\mu\nu} &= -2(D_\mu \xi_\nu + D_\nu \xi_\mu), \\ \delta B_{\mu\nu} &= D_\mu \Lambda_\nu - D_\nu \Lambda_\mu - 2B_{\mu\nu}{}^\rho \xi_\rho. \end{aligned} \tag{35}$$

The generator $\mathcal{G}$ of these transformations can be expressed as

$$\mathcal{G} = \int d\sigma \Big[2\xi^\mu \pi_\mu + 2\kappa \widetilde{\Lambda}_\mu x'^\mu \Big]. \tag{36}$$

with

$$\widetilde{\Lambda}_\nu = 2\xi^\mu B_{\mu\nu} + \Lambda^\mu G_{\mu\nu} = \Lambda_\nu - 2B_{\nu\mu}\xi^\mu. \tag{37}$$

The form of the generator makes it T-dual invariant, which will be discussed later.

3. The open string and its symmetries

From (35) and (37), we can conclude that the closed string is invariant under the general coordinate transformations

$$\begin{aligned} \delta_\xi G_{\mu\nu} &= -2(D_\mu \xi_\nu + D_\nu \xi_\mu), \\ \delta_\xi B_{\mu\nu} &= -2\xi^\rho B_{\rho\mu\nu} + 2(\partial_\mu b_\nu - \partial_\nu b_\mu), \quad b_\mu = B_{\mu\nu}\xi^\nu, \end{aligned} \tag{38}$$

with $D_\mu \xi_\nu = \partial_\mu \xi_\nu - \Gamma^\rho_{\mu\nu}\xi_\rho$, and the local gauge transformations

$$\begin{aligned} &\delta_\Lambda G_{\mu\nu} = 0, \\ &\delta_\Lambda B_{\mu\nu} = \partial_\mu \Lambda_\nu - \partial_\nu \Lambda_\mu, \end{aligned} \tag{39}$$

where we omit tilde in lambda ($\tilde{\Lambda}_\mu \to \Lambda_\mu$). These transformations are T-dual to each other, as can be seen comparing their generators. The action is invariant under these transformations in the closed string case, but is not invariant in the open string case. In paper [6], the open string action was proposed, which is invariant to the above transformations because of the additional surface term which is added to the standard action.

For the open string, beside the equations of motion the minimal action principle gives the boundary conditions on the string end-points

$$\begin{aligned} &\gamma_0^\mu \delta x^\mu \Big|_{\sigma=0,\pi} = 0, \\ &\gamma_0^\mu = x'^\mu - 2(G^{-1}B)^\mu{}_\nu \dot{x}^\nu. \end{aligned} \tag{40}$$

For each of the coordinates one can fulfill these conditions by choosing either the Neumann or the Dirichlet boundary condition. Let us mark the coordinates with the Neumann condition by x^a, $a = 0, 1, \cdots, p$ and the coordinates with the Dirichlet condition by x^i, $i = p+1, \cdots, D-1$.

The additional part of the open string action introduced in [6] is given in terms of the boundary conditions and it reads

$$S_{\partial\Sigma} = 2\int d\tau \Big[\kappa A_\mu(x)\dot{x}^\mu - \bar{A}_\mu(x)(G^{-1})^{\mu\nu}\gamma_\nu^{(0)}\Big]\Big|_{\sigma=0}^{\sigma=\pi}. \tag{41}$$

This term makes the open string theory invariant, taken that the introduced vector fields A_μ and $\bar{A}_\mu$ transform as

$$\begin{aligned} \delta_\Lambda A_\mu &= -\Lambda_\mu, \\ \delta_\xi \bar{A}_\mu &= -\xi_\mu. \end{aligned} \tag{42}$$

When the choice how to satisfy the boundary conditions is made, the surface term (41) reduces to

$$S_{\partial\Sigma} = 2\int d\tau \Big[\kappa A_a^N(x)\dot{x}^a - A_i^D(x)(G^{-1})^{ij}\gamma_j^{(0)}\Big]\Big|_{\sigma=0}^{\sigma=\pi}, \tag{43}$$

where A_a^N and A_i^D are $(p+1)$- and $(D-p-1)$-dimensional vector gauge fields, first living on the Dp-brane and the second orthogonal to Dp-brane.

It is well known that the Kalb-Ramond field term of the action

$$\int d^2\xi \varepsilon^{\alpha\beta} B_{\mu\nu}(x)\partial_\alpha x^\mu \partial_\beta x^\nu$$

can be regarded as an analogue of the coupling of the Maxwell field

$$\int d\tau A_\mu \dot{x}^\mu.$$

So, for the directions satisfying the Neumann boundary condition, the action on the boundary equals

$$S_{\partial\Sigma} = 2\kappa \int d\tau A_a^N(x)\dot{x}^a\Big|_{\sigma=0}^{\sigma=\pi}, \tag{44}$$

and it can be rewritten as

$$S = \kappa \int d^2\xi \, \mathcal{F}_{ab}^{(a)} \, \varepsilon^{\alpha\beta}\partial_\alpha x^a \partial_\beta x^b, \tag{45}$$

with

$$\mathcal{F}_{ab}^{(a)} = \partial_a A_b^N(x) - \partial_b A_a^N(x). \tag{46}$$

The complete surface term can be written in terms of the field strengths as

$$S_{\partial\Sigma} = \kappa \int d^2\xi \left(\mathcal{F}_{\mu\nu}^{(a)} \, \varepsilon^{\alpha\beta} + \frac{1}{2}\mathcal{F}_{\mu\nu}^{(s)} \, \eta^{\alpha\beta}\right)\partial_\alpha x^\mu \partial_\beta x^\nu, \tag{47}$$

with

$$\mathcal{F}^{(a)}_{\mu\nu} = \partial_\mu \mathcal{A}_{0\nu}(x) - \partial_\nu \mathcal{A}_{0\mu}(x),$$
$$\mathcal{F}^{(s)}_{\mu\nu} = 2(\partial_\mu \mathcal{A}_{1\nu}(x) + \partial_\nu \mathcal{A}_{1\mu}(x)), \tag{48}$$

where

$$\mathcal{A}_{0a} = A^N_a, \quad \mathcal{A}_{0i} = 2B_{ij}(G^{-1})^{jk} A^D_k, \tag{49}$$

and

$$\mathcal{A}_{1a} = 0, \quad \mathcal{A}_{1i} = -A^D_i. \tag{50}$$

Comparing the boundary actions (45) and (47) with the action (5), we conclude that the addition of the surface term has changed the background fields by

$$G_{\mu\nu} \to G_{\mu\nu} + \mathcal{F}^{(s)}_{\mu\nu} = \mathcal{G}_{\mu\nu}$$
$$B_{\mu\nu} \to B_{\mu\nu} + \mathcal{F}^{(a)}_{\mu\nu} = \mathcal{B}_{\mu\nu}. \tag{51}$$

3.1. The generators

Now that we have rewritten the open string action in the same form as the standard closed string action, we can determine the open string symmetry transformations and their generators in analogy to the closed string symmetry transformations and their generators. For the open string with Neumann boundary conditions the symmetry transformations are just (instead of (35))

$$\begin{aligned} \delta G_{\mu\nu} &= -2D_\mu \xi_\nu - 2D_\nu \xi_\mu, \\ \delta B_{\mu\nu} &= D_\mu \Lambda_\nu - D_\nu \Lambda_\mu + 2\mathcal{B}_\mu{}^\rho{}_\nu \xi_\rho, \end{aligned} \tag{52}$$

where $\mathcal{B}_\mu{}^\rho{}_\nu$ is a field strength of the changed Kalb-Ramond field $B_{\mu\nu} \to \mathcal{B}_{\mu\nu} = B_{\mu\nu} + \mathcal{F}^{(a)}_{\mu\nu}$, in comparison to the closed string case. However, the field strength of the additional part is zero and therefore $\mathcal{B}_{\mu\nu\rho} = B_{\mu\nu\rho}$. The generator of these transformations is

$$\mathcal{G} = \int d\sigma \Big[2\xi^\mu \pi_\mu + 2\kappa(2\xi^\mu \mathcal{B}_{\mu\nu} + \Lambda^\mu G_{\mu\nu}) x'^\nu \Big], \tag{53}$$

with

$$\pi_\mu = \kappa G_{\mu\nu}(x)\dot{x}^\nu - 2\kappa \mathcal{B}_{\mu\nu}(x) x'^\nu. \tag{54}$$

The open string symmetry transformations, for all the remaining choices for the boundary conditions are

$$\begin{aligned} \delta \mathcal{G}_{\mu\nu} &= -2(\mathbf{D}_\mu \xi_\nu + \mathbf{D}_\nu \xi_\mu), \\ \delta \mathcal{B}_{\mu\nu} &= \mathbf{D}_\mu \Lambda_\nu - \mathbf{D}_\nu \Lambda_\mu - 2\mathcal{B}_{\mu\nu}{}^\rho \xi_\rho, \end{aligned} \tag{55}$$

with the open string covariant derivative equal to

$$\mathbf{D}_\mu \Lambda^\nu = \partial_\mu \Lambda^\nu + \mathbf{\Gamma}^\nu_{\rho\mu} \Lambda^\rho,$$

where $\mathbf{\Gamma}^\nu_{\rho\mu}$ is a Christoffel symbol for the changed metric $\mathcal{G}_{\mu\nu} = G_{\mu\nu} + \mathcal{F}^{(s)}_{\mu\nu}$. Again the field strength of the changed Kalb-Ramond field $\mathcal{B}_{\mu\nu} = B_{\mu\nu} + \mathcal{F}^{(a)}_{\mu\nu}$ is just $\mathcal{B}_{\mu\nu\rho} = B_{\mu\nu\rho}$. The generator of the transformation is

$$\mathcal{G} = \int d\sigma \Big[2\xi^\mu \pi_\mu + 2\kappa \widetilde{\mathbf{\Lambda}}_\nu x'^\nu \Big], \tag{56}$$

with

$$\pi_\mu = \kappa \mathcal{G}_{\mu\nu}(x) \dot{x}^\nu - 2\kappa \mathcal{B}_{\mu\nu}(x) x'^\nu, \tag{57}$$

and

$$\widetilde{\mathbf{\Lambda}}_\nu = 2\xi^\mu \mathcal{B}_{\mu\nu} + \Lambda^\mu \mathcal{G}_{\mu\nu} = \Lambda_\nu - 2\mathcal{B}_{\nu\mu}\xi^\mu. \tag{58}$$

Using (48), we obtain the connection

$$\mathbf{\Gamma}_{\mu,\nu\rho} = \Gamma_{\mu,\nu\rho} - 2\partial_\nu \partial_\rho A^D_\mu. \tag{59}$$

So, if we chose A^D_μ linear in coordinate, the connection will remain the same as in the closed string case. Therefore, for such a choice, the symmetry transformations remain the same in all the cases considered. The generators however differ.

4. Conclusion

We considered the string theory of a closed and an open string, described by a standard string sigma model and the modified open string action introduced in [6]. We searched for the symmetry transformations of the space-time fields in which the strings move. We found these symmetries comparing the change in the energy-momentum tensor caused by a similarity transformation and the transformation of fields. These transformations are the symmetries because the conformal field theories corresponding to the initial and the transformed field configurations are isomorphic.

We obtained the explicit form of symmetry transformations for the closed string and the open string with the arbitrary choice of boundary conditions. It turned out that for the appropriate choice of the form of the vector fields on the boundary the symmetry transformations are the same in all cases considered. The generators of these transformations are of the following form

$$\mathcal{G} = 2 \int d\sigma \Big[\xi^\mu \pi_\mu + \widetilde{\Lambda}_\mu \kappa x'^\mu \Big] = \mathcal{G}_\xi + \mathcal{G}_\Lambda. \tag{60}$$

Let us at this point include the T-duality into the consideration, namely the complete T-dualization, i.e. the T-dualization along all initial coordinates.

The T-dualization procedure gives the T-duality coordinate transformation laws, which are connecting the coordinates of the initial string sigma model $S[x]$, defined in (5) and the totally T-dualized theory ${}^\star S[y]$, defined in (6). The coordinate transformation laws are obtained comparing the solutions of the equations of motion for the gauge fields of the auxiliary actions. They read

$$\partial_\pm x^\mu \cong -\kappa\, \Theta_\pm^{\mu\nu}[\Delta V^{(0)}]\partial_\pm y_\nu \mp 2\kappa \Theta_{0\pm}^{\mu\nu}\beta_\nu^\mp[V^{(0)}].$$

In the canonical form, in the zeroth order they reduce to

$$\pi_\mu \cong \kappa x'^\mu.$$

It is obvious, that in this case the components of the symmetry generator (60) are T-dual to each other

$$\mathcal{G}_\xi \cong \mathcal{G}_\Lambda.$$

Therefore, the generator of symmetries is T-dual to itself.

The broader presentation of these investigations will be presented elsewhere, as well as the consideration of the connection between generators in the more complicated backgrounds.

References

[1] K. Becker, M. Becker and J. H. Schwarz, *String Theory and M-Theory*, Cambridge University Press (2007).

[2] Lj. Davidović and B. Sazdović, *Phys. Rev.* **D 83** (2011) 066014; *JHEP* **11** (2015) 119.

[3] Lj. Davidović, B. Nikolić and B. Sazdović, *Eur. Phys. J.* **C 75** (2015) 576.

[4] Lj. Davidović and B. Sazdović, *Eur. Phys. J.* **C 74** (2014) 2683.

[5] Lj. Davidović, *Eur. Phys. J.* **C 76** (2016) 660.

[6] B. Sazdović, *From geometry to non-geometry via T-duality*, arXiv:1606.01938v2 [hep-th].

[7] M. Evans and B. Ovrut , *Phys. Rev.* **D 41** (1990), 3149; *Phys. Lett.* **B 231** (1989), 80; M. Evans and I. Giannakis, *Phys. Rev.* **D 44** (1991), 2467; M. Evans, I. Giannakis and D. Nanopoulos, *Phys. Rev.* **D 50** (1994), 4022.

[8] M. Evans, B.A. Ovrut, Phys. Rev. D 39 (1989) 3016; M. Evans, B.A. Ovrut, Phys. Rev. D 41 (1990) 10; M. Evans, B.A. Ovrut, Phys. Rev. D 44 (1991) 2467.

[9] B. Sazdović and D. Popović, *Mod. Phys. Lett.* **A 16** (2001) 589-602.

Tachyon scalar field in DBI and RSII cosmological context *

Dragoljub D. Dimitrijevic[†]
Faculty of Science and Mathematics, University of Nis, Serbia

Neven Bilic[‡]
Rudjer Boskovic Institute, Zagreb, Coratia

Goran S. Djordjevic[§]
Faculty of Science and Mathematics, University of Nis, Serbia

Milan Milosevic[¶]
Faculty of Science and Mathematics, University of Nis, Serbia

Abstract

The role tachyon fields may play in evolution of early universe is discussed. In particular, we discuss the inflationary scenario based on the tachyon field coupled with the radion of the second Randall-Sundrum model (RSII) and obtained results for the values of observable cosmological parameters, the tensor-to-scalar ratio r and the scalar spectral index n_s.

1. Introduction

The inflationary universe scenario in which the early universe undergoes a rapid expansion provides elegant mechanism to explain and solve horizon, flatness, etc. problems of the standard big-bang cosmology [1, 2, 3]. In a such inflationary universe primordial curvature perturbations, which seed cosmic microwave background temperature anisotropies and the structure formation are generated from vacuum fluctuations of the scalar field, inflaton.

General single field Lagrangian action of the inflaton field

$$S = \int d^4x\sqrt{-g}\mathcal{L}(X(\partial\phi), \phi), \tag{1}$$

* This work has been supported by the Serbian Ministry of Education, Science and Technological Development, Projects OI 174020 and OI 176021, as well as the SEENET-MTP Network under the ICTP grant NT-03.

† e-mail address: ddrag@pmf.ni.ac.rs
‡ e-mail address: bilic@irb.hr
§ e-mail address: gorandj@junis.ni.ac.rs
¶ e-mail address: mmilan@seenet-mtp.info

can be considered, where $X = \frac{1}{2}g^{\mu\nu}\partial_\mu\phi\partial_\nu\phi$ [4, 5]. For example, Lagrangian density of the standard form is

$$\mathcal{L}_{st}(\phi, \partial\phi) = X(\partial\phi) - V(\phi), \tag{2}$$

while popular class comprise tachyon scalar field models [6, 7, 8, 9, 10]

$$\mathcal{L}_{tach}(T, X) = -V(T)\sqrt{1 - 2X(\partial T)}. \tag{3}$$

These models are of particular interest as in these models inflation is driven by the tachyon field originating in string theory [11].

In cosmology, scalar fields can be connected with a perfect fluid which describes (dominant) matter in the Universe. Total action consists of term which describes gravity (Ricci scalar) plus term that describes cosmological fluid (scalar field Lagrangian)

$$S = \int d^4x\sqrt{-g}\left(-\frac{R}{16\pi G} + \mathcal{L}(X, T)\right), \tag{4}$$

from which Einstein equations are obtained

$$R_{\mu\nu} - \frac{1}{2}Rg_{\mu\nu} = 8\pi G T_{\mu\nu}. \tag{5}$$

From the expression of components of the energy-momentum tensor

$$T_{\mu\nu} = \frac{2}{\sqrt{-g}}\frac{\delta S}{\delta g^{\mu\nu}} = (P + \rho)u_\mu u_\nu - Pg_{\mu\nu}, \tag{6}$$

pressure, matter density and velocity 4-vector can be defined, respectively:

$$P(X, T) \equiv \mathcal{L}(X, T), \quad \rho(X, T) \equiv 2X\frac{\partial\mathcal{L}}{\partial X} - \mathcal{L}(X, T), \quad u_\mu \equiv \frac{\partial_\mu T}{\sqrt{2X}}. \tag{7}$$

Of course, this is model with scalar field minimally coupled to gravity.

2. Tachyon inflation

In this sectin we will focus on tachyon scalar field Lagrangian density

$$\mathcal{L}_{tach}(T, X) = -V(T)\sqrt{1 - g^{\mu\nu}\partial_\mu T\partial_\nu T}, \tag{8}$$

with the equation of motion

$$\left(g^{\mu\nu} - \frac{\partial^\mu T\partial^\nu T}{1 - g^{\rho\sigma}\partial_\rho T\partial_\sigma T}\right)\partial_\mu T\partial_\nu T = -\frac{1}{V(T)}\frac{dV}{dT}(1 - g^{\mu\nu}\partial_\mu T\partial_\nu T). \tag{9}$$

In the case of spatially homogenous scalar field the Friedmann equations are

$$H^2 \equiv \left(\frac{\dot{a}}{a}\right)^2 = \frac{8\pi G}{3} \frac{V}{\left(1-\dot{T}^2\right)^{1/2}}, \tag{10}$$

$$\frac{\ddot{T}}{1-\dot{T}^2} + 3H\dot{T} + \frac{V'}{V} = 0. \tag{11}$$

The slow-roll parameters are

$$\epsilon_{i+1} \equiv \frac{d\ln|\epsilon_i|}{dN}, \quad i \geq 0, \quad \epsilon_0 \equiv \frac{H_*}{H}, \tag{12}$$

$$\epsilon_1 = -\frac{\dot{H}}{H^2}, \quad \epsilon_2 = \frac{1}{H}\frac{\ddot{H}}{\dot{H}} + 2\epsilon_1, \tag{13}$$

where H_* is the Hubble parameter at some chosen time [9]. The number of e-folds is

$$N(t) = \int_{t_i}^{t_e} H(t)dt, \tag{14}$$

t_i and t_e correspond to beginning and end of inflation, respectively. In terms of the tachyon field, the slow-roll parameters can be expressed as

$$\epsilon_1 = \frac{3}{2}\dot{T}^2, \quad \epsilon_2 = \sqrt{\frac{2}{3\epsilon_1}}\frac{\epsilon_1'}{H} = 2\frac{\ddot{T}}{H\dot{T}}, \tag{15}$$

where $'$ denotes a derivative with respect to T [12, 13]. The tensor-to-scalar ratio r and the scalar spectral index n_s are defined as [9]

$$r = 16\epsilon_1, \quad n_s = 1 - 2\epsilon_1 - \epsilon_2, \tag{16}$$

and it is understood that ϵ_1 and ϵ_2 take their values at or close to the beginning of inflation. A simple tachyon model can be analyzed in the framework of the second Randall-Sundrum (RSII) model, which will be considered in the next section.

3. Tachyon inflation in an AdS braneworld

Randall-Sundrum model(s) imagine that the real world is a higher-dimensional universe described by warped geometry [14, 15, 16]. More concretely, our universe is a five-dimensional anti-de Sitter space and the elementary particles except for the graviton are localized on a (3 + 1)-dimensional brane(s). A simple cosmological model of this kind is based on the RSII model.

Cosmology on the brane is obtained by allowing the brane to move in the bulk. Equivalently, the brane is kept fixed at fifth coordinate $z = 0$

while making the metric in the bulk time dependent. The fluctuation of the interbrane distance implies the existence of a massless scalar field that causes a distortion of the bulk geometry, called radion.

The bulk spacetime of the extended RSII model

$$ds^2_{(5)} = G_{ab} dX^a dX^b, \tag{17}$$

in Fefferman-Graham coordinates are described by the metric [17]

$$ds^2_{(5)} = \frac{1}{k^2 z^2} \left[\left(1 + k^2 z^2 \eta(x)\right) g_{\mu\nu} dx^\mu dx^\nu - \frac{dz^2}{\left(1 + k^2 z^2 \eta(x)\right)^2} \right], \tag{18}$$

where $k = 1/l$ is inverse of the AdS curvature radius l and $\eta(x)$ is radion field. Now, if we add dynamical 3-brane, i.e. tachyon field (in terms of induced metric), the action, after integrating out fifth coordinate z becomes [10]:

$$\begin{aligned} S &= \int d^4x \sqrt{-g} \left(-\frac{R}{16\pi G} + \frac{1}{2} g^{\mu\nu} \Phi_{,\mu} \Phi_{,\nu} \right) - \\ &- \int d^4x \sqrt{-g} \frac{\sigma}{k^4 \Theta^4} (1 + k^2 \Theta^2 \eta)^2 \sqrt{1 - \frac{g^{\mu\nu} \Theta_{,\mu} \Theta_{,\nu}}{(1 + k^2 \Theta^2 \eta)^3}}, \end{aligned} \tag{19}$$

where Φ is now canonical radion field, $\eta = \sinh^2\left(\sqrt{4/3\pi G}\Phi\right)$ and Θ is tachyon field. Note that in the absence of radion we end up with tachyon condenzate:

$$S^{(0)}_{br} = -\int d^4x \sqrt{-g} \frac{\lambda}{\Theta^4} \sqrt{1 - g^{\mu\nu} \Theta_{,\mu} \Theta_{,\nu}}. \tag{20}$$

Going back, Lagrangian density (for brane-radion system) we are playing with is

$$\mathcal{L} = \frac{1}{2} g^{\mu\nu} \Phi_{,\mu} \Phi_{,\nu} - \frac{\lambda \psi^2}{\Theta^4} \sqrt{1 - \frac{g^{\mu\nu} \Theta_{,\mu} \Theta_{,\nu}}{\psi^3}}, \tag{21}$$

where $\psi = 1 + k^2 \Theta^2 \eta$ and $\lambda = \frac{\sigma}{k^4}$.

We will assume the spatially flat FRW spacetime on the observer brane

$$ds^2 = g_{\mu\nu} dx^\mu dx^\nu = dt^2 - a^2(t)(dr^2 + r^2 d\Omega^2). \tag{22}$$

Going from Lagrangian to Hamiltonian density $\mathcal{H}$

$$\mathcal{H} = \frac{1}{2} \Pi^2_\Phi + \frac{\lambda \psi^2}{\Theta^4} \sqrt{1 + \Pi^2_\Theta \Theta^8 / (\lambda^2 \psi)}, \tag{23}$$

Hamilton's equations become [18]

$$\dot{\Phi} = \frac{\partial \mathcal{H}}{\partial \Pi_\Phi}, \tag{24}$$

$$\dot{\Theta} = \frac{\partial \mathcal{H}}{\partial \Pi_\Theta}, \tag{25}$$

$$\dot{\Pi}_\Phi + 3H\Pi_\Phi = -\frac{\partial \mathcal{H}}{\partial \Phi}, \tag{26}$$

$$\dot{\Pi}_\Theta + 3H\Pi_\Theta = -\frac{\partial \mathcal{H}}{\partial \Theta}, \tag{27}$$

where $\Pi^\mu_\Phi = \frac{\partial \mathcal{L}}{\partial \Phi_{,\mu}}$, $\Pi^\mu_\Theta = \frac{\partial \mathcal{L}}{\partial \Theta_{,\mu}}$, and

$$\Pi_\Phi = \sqrt{g_{\mu\nu}\Pi^\mu_\Phi \Pi^\nu_\Phi}, \qquad \Pi_\Theta = \sqrt{g_{\mu\nu}\Pi^\mu_\Theta \Pi^\nu_\Theta}. \tag{28}$$

In the spatially flat Randall-Sundrum cosmology the Hubble expansion rate H is related to the Hamiltonian via the modified Friedmann equations [19, 20]

$$H \equiv \frac{\dot{a}}{a} = \sqrt{\frac{8\pi G}{3}\mathcal{H}\left(1 + \frac{2\pi G}{3k^2}\mathcal{H}\right)}, \tag{29}$$

$$\dot{H} = -4\pi G(\mathcal{H} + \mathcal{L})\left(1 + \frac{4\pi G}{3k^2}\mathcal{H}\right). \tag{30}$$

To solve this system of equations we rescale the time as $t = \tau/k$, expressing the system in terms of dimensionless quantities

$$h = H/k, \quad \phi = \Phi/(k\sqrt{\lambda}), \quad \pi_\phi = \Pi_\Phi/(k^2\sqrt{\lambda})), \tag{31}$$

$$\theta = k\Theta, \quad \pi_\theta = \Pi_\Theta/(k^4\lambda), \tag{32}$$

and obtain dimensionless pressure and energy density

$$\bar{p} \equiv \frac{\mathcal{L}}{k^4\lambda} = \frac{1}{2}\pi_\phi^2 - \frac{\psi^2}{\theta^4}\frac{1}{\sqrt{1 + \theta^8\pi_\theta^2/\psi}}, \tag{33}$$

$$\bar{\rho} \equiv \frac{\mathcal{H}}{k^4\lambda} = \frac{1}{2}\pi_\phi^2 + \frac{\psi^2}{\theta^4}\sqrt{1 + \theta^8\pi_\theta^2/\psi}. \tag{34}$$

Using $\kappa^2 = 8\pi\lambda G k^2$ [9], the Hamilton's dimensionless equations become

$$\dot{\phi} = \pi_\phi, \tag{35}$$

$$\dot{\theta} = \frac{\theta^4\psi\pi_\theta}{\sqrt{1 + \theta^8\pi_\theta^2/\psi}}, \tag{36}$$

$$\dot{\pi}_\phi = -3h\pi_\phi - \frac{\psi}{2\theta^2}\frac{4 + 3\theta^8\pi_\theta^2/\psi}{\sqrt{1+\theta^8\pi_\theta^2/\psi}}\eta', \tag{37}$$

$$\dot{\pi}_\theta = -3h\pi_\theta + \frac{\psi}{\theta^5}\frac{4 - 3\theta^{10}\eta\pi_\theta^2/\psi}{\sqrt{1+\theta^8\pi_\theta^2/\psi}}, \tag{38}$$

where the overdot now denotes a derivative with respect to τ and

$$h \equiv \frac{\dot{a}}{a} = \sqrt{\frac{\kappa^2}{3}\bar{\rho}\left(1+\frac{\kappa^2}{12}\bar{\rho}\right)}, \tag{39}$$

$$\psi = 1 + \theta^2\eta, \tag{40}$$

$$\eta = \sinh^2\left(\sqrt{\frac{\kappa^2}{6}}\phi\right), \tag{41}$$

$$\eta' = \frac{d\eta}{d\phi} = \sqrt{\frac{\kappa^2}{6}}\sinh\left(\sqrt{\frac{2\kappa^2}{3}}\phi\right). \tag{42}$$

We can add to this system two additional equations

$$\dot{h} = -\frac{\kappa^2}{2}(\bar{\rho}+\bar{p})\left(1+\frac{\kappa^2}{6}\bar{\rho}\right), \tag{43}$$

$$\dot{N} = h. \tag{44}$$

The first two slow-roll parameters are now

$$\epsilon_1 = -\frac{\dot{h}}{h^2}, \quad \epsilon_2 = 2\epsilon_1 + \frac{\ddot{h}}{h\dot{h}}. \tag{45}$$

The effect of the radion and the tachyon can be seen if we compare the slow-roll parameters for the full model with those for the model with inverse quartic tachyon potential. The number of e-folds is defined through Eq. (44) as

$$N = \int_{\tau_i}^{\tau_f} d\tau h. \tag{46}$$

Observable quantities, such as the tensor-to-scalar ratio r and the scalar spectral index n_s are related to the slow-roll parameters

$$r = \frac{\mathcal{P}_T}{\mathcal{P}_S}, \tag{47}$$

$$n_{\rm S} = \frac{d\ln \mathcal{P}_{\rm S}}{d\ln k}, \tag{48}$$

where $\mathcal{P}_{\rm S}$ and $\mathcal{P}_{\rm T}$ are the power spectra of scalar and tensor perturbations, respectively (evaluated at the horizon, i.e., for a wave-number satisfying $k = aH$). In this way (at the lowest order in slow-roll parameters)

$$\mathcal{P}_{\rm T} \simeq [1 - 2(1+C)\epsilon_1]\frac{16GH^2}{\pi}, \tag{49}$$

$$\mathcal{P}_{\rm S} \simeq [1 - 2(1+C-\alpha)\epsilon_1 - C\epsilon_2]\frac{GH^2}{\pi\epsilon_1}, \tag{50}$$

where $C = -2 + \ln 2 + \gamma \simeq -0.72$, α is a parameter related to the speed of sound expanded in ϵ_1

$$c_{\rm s} = 1 - 2\alpha\epsilon_1 + O(\epsilon_1^2), \tag{51}$$

and is equal to 1/12 [10] (for standard tachyon inflation $\alpha = 1/6$ [9]). To solve the system of equations (35)-(38) we need to choose initial conditions relevant for inflation.

4. Initial conditions

To find appropriate initial value of the field θ, we need to consider the pure tachyon model, with the slow-roll conditions [9]

$$\dot{\theta} \simeq \theta^4\pi_\theta \ll 1, \quad \dot{\pi}_\theta \ll 3h\pi_\theta. \tag{52}$$

The slow-roll parameters can be approximated as [10]

$$\epsilon_1 \simeq \frac{8\theta^2}{\kappa^2}\left(1 + \frac{\kappa^2}{6\theta^4}\right)\left(1 + \frac{\kappa^2}{12\theta^4}\right)^{-2}, \tag{53}$$

$$\epsilon_2 \simeq \frac{8\theta^2}{\kappa^2}\left(1 + \frac{\kappa^2}{12\theta^4}\right)^{-2}\left[1 + \frac{\kappa^2}{4\theta^4} - \frac{\kappa^2}{6\theta^4}\left(1 + \frac{\kappa^2}{6\theta^4}\right)^{-1}\right]. \tag{54}$$

In this regime $\kappa^2/\theta^4 \gg 1$ and corrections due to the RSII modification bring the equations in the following form

$$h \simeq \frac{1}{6}\frac{\kappa^2}{\theta^4}, \qquad \dot{\theta} \simeq 8\frac{\theta^3}{\kappa^2}, \qquad \ddot{\theta} \simeq 24\frac{\theta^2}{\kappa^2}\dot{\theta}, \tag{55}$$

$$\epsilon_1 \simeq 3\dot{\theta}^2 \simeq 192\frac{\theta^6}{\kappa^4}, \qquad \epsilon_2 \simeq 288\frac{\theta^6}{\kappa^4} \simeq \frac{3}{2}\epsilon_1. \tag{56}$$

So, in the slow-roll regime the tachyon inflation in the RSII modified cosmology proceeds in a quite distinct way compared with that in the standard

FRW cosmology. However, close to and at the end of inflation $\kappa^2/\theta_{\rm f}^4 \ll 1$ and we can safely neglect the RSII cosmology corrections, obtaining (for details see [10])

$$\epsilon_1(\theta_{\rm f}) \simeq \epsilon_2(\theta_{\rm f}) \simeq \frac{8\theta_{\rm f}^2}{\kappa^2} \simeq 1, \tag{57}$$

,

$$h(\theta_{\rm f}) \simeq \frac{8}{\sqrt{3}\kappa}, \tag{58}$$

$$N \simeq \frac{2}{3}\frac{1}{\epsilon_1(\theta_0)} - 1, \tag{59}$$

where θ_0 is initial value of the θ–field deep inside inflation (at the beginning of inflation). Note that in the standard tachyon inflation (for the potential $V = \lambda/\theta^4$) we would obtain for the number of e-folds

$$N_{\rm st.tach} \simeq \frac{\kappa^2}{8\theta_0^2} - 1 \simeq \frac{1}{\epsilon_1(\theta_0)} - 1. \tag{60}$$

The choice of initial values of the radion field is based on the natural scale dictated by observations, i.e., the scale between H and $G^{-1/2}$, and it would be of the order of k or a few orders larger, say in the range 10 to $1000k$. But dimensionless radion field ϕ is rescaled with respect to Φ/k by a factor of $1/\sqrt{\lambda} \simeq 10^{-4}$ so we can choose the initial value $\phi_0 = \phi(0)$ in the range from 0.001 to 0.5.

5. Values of couplings

It is necessary to find a phenomenologically acceptable choice of the couplings λ and κ. The evolution equations do not depend on λ, but its approximate value is needed for choosing appropriate initial conditions for the radion field. The value of λ may be estimated using the observational constraint on the amplitude of scalar perturbations. In this way, the condition

$$H\sqrt{G} \leq \sqrt{\pi A_s} \simeq 8.31 \times 10^{-5}, \tag{61}$$

must be satisfied close to and at the end of inflation, where $A_s \simeq 2.2 \times 10^{-9}$ is power spectrum amplitude measured by Planck [21], yielding

$$\lambda \geq 10^8. \tag{62}$$

Regarding estimation of coupling κ, the tension of a Dp-brane is given by [22]

$$\sigma = \frac{1}{(2\pi)^p \alpha'^{(p+1)/2} g_s}, \tag{63}$$

where g_s is the string coupling constant and $1/(2\pi\alpha')$ is the string tension. Using this (and $p = 3$, see [10]) we find a constraint

$$g_s \leq 10^{-12}\frac{M_s}{k}, \tag{64}$$

where $M_s = 1/\sqrt{\alpha'}$. Now, we can choose k and M_s such that the scale hierarchy $H < k < M_s < G^{-1/2}$ is satisfied. It follows that

$$\kappa \geq 8/\sqrt{3}. \tag{65}$$

6. Observational parameters

We are now able to obtain expression for r and n_s. Using (49) and (50) we find up to the second order [10]

$$r = 16\epsilon_1\left[1 - \frac{1}{6}\epsilon_1 + C\epsilon_2\right], \tag{66}$$

$$n_s = 1 - 2\epsilon_1 - \epsilon_2 - \left[2\epsilon_1^2 + \left(2C + \frac{17}{6}\right)\epsilon_1\epsilon_2 + C\epsilon_2\epsilon_3\right], \tag{67}$$

where it is understood that ϵ_1 and ϵ_2 take their values at or close to the beginning of inflation. Using these equations at linear order ($\epsilon_2 \simeq 3\epsilon_1/2$) we can find the approximate relations with the help of (59)

$$n_s \simeq 1 - \frac{7}{3}\frac{1}{N+1}, \tag{68}$$

$$r \simeq \frac{32}{7}(1 - n_s). \tag{69}$$

7. Numerical results

The system of equations (35)-(38) is evolved numerically starting from $\tau = 0$ up to some large τ of the order of 100. The initial values θ_0 and ϕ_0 are chosen as already described. The initial conjugate momenta are taken to be $\pi_{\theta 0} = \pi_{\phi 0} = 0$. The function $N(\tau)$ is solved simultaneously using (44) with $N(0) = 0$. The time evolution of the slow roll parameters ϵ_1 and ϵ_2 are obtained using (45). The inflation ends at a point τ_f at which $\epsilon_1(\tau_f) = 1$. The beginning of inflation at τ_i is then found by requiring

$$N(\tau_f) - N(\tau_i) = N. \tag{70}$$

More details can be found in [10]. Here, we only present in Fig. 1 the $n_s - r$ diagram with 10000 points superimposed on the observational constraints taken from the Planck Collaboration 2015 [21]. Each point in the diagram

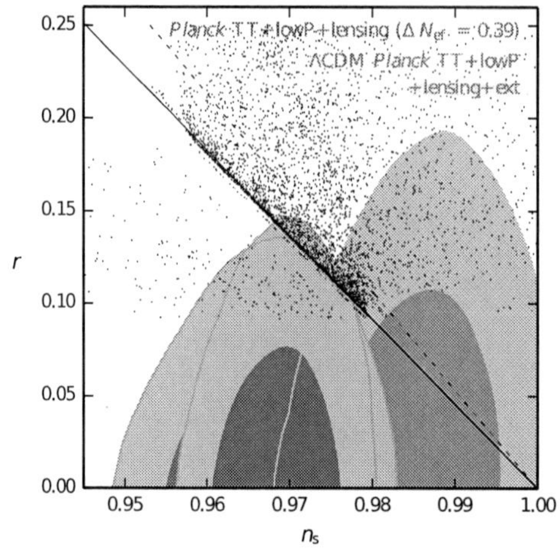

Figure 1: r versus $n_{\rm s}$ diagram with observational constraints from Ref. [21]. The dots represent the calculation in the tachyon-radion model for various N, κ and ϕ_0 chosen randomly in the range $60 \leq N \leq 120$, $1 \leq \kappa \leq 12$ and $0 \leq \phi_0 \leq 0.5$. The full line represents the slow-roll approximation (Eq. (69)) of the RSII model with no radion. The dashed line represents the slow-roll approximation of the standard tachyon model with inverse quartic potential.

corresponds to a set (N, κ, ϕ_0) chosen randomly in the range $60 \leq N \leq 120$, $1 \leq \kappa \leq 12$ and $0 \leq \phi_0 \leq 0.5$.

A still better agreement between our model and the observational data constraints is obtained for a very narrow range of parameters. The best fit is obtained for $\kappa = 1.25$, $\phi_0 = 0.05$ and $115 \leq N \leq 120$ [10].

8. Conclusions

We have investigated a model of inflation based on the dynamics of a $D3$-brane in the AdS5 bulk of the RSII model. The bulk metric is extended to include the back reaction of the radion excitations. The n_s/r relation here is substantially different from the standard one and is closer to the best observational value.

In general, the model is based on the brane dynamics which results in a definite potential with one free parameter only. We have analyzed the simplest tachyon model. In principle, the same mechanism could lead to a more general tachyon potential if the AdS_5 background metric is deformed by the presence of matter in the bulk. The agreement with observations is not ideal and it is fair to say that the present model is disfavored but not excluded.

In [10], we presented results in comparison to the $n_s - r$ diagram, Fig. 1. Here, Fig. 2 and Fig. 3 present histograms for the probability distribution for each parameter individually and compare them with the Planck results. P/P_{max} is the number of parameter values in a given interval divided by the maximal number of parameter values (normalized to unity). Solid lines represent the distribution of Planck's results, while a bar plots represent results for our model.

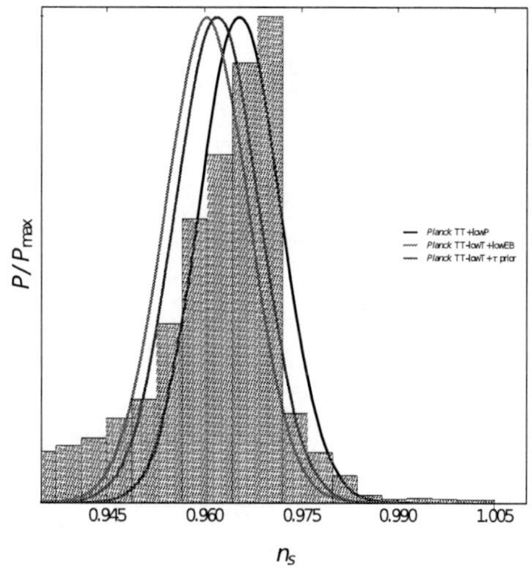

Figure 2: Histogram for density distribution for n_s.

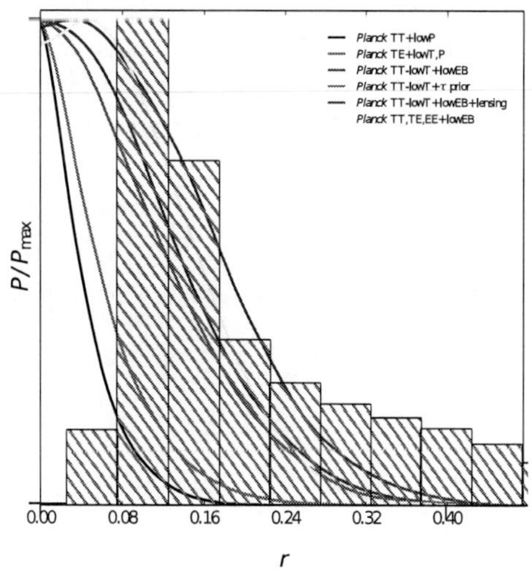

Figure 3: Histogram for density distribution for r.

This is in some sense a step backwards as far as the analysis of results are concerned, but it is easier to understand and analyze the statistics of the observational parameters separately rather then simultaneously. It should be clear in this way how to connect with the probability distribution for the Planck results that we used.

References

[1] A. A. Starobinsky, *JETP Lett.* **30** (1979) 682.

[2] A. H. Guth, *Phys. Rev. D* **23** (1981) 347.

[3] A. D. Linde, *Phys. Lett. B* **116** (1982) 335.

[4] J. Garriga and V. F. Mukhanov, *Phys. Lett. B* **458** (1999) 219.

[5] R. Bean, D. J. H. Chung and G. Geshnizjani, *Phys. Rev. D* **78** (2008) 023517.

[6] A. Feinstein, *Phys. Rev. D* **66** (2002) 063511.

[7] L. Kofman and A. D. Linde, *JHEP* **0207** (2002) 004.

[8] M. Sami, P. Chingangbam and T. Qureshi, *Phys. Rev. D* **66** (2002) 043530.

[9] D. A. Steer and F. Vernizzi, *Phys. Rev. D* **70** (2004) 043527.

[10] N. Bilic, D. D. Dimitrijevic, G. S. Djordjevic and M. Milosevic, *Int. J. Mod. Phys. A* **32** (2017) 1750039.

[11] A. Sen, *JHEP* **04** (2002) 048.

[12] N. Bilic, D. D. Dimitrijevic, G. S. Djordjevic, M. Milosevic, M. Stojanovic, *AIP Conf. Proc.* **1722** (2016) 050002.

[13] M. Milosevic, D. D. Dimitrijevic, G. S. Djordjevic, M. D. Stojanovic, *Serb. Astron. J.* **192** (2016) 1.

[14] L. Randall and R. Sundrum, *Phys. Rev. Lett.* **83** (1999) 4690.

[15] N. Bilic, S. Domazet and G. S. Djordjevic, *Phys. Rev. D* **96** (2017) 083518.

[16] N. Bilic, S. Domazet and G. S. Djordjevic, *Class. Quantum Grav.* **34** (2017) 165006.

[17] J. E. Kim, G. B. Tupper and R. D. Viollier, *Phys. Lett. B* **593** (2004) 209.

[18] N. Bilić and G. B. Tupper, *Central Eur. J. Phys.* **12** (2014) 147.

[19] R. Maartens and K. Koyama, *Living Rev. Rel.* **13** (2010) 5.

[20] M. C. Bento, O. Bertolami and A. A. Sen, *Phys. Rev. D* **67** (2003) 063511.

[21] P. A. R. Ade et al. [Planck Collaboration], arXiv:1502.01589 [astro-ph.CO].

[22] C. V. Johnson, *D-Branes*, Cambridge University Press, Cambridge, 2003.

Variation of a nonlocal modified Einstein gravity action*

Ivan Dimitrijevic†
Faculty of Mathematics, University of Belgrade

Jelena Stankovic‡
Teacher Education Faculty, University of Belgrade

Abstract

We consider nonlocal modified Einstein gravity without matter, where nonlocal term is of the form $P(R)\mathcal{F}(\Box)Q(R)$. Equations of motion are usually very complex. In this paper we present the derivation of the first and second variation of the action (1). The equations of motion are given. We give the expressions for perturbations in scalar, vector and tensor form.

1. Introduction

Although very successful, Einstein theory of gravity is not a final theory. There are many its modifications, which are motivated by quantum gravity, string theory, astrophysics and cosmology (for a review, see [2]). One of very promising directions of research is *nonlocal modified gravity* and its applications to cosmology (as a review, see [3] and [4]).

Under nonlocal modification of gravity we understand replacement of the scalar curvature R in the Einstein-Hilbert action by a suitable function $F(R, \Box, \Box^{-1}, R_{\mu\nu}R^{\mu\nu}, R_{\mu\nu\alpha\beta}R^{\mu\nu\alpha\beta}, \ldots)$, where $\Box = \nabla_\mu\nabla^\mu$ is d'Alembert operator and ∇_μ denotes the covariant derivative. As a review see [7, 6, 5] and references therein. Here, nonlocality means that Lagrangian contains an infinite number of space-time derivatives, i.e. derivatives up to an infinite order in the form of d'Alembert operator $\Box$ which is argument of an analytic function.

In the sequel we consider nonlocal modification of gravity where Einstein-Hilbertaction contains an additional nonlocal term of the form $P(R)\mathcal{F}(\Box)Q(R)$. In particular we consider a class of nonlocal gravity models without matter,

* Work partially supported by the Serbian Ministry of Education, Science and Technological Development, contract No. 174012.
† e-mail address: ivand@matf.bg.ac.rs
‡ e-mail address: jelenagg@gmail.com

given by the following action

$$S = \frac{1}{16\pi G}\int_M (R - 2\Lambda + P(R)\mathcal{F}(\Box)Q(R))\ \sqrt{-g}\ \mathrm{d}^4x, \tag{1}$$

where M is pseudo-Riemann manifold of signature $(1,3)$ with metric $(g_{\mu\nu})$, $\mathcal{F}(\Box) = \sum_{n=0}^{\infty} f_n \Box^n$, P and Q are differentiable functions of the scalar curvature R and Λ is cosmological constant. The corresponding Einstein equations of motion are complex. In this paper we will present their derivation. In order to obtain equations of motion for $g_{\mu\nu}$ we have to find the variation of the action (1) with respect to metric $g^{\mu\nu}$. In addition we also find the second variation of the action (1) and consider some cosmic perturbations.

2. Variation of curvature tensors

Let us start with a technical lemma:

Lemma 1. *The following relations hold*

$$\delta g = g g^{\mu\nu}\delta g_{\mu\nu} = -g g_{\mu\nu}\delta g^{\mu\nu}, \tag{2}$$

$$\delta\sqrt{-g} = -\frac{1}{2} g_{\mu\nu}\sqrt{-g}\delta g^{\mu\nu}, \tag{3}$$

$$\delta\Gamma^{\lambda}_{\mu\nu} = -\frac{1}{2}\left(g_{\nu\alpha}\nabla_\mu\delta g^{\lambda\alpha} + g_{\mu\alpha}\nabla_\nu\delta g^{\lambda\alpha} - g_{\mu\alpha}g_{\nu\beta}\nabla^\lambda\delta g^{\alpha\beta}\right), \tag{4}$$

where g is the determinant of the metric tensor.

Lemma 2. *The variation of Riemman tensor, Ricci tensor and scalar curvature satisfy the following relations*

$$\delta R^{\alpha}_{\mu\beta\nu} = \nabla_\beta\delta\Gamma^{\alpha}_{\mu\nu} - \nabla_\nu\delta\Gamma^{\alpha}_{\mu\beta}, \tag{5}$$

$$\delta R_{\mu\nu} = \nabla_\lambda\delta\Gamma^{\lambda}_{\mu\nu} - \nabla_\nu\delta\Gamma^{\lambda}_{\mu\lambda}, \tag{6}$$

$$\delta R = R_{\mu\nu}\delta g^{\mu\nu} - K_{\mu\nu}\delta g^{\mu\nu}, \tag{7}$$

$$\delta\nabla_\mu\nabla_\nu\psi = \nabla_\mu\nabla_\nu\delta\psi - \nabla_\lambda\psi\delta\Gamma^{\lambda}_{\mu\nu}, \tag{8}$$

where $K_{\mu\nu} = \nabla_\mu\nabla_\nu - g_{\mu\nu}\Box$.

Lemma 3. *Every scalar function $P(R)$ satisfies*

$$\int_M P K_{\mu\nu}\delta g^{\mu\nu}\sqrt{-g}\ \mathrm{d}^4x = \int_M K_{\mu\nu}P\ \delta g^{\mu\nu}\sqrt{-g}\ \mathrm{d}^4x. \tag{9}$$

Proof. At the beginning we prove $\int_M Pg_{\mu\nu}(\Box\delta g^{\mu\nu})\sqrt{-g}\,\mathrm{d}^4x = \int_M g_{\mu\nu}(\Box P)\delta g^{\mu\nu}\sqrt{-g}\,\mathrm{d}^4x$ by application of Stokes' theorem:

$$\begin{aligned}\int_M Pg_{\mu\nu}\Box\delta g^{\mu\nu}\sqrt{-g}\,\mathrm{d}^4x &= -\int_M \nabla_\alpha(Pg_{\mu\nu})\nabla^\alpha\delta g^{\mu\nu}\sqrt{-g}\,\mathrm{d}^4x \\ &= \int_M g_{\mu\nu}\Box P\,\delta g^{\mu\nu}\sqrt{-g}\,\mathrm{d}^4x.\end{aligned} \tag{10}$$

Let $Z^\mu = P\nabla_\nu\delta g^{\mu\nu} - \nabla_\nu P\delta g^{\mu\nu}$. Integration over M yields $\int_M \nabla_\mu Z^\mu\sqrt{-g}\,\mathrm{d}^4x = \int_{\partial M} Z^\mu n_\mu \mathrm{d}\partial M$, where n_μ is the unit normal to a hypersurface ∂M. Since the restriction $Z^\mu|_{\partial M}$ vanish, we obtain

$$\int_M P\nabla_\mu\nabla_\nu\delta g^{\mu\nu}\sqrt{-g}\,\mathrm{d}^4x = \int_M \nabla_\mu\nabla_\nu P\,\delta g^{\mu\nu}\sqrt{-g}\,\mathrm{d}^4x. \qquad \Box$$

Lemma 4. *Let $P(R)$ and $Q(R)$ be scalar functions. Then for all $n \in \mathbb{N}$*

$$\begin{aligned}\int_M P\delta\Box^n Q\sqrt{-g}\,\mathrm{d}^4x = &\frac{1}{2}\sum_{l=0}^{n-1}\int_M S_{\mu\nu}(\Box^l P, \Box^{n-1-l}Q)\delta g^{\mu\nu}\sqrt{-g}\,\mathrm{d}^4x \\ &+ \int_M \Box^n P\,\delta Q\sqrt{-g}\,\mathrm{d}^4x.\end{aligned} \tag{11}$$

Proof. The definition of the $\Box$ operator implies

$$\begin{aligned}I &= \int_M P\delta\Box^n Q\sqrt{-g}\,\mathrm{d}^4x = \int_M P\delta(g^{\mu\nu}\nabla_\mu\nabla_\nu\Box^{n-1}Q)\sqrt{-g}\,\mathrm{d}^4x \\ &= \int_M P\left(\nabla_\mu\nabla_\nu\,\Box^{n-1}Q\delta g^{\mu\nu} + g^{\mu\nu}\delta\nabla_\mu\nabla_\nu\Box^{n-1}Q\right)\sqrt{-g}\,\mathrm{d}^4x \\ &= \int_M P\left(\nabla_\mu\nabla_\nu\,\Box^{n-1}Q\delta g^{\mu\nu} + \Box\delta\Box^{n-1}Q - \nabla_\lambda\Box^{n-1}Q g^{\mu\nu}\delta\Gamma^\lambda_{\mu\nu}\right)\sqrt{-g}\,\mathrm{d}^4x.\end{aligned} \tag{12}$$

Moreover, from the Lemma 3 and Stokes' theorem we get

$$\begin{aligned}I = &\int_M P\Big(\nabla_\mu\nabla_\nu\Box^{n-1}Q\delta g^{\mu\nu} + \Box\delta\Box^{n-1}Q \\ &+\frac{1}{2}\nabla_\lambda\Box^{n-1}Q(2\nabla_\mu\delta g^{\lambda\mu} - g_{\mu\nu}\nabla^\lambda\delta g^{\mu\nu})\Big)\sqrt{-g}\,\mathrm{d}^4x \\ = &\int_M P\,\Box\delta\Box^{n-1}Q\sqrt{-g}\,\mathrm{d}^4x - \int_M \nabla_\mu P\nabla_\nu\Box^{n-1}Q\delta g^{\mu\nu}\sqrt{-g}\,\mathrm{d}^4x \\ &-\frac{1}{2}\int_M g_{\mu\nu}(\nabla^\lambda P\nabla_\lambda\Box^{n-1}Q + P\Box\Box^{n-1}Q)\delta g^{\mu\nu}\sqrt{-g}\,\mathrm{d}^4x \\ = &\int_M P\,\Box\delta\Box^{n-1}Q\sqrt{-g}\,\mathrm{d}^4x + \frac{1}{2}\int_M S_{\mu\nu}(P, \Box^{n-1}Q)\delta g^{\mu\nu}\sqrt{-g}\,\mathrm{d}^4x.\end{aligned} \tag{13}$$

Partial integration in the first term of the pervious formula completes the proof.

□

Theorem 1. *Let P and Q be scalar functions of scalar curvature, then*

$$\int_M P\delta(\sqrt{-g})\,\mathrm{d}^4x = -\frac{1}{2}\int_M g_{\mu\nu}P\delta g^{\mu\nu}\sqrt{-g}\,\mathrm{d}^4x, \tag{14}$$

$$\int_M P\delta R\sqrt{-g}\,\mathrm{d}^4x = \int_M (R_{\mu\nu}P - K_{\mu\nu}P)\,\delta g^{\mu\nu}\sqrt{-g}\,\mathrm{d}^4x, \tag{15}$$

$$\begin{aligned}\int_M P\delta(\mathcal{F}(\Box)Q)\sqrt{-g}\,\mathrm{d}^4x &= \int_M (R_{\mu\nu} - K_{\mu\nu})\left(Q'\mathcal{F}(\Box)P\right)\delta g^{\mu\nu}\sqrt{-g}\,\mathrm{d}^4x \\ &+ \frac{1}{2}\sum_{n=1}^{\infty} f_n \sum_{l=0}^{n-1}\int_M S_{\mu\nu}(\Box^l P, \Box^{n-1-l}Q)\delta g^{\mu\nu}\sqrt{-g}\,\mathrm{d}^4x,\end{aligned} \tag{16}$$

where $S_{\mu\nu}(A,B) = g_{\mu\nu}\nabla^\alpha A\nabla_\alpha B + g_{\mu\nu}A\Box B - 2\nabla_\mu A\nabla_\nu B.$

Proof. Equation (14) is a consequence of (3).
From Lemma 2 and Lemma 3 we get

$$\begin{aligned}\int_M P\delta R\sqrt{-g}\,\mathrm{d}^4x &= \int_M (R_{\mu\nu}P\delta g^{\mu\nu} - PK_{\mu\nu}\delta g^{\mu\nu})\sqrt{-g}\,\mathrm{d}^4x \\ &= \int_M (R_{\mu\nu}P - K_{\mu\nu}P)\,\delta g^{\mu\nu}\sqrt{-g}\,\mathrm{d}^4x.\end{aligned} \tag{17}$$

To prove (16) let us introduce the following notation $J_n = \int_M P\delta(\Box^n Q)\sqrt{-g}\,\mathrm{d}^4x$. Then $\int_M P\delta(\mathcal{F}(\Box)Q)\sqrt{-g}\,\mathrm{d}^4x = \sum_{n=0}^{\infty} f_n J_n$. The integral J_0 is calculated by applying (15). For $n > 0$ integral J_n is calculated by applying Lemma 4 and (15).

$$\begin{aligned}J_n = &\int_M \left(R_{\mu\nu}Q'\Box^n P - K_{\mu\nu}(Q'\Box^n P)\right)\delta g^{\mu\nu}\sqrt{-g}\,\mathrm{d}^4x \\ &+ \frac{1}{2}\sum_{l=0}^{n-1}\int_M S_{\mu\nu}(\Box^l P, \Box^{n-1-l}Q)\delta g^{\mu\nu}\sqrt{-g}\,\mathrm{d}^4x.\end{aligned} \tag{18}$$

Summation over n yields equation (16). □

3. Equations of motion

Let us consider the action (1). In order to calculate δS we introduce the following auxiliary actions

$$S_0 = \int_M (R - 2\Lambda)\,\sqrt{-g}\,\mathrm{d}^4x, \tag{19}$$

$$S_1 = \int_M P(R)\mathcal{F}(\Box)Q(R)\,\sqrt{-g}\,\mathrm{d}^4x. \tag{20}$$

Action S_0 is Einstein-Hilbert action and its variation is

$$\delta S_0 = \int_M (G_{\mu\nu} + \Lambda g_{\mu\nu})\,\delta g^{\mu\nu}\,\sqrt{-g}\,\mathrm{d}^4x. \tag{21}$$

Lemma 5. *Variation of the action S_1 is*

$$\begin{aligned}\delta S_1 = &-\frac{1}{2}\int_M g_{\mu\nu}P(R)\mathcal{F}(\Box)Q(R)\delta g^{\mu\nu}\,\sqrt{-g}\,\mathrm{d}^4x \\ &+\int_M (R_{\mu\nu}W - K_{\mu\nu}W)\,\delta g^{\mu\nu}\,\sqrt{-g}\,\mathrm{d}^4x \\ &+\frac{1}{2}\sum_{n=1}^{\infty} f_n \sum_{l=0}^{n-1}\int_M S_{\mu\nu}(\Box^l P(R), \Box^{n-1-l}Q(R))\delta g^{\mu\nu}\sqrt{-g}\,\mathrm{d}^4x, \end{aligned} \tag{22}$$

where $W = P'(R)\mathcal{F}(\Box)Q(R) + Q'(R)\mathcal{F}(\Box)P(R)$.

Theorem 2. *Variation of the action* (1) *is equal to zero iff*

$$\hat{G}_{\mu\nu} = G_{\mu\nu} + \Lambda g_{\mu\nu} - \frac{1}{2}g_{\mu\nu}P(R)\mathcal{F}(\Box)Q(R) + (R_{\mu\nu}W - K_{\mu\nu}W) + \frac{1}{2}\Omega_{\mu\nu} = 0, \tag{23}$$

where

$$W = P'(R)\mathcal{F}(\Box)Q(R) + Q'(R)\mathcal{F}(\Box)P(R), \tag{24}$$

$$\Omega_{\mu\nu} = \sum_{n=1}^{\infty} f_n \sum_{l=0}^{n-1} S_{\mu\nu}\left(\Box^l P(R), \Box^{n-1-l}Q(R)\right). \tag{25}$$

Proof. The proof of the Theorem is evident from the Lemma 5 and Theorem 1. □

4. Second variation of the action

In the following section we set $h_{\mu\nu} = \delta g_{\mu\nu}$. From Lemma 2 we see that $h^{\mu\nu} = -\delta g^{\mu\nu}$. Also let $h = g^{\mu\nu} h_{\mu\nu}$ be the trace of $h_{\mu\nu}$.

Operator $\delta\Box$ is defined by $(\delta\Box)V = \delta(\Box V) - \Box\delta V$. Then we can prove the following Lemma

Lemma 6. *Let U, V be scalar functions. Then*

$$(\delta\Box)V = -h^{\mu\nu}\nabla_\mu\nabla_\nu V - \nabla^\mu h_\mu^\lambda \nabla_\lambda V + \frac{1}{2}\nabla^\lambda h \nabla_\lambda V, \tag{26}$$

$$\int_M U(\delta\Box)V\sqrt{-g}\,\mathrm{d}^4x = \frac{1}{2}\int_M S_{\mu\nu}(U,V)\delta g^{\mu\nu}\sqrt{-g}\,\mathrm{d}^4x. \tag{27}$$

In the next lemma we find the variation of $\mathcal{F}(\Box)$.

Lemma 7. *Let U, V be scalar functions. Then,*

$$\int_M U\delta(\mathcal{F}(\Box))V\sqrt{-g}\,\mathrm{d}^4x = \sum_{n=1}^{\infty} f_n \sum_{l=0}^{n-1}\int_M S_{\mu\nu}(\Box^l U, \Box^{n-1-l}V)\delta g^{\mu\nu}\sqrt{-g}\,\mathrm{d}^4x \tag{28}$$

Proof. Note that $\delta\Box^n = \sum_{l=0}^{n-1}\Box^l(\delta\Box)\Box^{n-1-l}$ for $n > 0$ and $\delta\Box^0 = \delta\mathrm{Id} = 0$. Therefore sumation over n and integration yields

$$\int_M U\delta(\mathcal{F}(\Box))V\sqrt{-g}\,\mathrm{d}^4x = \sum_{n=1}^{\infty} f_n \sum_{l=0}^{n-1}\int_M U\Box^l(\delta\Box)\Box^{n-1-l}V\sqrt{-g}\,\mathrm{d}^4x \tag{29}$$

$$= \sum_{n=1}^{\infty} f_n \sum_{l=0}^{n-1}\int_M \Box^l U(\delta\Box)\Box^{n-1-l}V\sqrt{-g}\,\mathrm{d}^4x \tag{30}$$

$$= \sum_{n=1}^{\infty} f_n \sum_{l=0}^{n-1}\int_M S_{\mu\nu}(\Box^l U, \Box^{n-1-l}V)\delta g^{\mu\nu}\sqrt{-g}\,\mathrm{d}^4x. \tag{31}$$

□

Lemma 8. *Let U be scalar function. Then,*

$$\int_M U\delta W\sqrt{-g}\,\mathrm{d}^4x = \int_M (R_{\mu\nu}Y - K_{\mu\nu}Y + \frac{1}{2}\Psi_{\mu\nu})\delta g^{\mu\nu}\sqrt{-g}\,\mathrm{d}^4x, \tag{32}$$

$$Y = U(P''\mathcal{F}(\Box)Q + Q''\mathcal{F}(\Box)P) + (P'\mathcal{F}(\Box)(Q'U) + Q'\mathcal{F}(\Box)(P'U)), \tag{33}$$

$$\Psi_{\mu\nu} = \sum_{n=1}^{+\infty} f_n \sum_{l=0}^{n-1}\Big(S_{\mu\nu}(\Box^l(P'U), \Box^{n-1-l}Q) + S_{\mu\nu}(\Box^l(Q'U), \Box^{n-1-l}P)\Big) \tag{34}$$

Lemma 9. *Let A, B be scalar functions. Then,*

$$\int_M S_{\mu\nu}(\delta A, B)\delta g^{\mu\nu}\sqrt{-g}\,\mathrm{d}^4x = \int \sigma_1(B)\delta A\sqrt{-g}\,\mathrm{d}^4x, \tag{35}$$

$$\int_M S_{\mu\nu}(A, \delta B)\delta g^{\mu\nu}\sqrt{-g}\,\mathrm{d}^4x = \int \sigma_2(A)\delta B\sqrt{-g}\,\mathrm{d}^4x, \tag{36}$$

where

$$\sigma_1(B) = \nabla^\lambda h \nabla_\lambda B - 2\nabla_\mu h^{\mu\nu}\nabla_\nu B - 2h^{\mu\nu}\nabla_\mu\nabla_\nu B, \tag{37}$$

$$\sigma_2(A) = -\nabla^\lambda h \nabla_\lambda A - A\Box h - 2\nabla_\nu h^{\mu\nu}\nabla_\mu A - 2h^{\mu\nu}\nabla_\mu\nabla_\nu A. \tag{38}$$

Proof. To prove the first equation recall the definition of $S_{\mu\nu}(A, B)$

$$\int_M S_{\mu\nu}(\delta A, B)\delta g^{\mu\nu}\sqrt{-g}\,\mathrm{d}^4x \tag{39}$$

$$= \int_M \left(\nabla^\alpha(h\nabla_\alpha B) - h\Box B - 2\nabla_\mu(h^{\mu\nu}\nabla_\nu B)\right)\delta A\sqrt{-g}\,\mathrm{d}^4x \tag{40}$$

$$= \int_M \sigma_1(B)\delta A\sqrt{-g}\,\mathrm{d}^4x. \tag{41}$$

The proof of the second equation is similar. □

Lemma 10. *Let $\Omega_{\mu\nu} = \sum_{n=1}^{\infty} f_n \sum_{l=0}^{n-1} S_{\mu\nu}\left(\Box^l P(R), \Box^{n-1-l}Q(R)\right)$. Then,*

$$\begin{aligned}
&\int_M \delta\Omega_{\mu\nu}\delta g^{\mu\nu}\sqrt{-g}\,\mathrm{d}^4x \\
&= \int_M \sum_{n=1}^{+\infty} f_n \sum_{l=0}^{n-1}\Big(h_{\mu\nu}\nabla^\lambda\Box^l P\nabla_\lambda\Box^{n-1-l}Q + h\nabla_\mu\Box^l P\nabla_\nu\Box^{n-1-l}Q \\
&+ h_{\mu\nu}\Box^l P\Box^{n-l}Q - \frac{1}{2}S_{\mu\nu}(h\Box^l P, \Box^{n-1-l}Q) \\
&+ (R_{\mu\nu} - K_{\mu\nu})(P'\Box^l(\sigma_1(\Box^{n-1-l}Q)) + Q'\Box^l(\sigma_2(\Box^{n-1-l}P))) \\
&+ \frac{1}{2}\int_M \sum_{n=1}^{+\infty} f_n \sum_{l=1}^{n-1}\sum_{m=0}^{l-1}\Big(S_{\mu\nu}(\Box^m(\sigma_1(\Box^{n-1-l}Q)), \Box^{l-m-1}P) \\
&+ S_{\mu\nu}(\Box^m(\sigma_2(\Box^{n-1-l}P)), \Box^{l-m-1}Q)\Big)\delta g^{\mu\nu}\sqrt{-g}\,\mathrm{d}^4x
\end{aligned} \tag{42}$$

Proof. Note that

$$\delta\Omega_{\mu\nu} = \sum_{n=1}^{\infty} f_n \sum_{l=0}^{n-1}\delta S_{\mu\nu}\left(\Box^l P(R), \Box^{n-1-l}Q(R)\right). \tag{43}$$

Moreover,

$$\int_M \delta S_{\mu\nu}(A,B)\delta g^{\mu\nu}\sqrt{-g}\,\mathrm{d}^4x = \int_M \left(S_{\mu\nu}(\delta A,B)+S_{\mu\nu}(A,\delta B)\right)\delta g^{\mu\nu}\sqrt{-g}\,\mathrm{d}^4x$$
$$+\int_M \big(h_{\mu\nu}\nabla^\lambda A\nabla_\lambda B + h\nabla_\mu A\nabla_\nu B + h_{\mu\nu}A\Box B - \frac{1}{2}S_{\mu\nu}(hA,B)\big)\delta g^{\mu\nu}\sqrt{-g}\,\mathrm{d}^4x.$$

Using this formula for each term in $\delta\Omega_{\mu\nu}$ yields the result of the Lemma. □

Theorem 3. *The second variation of the action* (1) *is given by*

$$\delta^2 S = \frac{1}{16\pi G}\int_M \left(U_{\mu\nu} + R_{\mu\nu}X - K_{\mu\nu}X + \frac{1}{2}\chi_{\mu\nu} + \frac{1}{4}\Theta_{\mu\nu}\right)\delta g^{\mu\nu}\sqrt{-g}\,\mathrm{d}^4x, \tag{44}$$

where

$$U_{\mu\nu} = -\frac{1}{2}h_{\mu\nu}(R-2\Lambda+P\mathcal{F}(\Box)Q) + \delta R_{\mu\nu}(W+1) + \delta\Gamma^\lambda_{\mu\nu}\nabla_\lambda W$$
$$+ h_{\mu\nu}\Box W - \frac{1}{2}S_{\mu\nu}(h,W), \tag{45}$$

$$X = \frac{1}{2}(h + P'h\mathcal{F}(\Box)Q + Q'\mathcal{F}(\Box)(Ph)) + \Big(\delta R(P''\mathcal{F}(\Box)Q + Q''\mathcal{F}(\Box)P)$$
$$+ (P'\mathcal{F}(\Box)(Q'\delta R) + Q'\mathcal{F}(\Box)(P'\delta R))\Big), \tag{46}$$

$$\chi_{\mu\nu} = \frac{1}{2}\sum_{n=1}^{+\infty} f_n \sum_{l=0}^{n-1} S_{\mu\nu}(\Box^l(Ph),\Box^{n-l-1}Q)$$
$$-\sum_{n=1}^{+\infty} f_n \sum_{l=0}^{n-1}\Big(S_{\mu\nu}(\Box^l(P'M),\Box^{n-1-l}Q) + S_{\mu\nu}(\Box^l(Q'M),\Box^{n-1-l}P)\Big)$$
$$+\sum_{n=1}^{+\infty} f_n \sum_{l=0}^{n-1}\Big(h_{\mu\nu}\nabla^\lambda\Box^l P\nabla_\lambda\Box^{n-1-l}Q + h\nabla_\mu\Box^l P\nabla_\nu\Box^{n-1-l}Q$$
$$+ h_{\mu\nu}\Box^l P\Box^{n-l}Q - \frac{1}{2}S_{\mu\nu}(h\Box^l P,\Box^{n-1-l}Q)$$
$$+ (R_{\mu\nu}-K_{\mu\nu})(P'\Box^l(\sigma_1(\Box^{n-1-l}Q)) + Q'\Box^l(\sigma_2(\Box^{n-1-l}P)))\Big), \tag{47}$$

and

$$\Theta_{\mu\nu} = \sum_{n=1}^{+\infty} f_n \sum_{l=1}^{n-1}\sum_{m=0}^{l-1} \Big(S_{\mu\nu}(\Box^m(\sigma_1(\Box^{n-1-l}Q)), \Box^{l-m-1}P)$$
$$+ S_{\mu\nu}(\Box^m(\sigma_2(\Box^{n-1-l}P)), \Box^{l-m-1}Q)\Big), \quad (48)$$
$$\sigma_1(B) = \nabla^\lambda h \nabla_\lambda B - 2\nabla_\mu h^{\mu\nu}\nabla_\nu B - 2h^{\mu\nu}\nabla_\mu\nabla_\nu B, \quad (49)$$
$$\sigma_2(A) = -\nabla^\lambda h \nabla_\lambda A - A\Box h - 2\nabla_\nu h^{\mu\nu}\nabla_\mu A - 2h^{\mu\nu}\nabla_\mu\nabla_\nu A. \quad (50)$$

Proof. In the pervious section we calculated the first variation of the action (1)

$$\delta S = \frac{1}{16\pi G}\int_M \hat{G}_{\mu\nu}\delta g^{\mu\nu}\sqrt{-g}\,\mathrm{d}^4x. \quad (51)$$

Moreover the second variation $\delta^2 S$ is

$$\delta^2 S = \frac{1}{16\pi G}\int_M \left(\delta\hat{G}_{\mu\nu}\delta g^{\mu\nu} + \hat{G}_{\mu\nu}\delta^2 g^{\mu\nu} - \frac{1}{2}g_{\alpha\beta}\delta g^{\alpha\beta}\hat{G}_{\mu\nu}\delta g^{\mu\nu}\right)\sqrt{-g}\,\mathrm{d}^4x. \quad (52)$$

At the beginning note that

$$\int_M \delta\left(G_{\mu\nu} + \Lambda g_{\mu\nu}\right)\delta g^{\mu\nu}\sqrt{-g}\,\mathrm{d}^4x$$
$$= \int_M \left(\delta R_{\mu\nu} - \frac{1}{2}(R-2\Lambda)h_{\mu\nu} + \frac{1}{2}R_{\mu\nu}h - \frac{1}{2}K_{\mu\nu}h\right)\delta g^{\mu\nu}\sqrt{-g}\,\mathrm{d}^4x. \quad (53)$$

The next term is calculated by using Lemma 7

$$\int_M \delta\left(g_{\mu\nu}P\mathcal{F}(\Box)Q\right)\delta g^{\mu\nu}\sqrt{-g}\,\mathrm{d}^4x = \int_M h_{\mu\nu}P\mathcal{F}(\Box)Q\delta g^{\mu\nu}\sqrt{-g}\,\mathrm{d}^4x$$
$$-\frac{1}{2}\sum_{n=1}^{+\infty} f_n \sum_{l=0}^{n-1}\int S_{\mu\nu}(\Box^l(Ph), \Box^{n-l-1}Q)\delta g^{\mu\nu}\sqrt{-g}\,\mathrm{d}^4x$$
$$-\int_M (R_{\mu\nu} - K_{\mu\nu})\left(P'h\mathcal{F}(\Box)Q + Q'\mathcal{F}(\Box)(Ph)\right)\delta g^{\mu\nu}\sqrt{-g}\,\mathrm{d}^4x. \quad (54)$$

The third term is $\int_M \delta\left(R_{\mu\nu}W\right)\delta g^{\mu\nu}\sqrt{-g}\,\mathrm{d}^4x$ and it is equal to

$$\int_M \delta\left(R_{\mu\nu}W\right)\delta g^{\mu\nu}\sqrt{-g}\,\mathrm{d}^4x \quad (55)$$
$$= -\frac{1}{2}\int_M \left(\Box h_{\mu\nu} + \nabla_\mu\nabla_\nu h - 2\nabla_\lambda\nabla_\mu h_\nu^\lambda\right)W\delta g^{\mu\nu}\sqrt{-g}\,\mathrm{d}^4x$$
$$+ \int_M R_{\mu\nu}\delta W\delta g^{\mu\nu}\sqrt{-g}\,\mathrm{d}^4x \quad (56)$$

The last integral of the above formula is obtained by Lemma 8. Similarly, we obtain

$$\int_M \delta\left(K_{\mu\nu}W\right)\delta g^{\mu\nu}\sqrt{-g}\,\mathrm{d}^4x = \int_M \delta W K_{\mu\nu}\delta g^{\mu\nu}\sqrt{-g}\,\mathrm{d}^4x$$
$$-\int\left(\delta\Gamma^{\lambda}_{\mu\nu}\nabla_\lambda W + h_{\mu\nu}\Box W - \frac{1}{2}S_{\mu\nu}(h,W)\right)\delta g^{\mu\nu}\sqrt{-g}\,\mathrm{d}^4x \quad (57)$$

At the end the last term $\int_M \delta\Omega_{\mu\nu}\delta g^{\mu\nu}\sqrt{-g}\,\mathrm{d}^4x$ is calculated in Lemma 10.

□

5. Perturbations

5.1. Background

In this section we look for FRW metric with $k=0$ which can be written as

$$ds^2 = -dt^2 + a(t)^2(\mathrm{d}x^2 + \mathrm{d}y^2 + \mathrm{d}z^2) \quad (58)$$

Some relevant background quantities are

$$R = 12H^2 + 6\dot{H},\ \Gamma^0_{ij} = Hg_{ij},\ \Gamma^i_{j0} = H\delta^i_j,\ \Box = -\partial_t^2 - 3H\partial_t + \frac{\delta^{ij}\partial_i\partial_j}{a^2} \quad (59)$$

where the indexes i,j range as $1,2,3$. On the background all quantities are space homogeneous as the metric suggests.

For perturbations we employ the canonical ADM $(1+3)$ decomposition and introduce the conformal time τ such that

$$a\mathrm{d}\tau = dt$$

Then the general FRW metric (58) transforms

$$ds^2 = a(\tau)^2(-d\tau^2 + \mathrm{d}x^2 + \mathrm{d}y^2 + \mathrm{d}z^2). \quad (60)$$

5.2. Perturbations

The metric perturbations (see [1]) can be categorized in three types: scalar, vector and tensor perturbations. The component h_{00} is invariant under spatial rotations and translations and therefore

$$h_{00} = 2a(\tau)^2\phi. \quad (61)$$

The components h_{0i} are the sum of a spatial gradient of a function B and divergence free vector S_i.

$$h_{0i} = a(\tau)^2(\partial_i B + S_i). \quad (62)$$

Similarly, components h_{ij}, which transform as a tensor under 3− rotations are written as

$$h_{ij} = a(\tau)^2(2\psi\delta_{ij} + 2\partial^2_{ij}E + \partial_i F_j + \partial_j F_i + \varphi_{ij}), \tag{63}$$

where ψ and E are scalar functions, F_i is a vector with zero divergence and 3− tensor satisfies

$$\varphi^i_i = 0, \qquad \partial_i \varphi^i_j = 0. \tag{64}$$

Note that three are four scalar functions, two vectors with two independent components each and tensor s_{ij} has two independent components. Therefore, as expected we have total of ten functions.

The scalar perturbations are defined by scalar functions ϕ, ψ, B, E and perturbed metric around FRW b background is

$$ds^2 = a(\tau)^2\left[-(1-2\phi)d\tau^2 + \partial_i B \mathrm{d}\tau dx^i + ((1+2\psi)\delta_{ij} + 2\partial_i\partial_j E)dx^i dx^j\right], \tag{65}$$

The vector perturbations are defined by vectors S_i and F_i.

$$ds^2 = a(\tau)^2\left[-d\tau^2 + S_i B \mathrm{d}\tau dx^i + (\delta_{ij} + \partial_i F_j + \partial_j F_i)dx^i dx^j\right], \tag{66}$$

Tensor perturbations are defined by φ_{ij} and describe gravity waves, and have no analog in Newtonian theory.

$$ds^2 = a(\tau)^2\left[-d\tau^2 + (\delta_{ij} + \varphi_{ij})dx^i dx^j\right], \tag{67}$$

Each of the types of perturbations can be studied separately. In this form of perturbations we have

$$h_{\mu\nu} = a(\tau)^2\begin{pmatrix} 2\phi & \partial_i B + S^i \\ \partial_i B + S^i & 2\psi \mathrm{Id} + 2\,\mathrm{Hess}\,E + \partial_j F_i + \partial_i F_j + \varphi_{ij}\end{pmatrix}, \tag{68}$$

$$h = -2\phi + 6\psi + 2\triangle E, \tag{69}$$

$$\begin{aligned}\delta R = &\frac{2}{a^2}(6\frac{a''}{a}\phi + \triangle(\phi - 2\psi) + 3\frac{a'}{a}\triangle(B + E') \\ &+ 3\frac{a'}{a}(\phi' + 3\psi') + \triangle(B' + E'') + 3\psi'').\end{aligned} \tag{70}$$

Moreover, let $A^\mu = \nabla_\nu h^{\mu\nu}$, then

$$A^0 = 6\frac{a'}{a^3}(\phi + \psi) + \frac{1}{a^2}\triangle B + 2\frac{a'}{a^3}\triangle E + \frac{2}{a^2}\phi', \tag{71}$$

$$A^i = 4\frac{a'}{a^3}(\partial_i B + S^i) + \frac{2}{a^2}(\partial_i\psi + \triangle\partial_i E) + \frac{1}{a^2}(\partial_i B' + S'^i + \triangle F^i). \tag{72}$$

Let $r_{\mu\nu} = \delta R_{\mu\nu}$, then

$$r_{00} = -\frac{1}{a}\Big(a'(\triangle(B+E') + 3(\phi' + \psi')) + a(\triangle(\phi + B' + E'') + 3\psi'')\Big), \tag{73}$$

$$r_{i0} = -\frac{1}{a^2}\Big((a'^2 + aa'')\partial_i B + 2aa'\partial_i\phi\Big) - 2\partial_i\psi', \tag{74}$$

$$r_{ii} = \frac{1}{a^2}\Big(2(a'^2 + aa'')(\phi + \psi + \partial_{ii}^2 E) + aa'(\triangle(B+E') + 2\partial_{ii}^2(B+E') + \phi' + 5\psi') + a^2(\partial_{ii}^2(\phi + B' + E'' - \psi) + \psi'' - \triangle\psi)\Big), \tag{75}$$

$$r_{ij} = \frac{1}{a^2}\Big(2\partial_{ij}^2 E(a'^2 + aa'') + 2aa'\partial_{ij}^2(B+E') + a^2\partial_{ij}^2(\phi - \psi + B' + E'')\Big),\ i \neq j. \tag{76}$$

Out of 4 scalar modes only 2 are gauge invariant. The convenient gauge invariant variables (Bardeen potentials) are introduced as

$$\Phi = \phi - \frac{1}{a}(a(B-E'))', \qquad \Psi = \psi + \frac{a'}{a}(B-E'). \tag{77}$$

The prime denotes the differentiation with respect to the conformal time τ and the dot as before w.r.t. the cosmic time t.

The $(1+3)$ structure suggests to represent the perturbation quantities (which can depend on all 4 coordinates) as

$$f(\tau, \vec{x}) = f(\tau, k)Y(k, \vec{x}) \tag{78}$$

where $\vec{x} = (x, y, z)$ and $k = |\vec{k}|$ comes from the definition of the Y-functions as spatial Fourier modes

$$\delta^{ij}\partial_i\partial_j Y = -k^2 Y \tag{79}$$

Obviously

$$Y = Y_0 e^{\pm i\vec{k}\vec{x}} \tag{80}$$

The relevant expressions for the d'Alembertian operator are

$$\Box = -\frac{1}{a^2}\partial_\tau^2 - 2\frac{a'}{a^3}\partial_\tau - \frac{k^2}{a^2} = -\partial_t^2 - 3H\partial_t - \frac{k^2}{a^2} \tag{81}$$

Recall that considering the de Sitter background we do not need the perturbed d'Alembertian operator to the linear order in perturbation analysis. However, all the expression in this subsection are valid for a generic scale factor a.

6. Concluding Remarks

In this paper we have considered a nonlocal gravity model without matter given by the action in the form

$$S = \frac{1}{16\pi G} \int_M (R - 2\Lambda + P(R)\mathcal{F}(\Box)Q(R)) \sqrt{-g}\, \mathrm{d}^4x. \tag{82}$$

We have derived the equations of motion for this action. We also have presented the second variation of the previous action.

In many research papers there are equations of motion which are a special case of our equations. In the case $P(R) = Q(R) = R$ we obtain

$$S = \frac{1}{16\pi G} \int_M (R - 2\Lambda + R\mathcal{F}(\Box)R) \sqrt{-g}\, \mathrm{d}^4x.$$

This nonlocal model is elaborated is the series of papers [8, 9, 10, 11, 12, 13, 14, 15].

The action (82) for $P(R) = R^{-1}$ and $Q(R) = R$ was introduced in [16] as a new approach to nonlocal gravity. This model one can also find in [17].

The case $P(R) = R^p$ and $Q(R) = R^q$ we analyze in [19, 20].

For the case $P(R) = (R + R_0)^m$ and $Q(R) = (R + R_0)^m$ see [18].

In the last case we have $R =$ const. Studies of this model can be found in [21, 22, 19].

References

[1] V. Mukhanov, *Physical foundations of cosmology*, (Cambridge, 2005)

[2] T. Clifton, P. G. Ferreira, A. Padilla, C. Skordis, Modified gravity and cosmology, Physics Reports **513** (2012) 1–189; [arXiv:1106.2476v2 [astro-ph.CO]].

[3] S. Nojiri, S. D. Odintsov, Unified cosmic history in modified gravity: from $F(R)$ theory to Lorentz non-invariant models, Physics Reports **505** (2011) 59–144; [arXiv:1011.0544v4 [gr-qc]].

[4] R. P. Woodard, Nonlocal models of cosmic acceleration, [arXiv:1401.0254 [astro-ph.CO]] (2014).

[5] L. Buoninfante, A. S. Koshelev, G. Lambiase and A.Mazumdar, Classical properties of non-local, ghost- and singularity-free gravity, [arXiv:1802.00399 [gr-qc]].

[6] E. Elizalde, E. O. Pozdeeva, and S. Yu. Vernov. Stability of de Sitter So- lutions in Non-local Cosmological Models. PoS, QFTHEP2011:038, 2013, arXiv:1202.0178.

[7] B. Dragovich, On Nonlocal modified gravity and cosmology, Lie Theory and Its Applications in Physics, volume 111 of Springer Proceedings in Mathematics and Statistics, pages 251-262, 2014.

[8] T. Biswas, A. Mazumdar, , W. Siegel Bouncing universes in string-inspired gravity. J. Cosmology Astropart. Phys. **0603**, 009 (2006) [arXiv:hep-th/0508194]

[9] T. Biswas, T. Koivisto, A. Mazumdar, Towards a resolution of the cosmological singularity in non-local higher derivative theories of gravity. J. Cosmology Astropart. Phys. **1011**, 008 (2010) [arXiv:1005.0590v2 [hep-th]].

[10] T. Biswas, E. Gerwick, Koivisto, A. Mazumdar, Towards singularity and ghost free theories of gravity. Phys. Rev. Lett. **108**, 031101 (2012) [arXiv:1110.5249v2 [gr-qc]]

[11] A. S. Koshelev, S. Yu. Vernov, On bouncing solutions in non-local gravity, Phys. Part. Nuclei **43**, 666668 (2012) [arXiv:1202.1289v1 [hep-th]].

[12] I. Dimitrijevic, B. Dragovich, J. Grujic , Z. Rakic, On modified gravity. Springer Proceedings in Mathematics and Statistics **36**, 251–259 (2013) [arXiv:1202.2352 [hep-th]]

[13] I. Dimitrijevic, B. Dragovich, J. Grujic Z. Rakic, New cosmological solutions in non-local modified gravity. Rom. Journ. Phys. **58** (5-6), 550–559 (2013) [arXiv:1302.2794 [gr-qc]]

[14] T. Biswas, A. Conroy, A.S. Koshelev, A. Mazumdar Generalized gost-free quadratic curvature gravity. [arXiv:1308.2319 [hep-th]]

[15] T. Biswas, A. S. Koshelev, A. Mazumdar, S. Yu. Vernov, Stable bounce and inflation in non-local higher derivative cosmology, JCAP **08** (2012) 024, [arXiv:1206.6374v2 [astro-ph.CO]].

[16] I. Dimitrijevic, B. Dragovich, J. Grujic and Z. Rakic, A new model of nonlocal modified gravity, Publications de l'Institut Mathematique **94** (2013), 187–196.

[17] I. Dimitrijevic, B. Dragovich, J. Grujic and Z. Rakic, Some power-law cosmological solutions in nonlocal modified gravity in: Lie Theory and Its Applications in Physics, *Springer Proceedings in Mathematics and Statistics*, 2014, pp. 241–250.

[18] I. Dimitrijevic, B. Dragovich, J. Stankovic, A. S. Koshelev and Z. Rakic, On Nonlocal Modified Gravity and Its Cosmological Solutions, Springer Proceedings in Mathematics & Statistics **191** (2016) 35–51. arXiv:1701.02090 [hep-th].

[19] I. Dimitrijevic, B. Dragovich, J. Grujic and Z. Rakic, Some Cosmological Solutions of a Nonlocal Modified Gravity, Filomat 29 (3), (2015) 619–628, arXiv:1508.05583 [hep-th].

[20] I. Dimitrijevic, Cosmological solutions in modified gravity with monomial nonlocality, Applied Mathematics and Computation, 285 (3), (2016) 195–203.

[21] I. Dimitrijevic, B. Dragovich, J. Grujic and Z. Rakic, Constant Curvature Cosmological Solutions in Nonlocal Gravity, AIP Conference Proceedings **1634**, (2014) 18–23.

[22] I. Dimitrijevic, B. Dragovich, Z. Rakic and J. Stankovic, On Nonlocal Gravity with Constant Scalar Curvature, Publications de l'Institut Mathematique (2017), to be published

The noncommutative $SO(2,3)_\star$ gravity model

Marija Dimitrijević Ćirić*
Faculty of Physics, University of Belgrade
Dragoljub Gočanin†
Faculty of Physics, University of Belgrade
Nikola Konjik‡
Faculty of Physics, University of Belgrade
Voja Radovanović§
Faculty of Physics, University of Belgrade

Abstract

In this review, noncommutative gravity is treated as a gauge theory of the noncommutative $SO(2,3)_\star$ group. We assume that the spacetime deformation is canonical. The Seiberg-Witten map is used to express noncommutative fields in terms of the corresponding commutative fields. In addition to pure gravity, we consider couplings to matter fields, in particular, fermion and $U(1)$ gauge field. The analysis can be extended to non Abelian gauge fields and scalar fields.

1. Introduction

Quantum Field Theory (QFT) and General Relativity (GR) are widely regarded as the two pillars of modern theoretical physics. Although these theories have been tested to an excellent degree of accuracy in their respective areas of applicability, occurrence of singularities in both of them strongly indicates that they are incomplete. GR, as a classical theory of gravitation, describes large-scale geometric structure of spacetime and its relation to the distribution of matter. On the other hand, QFT, standing on the principles of Quantum Mechanics and Special Relativity, provides us with the Standard Model of elementary particles which successfully utilizes the concept of gauge (local) symmetry to describe the fundamental

* e-mail address: dmarija@ipb.ac.rs
† e-mail address: dgocanin@ipb.ac.rs
‡ e-mail address: konjik@ipb.ac.rs
§ e-mail address: rvoja@ipb.ac.rs

particle interactions (electro-weak and strong). In spite of being fundamentally inconsistent with each other[1], both theories (QFT and GR) rely on the concepts of continuous spacetime and that of point particle. This is indeed appropriate, but, in the case of GR, only at large distances and only approximately. In order to establish the, still elusive, theory of "Quantum Gravity", i.e. the theory of the quantum structure of spacetime, at very short distances (very high energies) we must go beyond the usual concepts of QFT and GR to which we are accustomed. Variety of ways of treating the problem has been proposed so far, stemming from String Theory, Loop Quantum Gravity, Noncommutative (NC) Field Theory, etc.

Recently, a lot of attention has been devoted to NC gravity, as a special case of NC Field Theory that doesn't include matter fields, and many different approaches have been developed. In [1, 2, 3] a deformation of pure Einstein gravity via Seiberg-Witten (SW) map is proposed. Twist approach was explored in [4, 5, 6, 7]. Some other proposals are given in [8, 9, 10, 11, 12, 13, 14, 15, 16, 17, 18, 19, 20]. The connection to Supergravity was established in [21, 22]. Finally, the most relevant for our work is the approach developed by the members of our group [23, 24, 25, 26]. In this approach NC gravity is treated as a gauge theory of canonically deformed Anti de Sitter (AdS) group $SO(2,3)_\star$ (NC version of ordinary $SO(2,3)$ group).

A next natural step is to tackle the problem of introducing matter fields on NC spacetime and their interaction with NC gravity. Our theoretical model is the extension of the previously established $SO(2,3)_\star$ model of pure NC gravity [23, 24, 25, 26]. During the development of this model, the subject was treated from a different point of view by Aschieri and Castellani [27, 28, 29, 30]. Their model, based on the deformed $SO(1,3)$ symmetry, is significantly different from the one that we propose. On the formal side, in their $SO(1,3)_\star$ model, there is a problem of vierbein not being a well-defined gauge field. This is not an issue if we use $SO(2,3)$ gauge group. More importantly, our model leads to some very definite physical predictions and allows us to actually derive them in detail, while on the other hand no elaboration concerning any potential physical effects has been given by the upper mentioned authors. Also, we should emphasize that the differences between the two models revealed themselves already in the case of pure gravity, namely, the canonical NC deformation of Minkowski space actually is obtained in $SO(2,3)_\star$ model [23].

[1]Except in a certain narrow range of parameters where semi-classical discipline of QFT in curved spacetime background provides us with some reliable predictions, like the well-known Hawking radiation.

2. Introducing matter fields on NC spacetime

2.1. Deformation quantization

The construction of a NC Field Theory, i.e. Field Theory on NC spacetime, is based on the general method of deformation quantization via NC $\star$-products developed (mainly) by Flato, Sternheimer and Kontsevich [31, 32, 33]. One speaks of a *deformation* of an object/structure whenever there is a family of similar objects/structures for which we can parametrise their "distortion" from the original, "undeformed" one. In physics, this so called *deformation parameter* appears as some fundamental constant of nature that measures the deviation from the "classical" (undeformed) theory. When it is zero, the classical theory is restored. We want to deform the structure of continuous spacetime. This way of "quantizing" spacetime is essentially different from the standard QFT quantization procedure for matter fields. The spacetime coordinates (ordinary $3+1$) are proclaimed to be mutually *incompatible.* Analogously to the Heisenberg's uncertainty relations for a conjugate coordinate-momentum pair of a particle, there exist a non-zero lower bound for the product of uncertainties $\Delta x^\mu \Delta x^\nu$ for a pair of two different coordinates. In order to capture this "fuzziness" of spacetime, an abstract algebra of *NC coordinates* is introduced as a deformation of the ordinary structure. These NC coordinates, denoted by $\hat{x}^\mu$, satisfy some non-trivial commutation relations, and so, it is no longer the case that $\hat{x}^\mu \hat{x}^\nu = \hat{x}^\nu \hat{x}^\mu$. Abandoning this basic property of spacetime leads to various new physical effects that were not present in the theory based on classical spacetime. The simplest case of noncommutativity is the so called *canonical noncommutativity,* defined by

$$[\hat{x}^\mu, \hat{x}^\nu] = i\theta^{\mu\nu} \ , \tag{2.1}$$

where $\theta^{\mu\nu}$ are components of a *constant* antisymmetric matrix. Other important choices include Lie algebra-like deformation and κ-deformation.

Instead of deforming abstract algebra of coordinates one can take alternative, but equivalent, approach in which noncommutativity appears in the form of NC products of functions (NC fields) of *commutative* coordinates. These products are called *star products* ($\star$-products). Specifically, to establish *canonical noncommutativity,* we use NC Moyal-Weyl $\star$-product,

$$(\hat{f} \star \hat{g})(x) = e^{\frac{i}{2}\theta^{\alpha\beta}\frac{\partial}{\partial x^\alpha}\frac{\partial}{\partial y^\beta}} f(x)g(y)|_{y\to x} \ . \tag{2.2}$$

The first term in the expansion of the exponential is the ordinary commutative point-wise multiplication of functions. Using this definition one an easily check that

$$[x^\mu \stackrel{\star}{,} x^\nu] = i\theta^{\mu\nu} \ , \tag{2.3}$$

which means that the Moyal-Weyl $\star$-product (2.2) provides a suitable representation of the canonical noncommutativity. The quantity $\theta^{\mu\nu}$ is a "small" *deformation parameter* that has dimensions of $(length)^2$. It is a fundamental constant like the Planck length or the speed of light.

2.2. Enveloping algebra approach and the Seiberg-Witten map

In classical (undeformed) gauge field theories, generators T_A $(A = 1, 2, ..., N)$ of a gauge group (generally non-Abelian) satisfy some N-dimensional Lie algebra commutation relations:

$$[T_A, T_B] = f^C{}_{AB} T_C \ , \tag{2.4}$$

with structure constants $f^C{}_{AB}$. Variation of the matter field ψ under infinitesimal gauge transformation is given by:

$$\delta_\alpha \psi = i\alpha\psi \ , \tag{2.5}$$

where infinitesimal gauge parameter $\alpha(x) = \alpha^A(x)T_A$ belongs to the Lie algebra of the gauge group and depends on the spacetime coordinates. These transformations close in the algebra:

$$[\delta_\alpha, \delta_\beta] = \delta_{-i[\alpha,\beta]} \ . \tag{2.6}$$

Covariant derivative of the ψ field is

$$D_\mu \psi = \partial_\mu - i\omega_\mu \psi \ , \tag{2.7}$$

where $\omega_\mu(x) = \omega_\mu^A(x)T_A$ is the Lie algebra-valued gauge potential.

Analogously, a variation of a NC matter field $\widehat{\psi}$ under deformed infinitesimal gauge transformation with NC gauge parameter $\widehat{\Lambda}(x)$ is defined as:

$$\delta^\star_{\widehat{\Lambda}} \widehat{\psi} = i\widehat{\Lambda} \star \widehat{\psi} \ . \tag{2.8}$$

However, there is a problem concerning the closure condition for NC gauge transformations. If the parameter $\widehat{\Lambda}$ is supposed to be Lie algebra-valued, $\widehat{\Lambda}(x) = \widehat{\Lambda}^A(x)T_A$, it follows that

$$\begin{aligned} [\delta^\star_{\widehat{\Lambda}_1} \overset{\star}{,} \delta^\star_{\widehat{\Lambda}_2}]\widehat{\psi} &= (\widehat{\Lambda}_1 \star \widehat{\Lambda}_2 - \widehat{\Lambda}_2 \star \widehat{\Lambda}_1) \star \widehat{\psi} \\ &= \frac{1}{2}\Big([\widehat{\Lambda}_1^A \overset{\star}{,} \widehat{\Lambda}_2^B]\{T_A, T_B\} + \{\widehat{\Lambda}_1^A \overset{\star}{,} \widehat{\Lambda}_2^B\}[T_A, T_B]\Big) \star \widehat{\psi} \end{aligned} \tag{2.9}$$

The commutator of two NC gauge transformations does not generally close in the Lie algebra of gauge group because the anti-commutator $\{T_A, T_B\}$ does not in general belong to this algebra. To overcome this difficulty, we apply the *enveloping algebra approach.* The enveloping algebra is "large enough" to ensure the closure property of NC gauge transformations if NC gauge parameter $\widehat{\Lambda}$ takes values in it. The covariant derivative of NC field $\widehat{\psi}$ is defined by:

$$D_\mu \widehat{\psi} = \partial_\mu \widehat{\psi} - i\widehat{\omega}_\mu \star \widehat{\psi} \ . \tag{2.10}$$

Here we introduced NC gauge potential $\widehat{\omega}_\mu$ that also belongs to the enveloping algebra and can be represented in its basis. But, enveloping algebra has an infinite basis and it seems that by invoking it we are actually introducing infinite number of new degrees of freedom (new fields) in the NC-deformed theory. The solution to the problem is provided by the Seiberg-Witten map [34, 35]. It relies on the fact that NC quantities can be represented as a perturbation series in powers of the deformation parameter $\theta^{\alpha\beta}$ with expansion coefficients built out of the commutative fields, e.g. NC spinor $\widehat{\psi}$ field can be represented as:

$$\widehat{\psi} = \psi - \frac{1}{4}\theta^{\alpha\beta}\omega_\alpha(\partial_\beta + D_\beta)\psi + \mathcal{O}(\theta^2) \ . \tag{2.11}$$

where ω_α is the ordinary gauge potential. It is clear that at the zeroth perturbative order NC fields consistently reduce to their undeformed counterparts.

2.3. Commutative model and its NC deformation

The first step in our method is to construct a well-defined "classical" (undeformed) *commutative model* of an action that will subsequently be *deformed* by substituting ordinary commutative point-wise multiplication with NC Moyal-Weyl $\star$-product and replacing ordinary commutative fields by their NC counterparts. Classical action is built out of the commutative fields on *Minkowski space* that transform either in the fundamental or the adjoint representation of the gauge symmetry group. For pure gravity this will be $SO(2,3)$ "gravitational" gauge symmetry but if we want to include interacting Dirac fermions in the theory we must upgrade it to $SO(2,3)\otimes U(1)$. The SW map than ensures that NC-deformed action possesses deformed gauge symmetry, i.e. $SO(2,3)_\star \otimes U(1)_\star$. Gravity emerges only after suitable *gauge fixing*, i.e. gauge symmetry breaking, and there is no need for introducing metric tensor a priori. The commutative action must be consistent with the action for *classical electrodynamics in curved spacetime* with the usual $SO(1,3)\otimes U(1)$ gauge symmetry. Therefore we must break the $SO(2,3)$ gauge symmetry down to the local Lorentz $SO(1,3)$ symmetry. We accomplish this by introducing an *auxiliary field* ϕ and fixing its value. Assuming that deformation parameter is "small", we *perturbatively expend* the NC action in powers of the deformation parameter via SW map. In this way we ensure the invariance of the NC expansion under original udeformed gauge transformations order-by-order. The *first order term* in the NC expansion contains the "strongest" NC effects and for that reason a model that predicts non vanishing first order NC correction after the symmetry breaking would be preferred.

2.4. $SO(2,3)_\star$ model of NC gravity

The $SO(2,3)_\star$ model of NC gravity is established in [23, 24, 25, 26]. In this approach, NC gravity is treated as as a *gauge theory* of *canonically deformed* Anti de Sitter (AdS) gauge group $SO(2,3)_\star$. The generators of $SO(2,3)$ group are denoted by M_{AB} (group indices $A, B, ...$ taking values: $0,1,2,3,5$) and they satisfy the AdS algebra:

$$[M_{AB}, M_{CD}] = i(\eta_{AD}M_{BC} + \eta_{BC}M_{AD} - \eta_{AC}M_{BD} - \eta_{BD}M_{AC}) \ , \quad (2.12)$$

where η_{AB} is the $5D$ flat metric with signature $(+,-,-,-,+)$. By introduce *momenta generators* as $P_a = \frac{1}{l}M_{a5}$, where l is a constant length scale, we can recast the AdS algebra (2.12) in the following form:

$$\begin{aligned} [M_{ab}, M_{cd}] &= i(\eta_{ad}M_{bc} + \eta_{bc}M_{ad} - \eta_{ac}M_{bd} - \eta_{bd}M_{ac}) \ , && (2.13) \\ [M_{ab}, P_c] &= i(\eta_{bc}P_a - \eta_{ac}P_b) \ , && (2.14) \\ [P_a, P_b] &= -\frac{i}{2l^2}M_{ab} \ . && (2.15) \end{aligned}$$

In the limit $l \to \infty$ the AdS algebra reduces to the Poincare algebra. This is known as the Wigner-Inonu contraction. A realisation of (2.12) algebra can be obtained from $5D$ gamma matrices Γ^A that satisfy Clifford algebra: $\{\Gamma^A, \Gamma^B\} = 2\eta^{AB}$; the generators are given by $M_{AB} = \frac{i}{4}[\Gamma_A, \Gamma_B]$. One choice of $5D$ gamma matrices is $\Gamma_A = (i\gamma_a\gamma_5, \gamma_5)$, where γ_a are the usual $4D$ gamma matrices. The local Lorentz indices $a, b, ...$ take values: $0,1,2,3$. In this particular representation, $SO(2,3)$ generators are: $M_{ab} = \frac{i}{4}[\gamma_a, \gamma_b] = \frac{1}{2}\sigma_{ab}$ and $M_{5a} = \frac{1}{2}\gamma_a$. The gauge potential ω_μ of $SO(2,3)$ gauge group gives us the spin-connection $\omega_\mu{}^{ab}$ and the vierbein e_μ^a field:

$$\omega_\mu = \frac{1}{2}\omega_\mu{}^{AB}M_{AB} = \frac{1}{4}\omega_\mu{}^{ab}\sigma_{ab} - \frac{1}{2l}e_\mu^a\gamma_a \ . \quad (2.16)$$

Note that in this framework the vierbein field e_μ^a is treated as an additional gauge field, standing on equal footing with the spin-connection (which is the gravitational gauge field for the Lorentz group $SO(1,3)$). This is an important advantage of the theory with $SO(2,3)$ gauge symmetry. Vierbein is related to the metric tensor by $\eta_{ab}e_\mu^a e_\nu^b = g_{\mu\nu}$ and $e = \det(e_\mu^a) = \sqrt{-g}$.

The "gravitational" field strength associated with the gauge potential ω_μ is

$$F_{\mu\nu} = 2\partial_{(\mu}\omega_{\nu)} - i[\omega_\mu, \omega_\nu] = \left(R_{\mu\nu}{}^{ab} - \frac{2}{l^2}e^a_{(\mu}e^b_{\nu)}\right)\frac{\sigma_{ab}}{4} - \frac{1}{l}T_{\mu\nu}{}^a\frac{\gamma_a}{2} \ , \quad (2.17)$$

where $R_{\mu\nu}{}^{ab}$ is the *curvature tensor* and $T_{\mu\nu}{}^a$ the *torsion*:

$$\begin{aligned} R_{\mu\nu}{}^{ab} &= \partial_\mu\omega_\nu{}^{ab} - \partial_\nu\omega_\mu{}^{ab} + \omega_\mu{}^{ac}\omega_\nu{}^{cb} - \omega_\mu{}^{bc}\omega_\nu{}^{ca} \ , && (2.18) \\ T_{\mu\nu}{}^a &= \nabla_\mu e_\nu^a - \nabla_\nu e_\mu^a \ . && (2.19) \end{aligned}$$

A necessary step in obtaining GR from $SO(2,3)$ model is the *gauge fixing*, i.e. symmetry breaking from $SO(2,3)$ down to $SO(1,3)$. For that reason one usually introduces an *auxiliary field* $\phi = \phi^A\Gamma_A$ as in [36, 37]. We break the symmetry by fixing the value of the auxiliary field, in particular, by setting $\phi^a = 0$ and $\phi^5 = l$. This field is a spacetime scalar and an internal-space vector and satisfies the constraint: $\phi^A\phi_A = l^2$. It transforms in the adjoint representation of $SO(2,3)$ and its covariant derivative is given by

$$D_\mu\phi = \partial_\mu\phi - i[\omega_\mu, \phi] \,. \tag{2.20}$$

After the gauge fixing, the components of $D_\mu\phi$ reduce to $(D_\mu\phi)^a = e_\mu^a$ and $(D_\mu\phi)^5 = 0$. In this way the vierbein field emerges from the auxiliary field ϕ after the symmetry breaking.

In the papers of Stelle, West and Wilczek [36, 37] a commutative action for *pure gravity* with $SO(2,3)$ gauge symmetry was constructed. Also, in the papers of Chamseddine and Mukhanov [38, 39], GR is formulated by gauging $SO(1,4)$ or, more suitable for SUGRA, $SO(2,3)$ group. Building on their work, in [24] the $SO(2,3)$ model of pure gravity action and its NC deformation were analysed. The commutative action consists of three parts:

$$S_1 = \frac{ilc_1}{64\pi G_N}\mathrm{Tr}\int \mathrm{d}^4x\; \epsilon^{\mu\nu\rho\sigma}F_{\mu\nu}F_{\rho\sigma}\phi \,, \tag{2.21}$$

$$S_2 = \frac{c_2}{128\pi G_N l}\mathrm{Tr}\int d^4x\; \epsilon^{\mu\nu\rho\sigma}F_{\mu\nu}D_\rho\phi D_\sigma\phi\phi + h.c. \,, \tag{2.22}$$

$$S_3 = -\frac{ic_3}{128\pi G_N l}\mathrm{Tr}\int d^4x\; \epsilon^{\mu\nu\rho\sigma}D_\mu\phi D_\nu\phi D_\rho\phi D_\sigma\phi\phi \,, \tag{2.23}$$

After the gauge fixing we obtain:

$$S = -\frac{1}{16\pi G_N}\int d^4x\left(\frac{c_1 l^2}{16}\epsilon^{\mu\nu\rho\sigma}\epsilon_{abcd}R_{\mu\nu}{}^{ab}R_{\rho\sigma}{}^{cd} + \sqrt{-g}\Big((c_1+c_2)R - \frac{6}{l^2}(c_1+2c_2+2c_3)\Big)\right) . \tag{2.24}$$

For the sake of generality, three a priori undetermined dimensionless constants are introduced. These can be fixed by some consistency conditions. The first part is the topological Gauss-Bonnet term which doesn't effect the equations of motion, and so, we can set $c_1 = 0$. The Einstein-Hilbert term requires $c_1 + c_2 = 1$, while the absence of the cosmological constant is provided with $c_1 + 2c_2 + 2c_3 = 0$.

To construct a NC gravity model, we start from a general action of the form

$$\widehat{S} = c_1\widehat{S}_1 + c_2\widehat{S}_2 + c_3\widehat{S}_3 \,, \tag{2.25}$$

with

$$\begin{aligned}
\widehat{S}_1 &= \frac{il}{64\pi G_N}\mathrm{Tr}\int \mathrm{d}^4x\; \epsilon^{\mu\nu\rho\sigma}\hat{F}_{\mu\nu}\star\hat{F}_{\rho\sigma}\star\hat{\phi}\,,\\
\widehat{S}_2 &= \frac{1}{64\pi G_N l}\mathrm{Tr}\int \mathrm{d}^4x\; \epsilon^{\mu\nu\rho\sigma}\hat{\phi}\star\hat{F}_{\mu\nu}\star\hat{D}_\rho\hat{\phi}\star\hat{D}_\sigma\hat{\phi}+\text{c.c.}\,,\\
\widehat{S}_3 &= -\frac{i}{128\pi G_N l}\mathrm{Tr}\int \mathrm{d}^4x\,\varepsilon^{\mu\nu\rho\sigma}D_\mu\hat{\phi}\star D_\nu\hat{\phi}\star\hat{D}_\rho\hat{\phi}\star\hat{D}_\sigma\hat{\phi}\star\hat{\phi}\,.
\end{aligned}$$

The action is formulated in the 4-dimensional Minkowski space as an ordinary NC gauge theory. It is invariant under the NC $SO(2,3)_\star$ gauge transformations and the SW map ensures that after the perturbative expansion in powers of $\theta^{\alpha\beta}$, it possesses the ordinary commutative $SO(2,3)$ gauge symmetry order-by-order in $\theta^{\alpha\beta}$. It was found that the first order NC correction to the commutative action equals zero. This result was also obtained in [40]. The first non-vanishing correction is *quadratic* in the NC parameter; it is long and difficult to calculate. However, different limits of the second order correction can be analyzed. We are interested in the low energy expansion: we keep only the terms of the zeroth, the first and the second order in the derivatives of the vierbeins (linear in $R_{\alpha\beta\gamma\delta}$, quadratic in $T_{\alpha\beta}{}^a$). Then, the expanded action is given by

$$\begin{aligned}
S_{NC} = &-\frac{1}{16\pi G_N}\int \mathrm{d}^4x\, e\left(R-\frac{6}{l^2}(1+c_2+2c_3)\right)\\
&+\frac{1}{128\pi G_N l^4}\int \mathrm{d}^4x\, e\,\theta^{\alpha\beta}\theta^{\gamma\delta}\Big((-2+12c_2+38c_3)R_{\alpha\beta\gamma\delta}\\
&+(4-18c_2-44c_3)R_{\alpha\gamma\beta\delta}-(6+22c_2+36c_3)g_{\beta\delta}R_{\alpha\mu\gamma}{}^\mu+\frac{6+28c_2+56c_3}{l^2}g_{\alpha\gamma}g_{\beta\delta}\\
&+(5-\frac{9}{2}c_2-7c_3)T^a_{\alpha\beta}T_{\gamma\delta a}+(-10+\frac{9}{2}c_2+14c_3)T^a_{\alpha\gamma}T_{\beta\delta a}+(3-3c_2-2c_3)T_{\alpha\beta\gamma}T_{\delta\mu}{}^\mu\\
&+(1+2c_2)T_{\alpha\beta\rho}T^\rho{}_{\gamma\delta}+8T_{\alpha\gamma\delta}T_{\beta\mu}{}^\mu-(2c_2+4c_3)T_{\alpha\gamma\rho}T^\rho{}_{\delta\beta}\\
&+(2c_2+4c_3)g_{\beta\delta}T_{\gamma\sigma}{}^\sigma T_{\alpha\rho}{}^\rho-(2c_2+4c_3)T_{\alpha\rho\sigma}T_\gamma{}^{\sigma\rho}g_{\beta\delta}+(-2+4c_2+18c_3)T_{\alpha\beta\gamma}e^\rho_a\nabla_\delta e^a_\rho\\
&+(6-8c_2-8c_3)T_{\alpha\gamma\beta}e^\rho_a\nabla_\delta e^a_\rho+(2+4c_2+12c_3)T_{\alpha\gamma}{}^\mu e^a_\beta\nabla_\delta e^a_\mu-T_{\alpha\beta}{}^\mu e^a_\delta\nabla_\gamma e^a_\mu\\
&+(-6-8c_2-16c_3)T_{\delta\rho\beta}e^\rho_a\nabla_\alpha e^a_\gamma-(2c_2+4c_3)g_{\alpha\gamma}T_{\mu\beta}{}^\mu e^\rho_a\nabla_\delta e^a_\rho-(2c_2+4c_3)g_{\beta\delta}T_{\alpha\rho}{}^\sigma e^\rho_a\nabla_\gamma e^a_\sigma\\
&-(4+16c_2+32c_3)e^\mu_a e_{b\beta}\nabla_\gamma e^a_\alpha\nabla_\delta e^b_\mu+(4+12c_2+32c_3)e_{\delta a}e^\mu_b\nabla_\alpha e^a_\gamma\nabla_\beta e^b_\mu\\
&-(2+4c_2+8c_3)g_{\beta\delta}e^\mu_a e^\nu_b\nabla_\gamma e^a_\mu\nabla_\alpha e^b_\nu+(2+4c_2+8c_3)g_{\beta\delta}e^\mu_a e^\rho_c\nabla_\alpha e^a_\rho\nabla_\gamma e^c_\mu\Big)\,.
\end{aligned} \tag{2.26}$$

Equations of motion are obtained by varying with respect to the vierbeins and the spin connection independently:

$$\begin{aligned}
\delta e^a_\mu:\quad & R_{\alpha\gamma}{}^{cd}e^\gamma_d e^\alpha_a e^\mu_c-\frac{1}{2}e^\mu_a R+\frac{3}{l^2}(1+c_2+2c_3)e^\mu_a=\tau_a{}^\mu\\
&=-\frac{8\pi G_N}{e}\frac{\delta S^{(2)}_{NC}}{\delta e^a_\mu}\,,
\end{aligned} \tag{2.27}$$

$$\delta\omega_\mu{}^{ab}:\quad T_{ac}{}^c e^\mu_b-T_{bc}{}^c e^\mu_a-T_{ab}{}^\mu=S_{ab}{}^\mu=-\frac{16\pi G_N}{e}\frac{\delta S^{(2)}_{NC}}{\delta\omega_\mu{}^{ab}}\,. \tag{2.28}$$

The effective energy-momentum tensor $\tau_a{}^\mu$ and the effective spin-tensor $S_{ab}{}^\mu$ depend on $\theta^{\alpha\beta}$ and we can conclude that *noncommutativity is a source of curvature and torsion.* Using equation (2.27) it can be shown [23] that a NC correction to Minkowski metric is of the form:

$$\begin{aligned} g_{00} &= 1-\frac{11}{2l^6}\theta^0{}_m\theta^0{}_n x^m x^n - \frac{11}{8l^6}\theta^2 r^2 = 1 - R_{0m0n}x^m x^n\,,\\ g_{0i} &= -\frac{11}{3l^6}\theta^0{}_m\theta^i{}_n x^m x^n = -\frac{2}{3}R_{0min}x^m x^n\,,\\ g_{ij} &= -\delta_{ij} - \frac{11}{6l^6}\theta^i{}_m\theta^j{}_n x^m x^n + \frac{11}{24l^6}\delta^{ij}\theta^2 r^2 - \frac{11}{24l^6}\theta^2 x^i x^j\\ &= -\delta_{ij} - \frac{1}{3}R_{imjn}x^m x^n\,, \end{aligned} \tag{2.29}$$

where $R_{\mu\nu\rho\sigma}$ are the components of the Reimann tensor for this solution.

Thus, in $SO(2,3)_\star$ model, there actually exists a *canonical NC deformation of Minkowski space.* This result suggests that the coordinates x^μ we started with are actually *Fermi normal coordinates.* These are inertial coordinates of a local observer moving along a geodesic and can be constructed in a small neighbourhood along the geodesic (inside a small cylinder surrounding the geodesic) [41, 42, 43]. The breaking of diffeomorphism symmetry due to canonical noncommutivity we understand as a consequence of working in a preferred reference system given by the Fermi normal coordinates. A local observer moving along the geodesic measures $\theta^{\alpha\beta}$ to be constant. In any other reference frame this will not be the case.

3. Commutative actions for matter fields and their NC deformation

First, we will consider *non-interacting fermions* in canonically deformed spacetime. In our recant paper entitled *Dirac field and gravity in NC $SO(2,3)_\star$ model* [44] we have proposed the following kinetic-type action for the Dirac spinor field, invariant under local $SO(2,3)$ transformations:

$$S_{\psi,kin} = \frac{i}{12}\int d^4x\; \varepsilon^{\mu\nu\rho\sigma}\Big[\bar\psi D_\mu\phi D_\nu\phi D_\rho\phi D_\sigma\psi - D_\sigma\bar\psi D_\mu\phi D_\nu\phi D_\rho\phi\psi\Big]\,. \tag{3.30}$$

After the symmetry braking it becomes:

$$S_{\psi,kin} = \frac{i}{2}\int d^4x\; e\;\left[\bar\psi\gamma^\sigma\nabla_\sigma\psi - \nabla_\sigma\bar\psi\gamma^\sigma\psi\right] - \frac{2}{l}\int d^4x\; e\;\bar\psi\psi\,, \tag{3.31}$$

which is exactly the Dirac action in curved spacetime for spinors of mass $2/l$. To obtain fermions with arbitrary mass m, not just $2/l$, we have to include the following additional "mass terms" (terms of the type $\bar\psi ... \psi$) in the action:

$$S^{(1)}_{\psi,m} = \frac{i}{2}c_1\Big(\frac{m}{l}-\frac{2}{l^2}\Big)\int d^4x\, \varepsilon^{\mu\nu\rho\sigma}\Big[\bar\psi D_\mu\phi D_\nu\phi D_\rho\phi D_\sigma\phi\phi\psi + \bar\psi\phi D_\mu\phi D_\nu\phi D_\rho\phi D_\sigma\phi\psi\Big]\,,$$
$$S^{(2)}_{\psi,m} = \frac{i}{2}c_2\Big(\frac{m}{l}-\frac{2}{l^2}\Big)\int d^4x\, \varepsilon^{\mu\nu\rho\sigma}\Big[\bar\psi D_\mu\phi D_\nu\phi D_\rho\phi\phi D_\sigma\phi\psi + \bar\psi D_\mu\phi\phi D_\nu\phi D_\rho\phi D_\sigma\phi\psi\Big]\,,$$
$$S^{(3)}_{\psi,m} = i\, c_3\Big(\frac{m}{l}-\frac{2}{l^2}\Big)\int d^4x\, \varepsilon^{\mu\nu\rho\sigma}\ \bar\psi D_\mu\phi D_\nu\phi\phi D_\rho\phi D_\sigma\phi\psi\,. \tag{3.32}$$

If we demand that the a priori undetermined dimensionless coefficients c_1, c_2, and c_3 satisfy the constraint: $c_2 - c_1 - c_3 = -\frac{1}{24}$, after the symmetry breaking, the sum of the three terms in (3.32) becomes:

$$S_{\psi,m} = -\left(m - \frac{2}{l}\right)\int d^4x\; e\,\bar\psi\psi\,, \tag{3.33}$$

and the total action, $S_\psi = S_{\psi,kin} + S_{\psi,m}$, is exactly equal to the Dirac action for spinors of mass m in curved spacetime,

$$S_\psi = \frac{i}{2}\int d^4x\; e\ [\bar\psi\gamma^\sigma\nabla_\sigma\psi - \nabla_\sigma\bar\psi\gamma^\sigma\psi] - m\int d^4x\; e\,\bar\psi\psi\,. \tag{3.34}$$

NC-deformed version of this spinorial action (denoted by a "hat" symbol) is obtained by replacing the ordinary field multiplication with NC $\star$-product of NC-deformed fields; for example, kinetic term (3.30) becomes:

$$\widehat S_{\psi,kin} = \frac{i}{12}\int d^4x\; \varepsilon^{\mu\nu\rho\sigma}\Big[\widehat{\bar\psi}\star(D_\mu\widehat\phi)\star(D_\nu\widehat\phi)\star(D_\rho\widehat\phi)\star(D_\sigma\widehat\psi)$$
$$-(D_\sigma\widehat{\bar\psi})\star(D_\mu\widehat\phi)\star(D_\nu\widehat\phi)\star(D_\rho\widehat\phi)\star\widehat\psi\Big]\,. \tag{3.35}$$

This action is hermitian (up to a surface term that vanishes) and it is endowed with deformed $SO(2,3)_\star$ gauge symmetry. It turns out that the leading term in the NC expansion does not vanish after the symmetry breaking and so we obtain linear NC correction to the classical Dirac action in curved spacetime. The calculation is long and tedious and we will not present the details here. Schematically, the spinorial piece is given by:

$$\widehat S^{(1)}_\psi = \theta^{\alpha\beta}\int d^4x\; e\,\bar\psi\Big(\mathcal{A}_{\alpha\beta}{}^{\rho\sigma}\nabla_\rho\nabla_\sigma + \mathcal{B}_{\alpha\beta}{}^{\sigma}\nabla_\sigma + \mathcal{C}_\alpha\nabla_\beta + \mathcal{D}_{\alpha\beta}\Big)\psi\,. \tag{3.36}$$

Objects $\mathcal{A},\mathcal{B},\mathcal{C},\mathcal{D}$ are complicated functions of geometric quantities, e.g. there appear new interactions like: $\bar\psi\sigma_\alpha{}^\sigma\nabla_\beta\nabla_\sigma\psi$, $\bar\psi R_{\alpha\beta}{}^{\rho\sigma}\gamma_\rho\nabla_\sigma\psi$, $\bar\psi T_{\alpha\beta}{}^{\sigma}\nabla_\sigma\psi$, $\bar\psi\sigma_{\alpha\beta}\psi$, etc. More importantly, we want to emphasize the fact that this θ-linear NC correction pertains even in the limit of *flat spacetime*. This enables us to derive some tangible phenomenological consequences of our model that could potentially be tested experimentally in a not to far future.

3.1. Free electron in flat NC spacetime

One peculiar property of our model is that noncommutativity effects pertain even in flat spacetime. The NC-deformed Dirac equation for an electron in Minkowski space can be derived by varying the action (3.36) with respect to $\bar{\psi}$. In particular, assuming for simplicity that $\theta^{12} = -\theta^{21} = \theta \neq 0$ and all others equal to zero, we obtain:

$$\Big[i\partial\!\!\!/ - m - \frac{\theta}{2l}({\sigma_1}^{\sigma}\partial_2\partial_\sigma - {\sigma_2}^{\sigma}\partial_1\partial_\sigma) + \frac{7i\theta}{12l^2}(\gamma_0\gamma_5\partial_3 - \gamma_3\gamma_5\partial_0) \\ -\theta\left(\frac{m}{2l^2} + \frac{1}{3l^3}\right)\sigma_{12}\Big]\psi = 0\ . \qquad (3.37)$$

Non trivial solutions of this homogeneous matrix equation exist, if and only if, the determinant of the matrix that acts on ψ equals zero. This condition will give us the *dispersion relation* for an electron. Specifically, for an electron moving along the z-direction, i.e. in the direction orthogonal to the noncommutative x, y-plane there are four different solutions for the energy function:

$$\begin{aligned} E_{1,2}(\mathbf{p}) &= E_{\mathbf{p}} \mp \left[\frac{m^2}{12l^2} - \frac{m}{3l^3}\right]\frac{\theta}{E_{\mathbf{p}}} + \mathcal{O}(\theta^2)\ , \\ E_{3,4}(\mathbf{p}) &= -E_{\mathbf{p}} \pm \left[\frac{m^2}{12l^2} - \frac{m}{3l^3}\right]\frac{\theta}{E_{\mathbf{p}}} + \mathcal{O}(\theta^2)\ , \end{aligned} \qquad (3.38)$$

with $E_{\mathbf{p}} = \sqrt{m^2 + p_z^2}$. This is reminiscent of the well known *Zeeman effect*. The deformation parameter θ plays the role of a constant background magnetic field that causes the splitting of electron energy levels.

Working in the rest frame ($\mathbf{p} = 0$), from (3.38) we conclude that electron's mass gets modified due to noncommutativity of the background spacetime and the correction is *linear* in the deformation parameter:

$$\begin{aligned} E_{1,2}(0) &= m \mp \left[\frac{m}{12l^2} - \frac{1}{3l^3}\right]\theta + \mathcal{O}(\theta^2)\ , \\ E_{3,4}(0) &= -m \pm \left[\frac{m}{12l^2} - \frac{1}{3l^3}\right]\theta + \mathcal{O}(\theta^2)\ . \end{aligned} \qquad (3.39)$$

Another important prediction of $SO(2,3)_\star$ model is that NC energy levels are *helicity-dependent*. The electron states (presented in [44]) with different helicity have different energies. This means that NC space effectively acts as a *birefringent medium* for electrons propagating in it.

Such NC modification of the *non-interacting theory* could not have be achieved by directly introducing noncommutativity into the free Dirac action, i.e. by minimal substitution,

$$\widehat{S} = \int d^4x \, \widehat{\bar{\psi}} \star (i\gamma^\mu \partial_\mu - m)\widehat{\psi} \; , \tag{3.40}$$

because generally $\int d^4x \, \widehat{f} \star \widehat{g} = \int d^4x \, fg$, and so, non-interacting theories are not modified due to noncommutativity since they include only kinetic term which is quadratic.

3.2. Interacting Dirac fermions; U(1) gauge field

In order to have interacting Dirac fermions we need to incorporate $U(1)$ gauge field into our framework. Accordingly, the gauge group must be upgraded from $SO(2,3)$ to $SO(2,3) \otimes U(1)$. In the paper entitled *Noncommutative Electrodynamics from $SO(2,3)_\star$ model of Noncommutative Gravity* [45] (under revision), we propose the following action for the $U(1)$ gauge field:

$$S_A = -\frac{1}{16l} Tr \int d^4x \, \varepsilon^{\mu\nu\rho\sigma} \Big\{ f\mathbb{F}_{\mu\nu} + \frac{i}{3!} f^2 D_\mu\phi D_\nu\phi \Big\} D_\rho\phi D_\sigma\phi\phi + h.c. \tag{3.41}$$

It includes an additional auxiliary field $f = \frac{1}{2} f^{AB} M_{AB}$. Like ϕ, this field transforms in the adjoint representation of $SO(2,3)$ and it is invariant under $U(1)$ (i.e. not charged). Its role is to produce the canonical kinetic term for $U(1)$ gauge field in curved spacetime since we cannot define the Hodge dual operation without prior knowledge of the metric tensor.

After the gauge fixing, the purely gravitational part of the action (3.41) vanishes and we obtain:

$$S_A = \frac{1}{2} \int d^4x \; e \; f^{ab} e_a^\mu e_b^\nu \mathcal{F}_{\mu\nu} + \frac{1}{4} \int d^4x \; e \; (f^{ab} f_{ab} + 2 f^{a5} f_a{}^5) \; . \tag{3.42}$$

We use equations of motion for the components of the auxiliary field f,

$$f_{a5} = 0 \; , \quad f_{ab} = -e_a^\mu e_b^\nu \mathcal{F}_{\mu\nu} \; . \tag{3.43}$$

to eliminate the auxiliary field in the action (3.42) and this leaves us with the canonical kinetic term for pure $U(1)$ gauge field in curved spacetime:

$$S_A = -\frac{1}{4} \int d^4x \; e \; \mathcal{F}_{\mu\nu} \mathcal{F}^{\mu\nu} \; . \tag{3.44}$$

3.3. Electron in background magnetic field

With the $U(1)$ gauge field include, we can investigate the behaviour of an electron in background electromagnetic field, see [45] for detail. The action for NC electrodynamics in Minkowski space up to the first order in $\theta^{\alpha\beta}$ is given by:

$$\begin{aligned}
\widehat{S}_{flat} &= \int d^4x\, \bar\psi(i\mathcal{D}\!\!\!/ - m)\psi - \frac{1}{4}\int d^4x\, \mathcal{F}_{\mu\nu}\mathcal{F}^{\mu\nu} \\
&\quad -\theta^{\alpha\beta}\int d^4x \left[\frac{1}{8}\mathcal{F}_{\alpha\beta}\mathcal{F}^{\mu\nu}\mathcal{F}_{\mu\nu} + \frac{1}{2}\mathcal{F}_{\alpha\mu}\mathcal{F}_{\beta\nu}\mathcal{F}^{\mu\nu}\right] \\
&\quad +\theta^{\alpha\beta}\int d^4x\, \bar\psi\Bigg[-\frac{1}{2l}\sigma_\alpha{}^\sigma\mathcal{D}_\beta\mathcal{D}_\sigma + \frac{7i}{24l^2}\varepsilon_{\alpha\beta}{}^{\rho\sigma}\gamma_\rho\gamma_5\mathcal{D}_\sigma - \left(\frac{m}{4l^2} + \frac{1}{6l^3}\right)\sigma_{\alpha\beta} \\
&\quad +\frac{3i}{4}\mathcal{F}_{\alpha\beta}\mathcal{D}\!\!\!/ - \frac{i}{2}\mathcal{F}_{\alpha\mu}\gamma^\mu\mathcal{D}_\beta - \left(\frac{3m}{4} - \frac{1}{4l}\right)\mathcal{F}_{\alpha\beta}\Bigg]\psi\ . \qquad (3.45)
\end{aligned}$$

where we introduced the flat spacetime covariant derivative $\mathcal{D}_\mu = \partial_\mu - iA_\mu$ (for electron $q = -1$). We notice immediately that this action is different from the actions for NC electrodynamics already present in the literature [46, 47, 48]. The new interaction terms, specific to the $SO(2,3)_\star$ model, appear as residual from the gravitational interaction and they lead to some non trivial physical consequences, such as, for instance, a deformation of electron's dispersion relation.

By varying NC action (3.45) with respect to $\bar\psi$ we obtain NC-deformed Dirac equation for electron coupled to some background electromagnetic field A_μ:

$$\left(i\partial\!\!\!/ - m + A\!\!\!/ + \theta^{\alpha\beta}\mathcal{M}_{\alpha\beta}\right)\psi = 0\ . \qquad (3.46)$$

In [45] we investigate a special case of an electron propagating in constant background magnetic field $\mathbf{B} = B\mathbf{e}_z$ in order to see how noncommutativity deforms its energy levels.

Classical (undeformed) energy levels for a *relativistic electron* are given by:

$$E^{(0)}_{n,s} = \sqrt{p_z^2 + m^2 + (2n+s+1)B}\ . \qquad (3.47)$$

where $s = \pm 1$ is the σ_z eigenvalue for the spin-state of an electron. We are looking for linear NC correction $E^{(1)}_{n,s} \sim \theta$ of the energy levels (3.47). If we assume that only two spatial coordinates are mutually incompatible, e.g. $[x^1, x^2] = i\theta^{12}$, than we have $\theta^{12} = -\theta^{21} =: \theta \neq 0$ and all other components of $\theta^{\alpha\beta}$ equal to zero. In this case we obtain:

$$\begin{aligned}
E^{(1)}_{n,s} &= -\frac{\theta s}{E^{(0)}_{n,s}}\left[\frac{m^2}{12l^2} - \frac{m}{3l^3}\right] - \frac{\theta B s}{E^{(0)}_{n,s}(E^{(0)}_{n,s} + m)}\left[\frac{m}{12l^2} - \frac{1}{3l^3}\right](2n+s+1) \\
&\quad + \frac{\theta B^2}{2E^{(0)}_{n,s}}(2n+s+1)\ . \qquad (3.48)
\end{aligned}$$

Non-relativistic limit of NC energy levels is obtained by expanding undeformed energy function $E^{(0)}_{n,s}$ assuming that p_z^2, $B \ll m^2$; for a *non-relativistic electron* constrained to move in NC x, y-plane ($p_z = 0$) expending (3.48) we obtain:

$$E_{n,s} = \left[m - s\theta\left(\frac{m}{12l^2} - \frac{1}{3l^3}\right)\right] + \frac{2n+s+1}{2m}B_{eff} - \frac{(2n+s+1)^2}{8m^3}B^2_{eff}\ , \quad (3.49)$$

where we introduced $B_{eff} = (B + \theta B^2)$ as an *effective magnetic field.*

If we compare the expression for NC-deformed energy levels (3.49) with the one for undeformed energy levels, we conclude that the only effect of constant spacetime noncommutaivity is to modify the mass of an electron and the value of the background magnetic field. This interpretation is in accord with String Theory. In the famous paper of Seiberg and Witten [35] it is argued that coordinate functions of the endpoints of an open string constrained to a D-brane in the presence of a constant Neveu-Schwarz B-field (equivalent to a constant magnetic field on the brane-world) satisfy constant noncommutativity algebra. The implication is that a relativistic field theory on NC spacetime can be interpreted as a low energy limit of the theory of open strings.

From energy function (3.49) we can derive NC-deformed *magnetic moment of an electron* in the state (n, s) for weak magnetic field:

$$\mu_{n,s} = -\frac{\partial E_{n,s}}{\partial B} = -\mu_B(2n+s+1)(1+\theta B), \quad (3.50)$$

where, if we restore the units, $\mu_B = \frac{e\hbar}{2mc}$ is the Bohr magneton. We recognise $-(2n+1)\mu_B$ as the *diamagnetic moment* of an electron and $-s\mu_B$ as the spin magnetic moment. The θB-term is another potentially observable phenomenological prediction of our model. It is a linear NC correction to the electron's dipole moment.

4. Conclusion

In this article we discussed coupling of matter fields with gravity in the framework of NC $SO(2,3)_\star$ gauge theory of gravity. Taking the enveloping algebra approach, along with the Seiberg-Witten map, we constructed the perturbative actions and derived the *deformed* equations of motion for the Dirac field coupled to $U(1)$ gauge field and gravity. In this way we have established the NC Electrodynamics in curved spacetime induced by NC $SO(2,3)_\star$ gravity. The fact that NC effects pertain even in flat spacetime limit of this model enables one to study behaviour of an electron in a background electromagnetic field. Corrections to the relativistic Landau levels of an electron in a constant magnetic field are given by (3.48) and their non-relativistic limit is (3.49). The NC corrections are different for different Landau levels. It is well known that the physics of the Lowest Landau Level (LLL) is closely related to the physics of Quantum Hall Effect (QHE). Using the obtained results, we plan to investigate NC corrections to the QHE. In this way, together with the induced NC magnetic moment (3.50) and the NC-induced magnetization in materials we hope to obtain some constraints on noncommutaivity parameter from condensed matter experiments.

Starting from (3.45) one can check renormalizability of the model. It is known that, the so-called Minimal NC Electrodynamics, a theory obtained by directly introducing NC Moyal-Weyl $\star$-product in the classical Dirac action for fermions coupled with $U(1)$ gauge field in Minkowski space,

$$\widehat{S} = \int d^4x\, \widehat{\bar{\psi}} \star (i\gamma^\mu D_\mu - m)\widehat{\psi} - \frac{1}{4}\int d^4x\, \widehat{F}_{\mu\nu} \star \widehat{F}^{\mu\nu}\,,$$

is not a renormalizabile theory because of the fermionic loop contributions [46, 47, 48]. It would be interesting to see if additional terms present in the NC $SO(2,3)_\star$ gravity induced Electrodynamics (3.45) can improve this behaviour.

The NC $SO(2,3)_\star$ gravity model also enables one to introduce coupling with scalar and non-Abelian gauge fields. In this way, it is possible to progress towards generalizing Standard Model to a NC spacetime using the setup we described in this article.

References

[1] A. H. Chamseeddine, *Deforming Einstein's gravity*, Phys. Lett. B **504** 33 (2001), [hep-th/0009153].

[2] A. H. Chamseeddine *$SL(2,C)$ gravity with a complex vierbein and its noncommutative extension*, Phys. Rev. D **69**, 024015 (2004)

[3] M. A. Cardella and D. Zanon, *Noncommutative deformation of four-dimensional gravity*, Class. Quant. Grav. **20**, L95 (2003), [hep-th/0212071].

[4] P.Aschieri, C. Blohmann, M. Dimitrijević, F. Meyer, P. Schupp and J. Wess, *A Gravity Theory on Noncommutative Spaces*, Class. Quant. Grav. **22**, 3511 (2005), [hep-th/0504183].

[5] P. Aschieri, M. Dimitrijević, F. Meyer and J. Wess, *Noncommutative Geometry and Gravity*, Class. Quant. Grav. **23**, 1883 (2006), [hep-th/0510059].

[6] T. Ohl and A. Schenckel, *Cosmological and Black Hole Spacetimes in Twisted Noncommutative Gravity*, JHEP **0910** (2009) 052, [arXiv: 0906.2730].

[7] P. Aschieri and L. Castellani, *Noncommutative Gravity Solutions*, J. Geom. Phys. **60**, 375-393 (2010), [arXiv:0906.2774].

[8] H. S. Yang, *Emergent gravity from noncommutative spacetime*, Int. J. Mod. Phys. **A24**, 4473 (2009), [hep-th/0611174].

[9] H. Steinacker, *Emergent Geometry and Gravity from Matrix Models: an Introduction*, Class. Quant. Grav. **27**, 133001 (2010), [arXiv:1003.4134].

[10] M. Burić and J. Madore, *Spherically Symmetric Noncommutative Space: $d = 4$*, Eur. Phys. J. **C58**, 347 (2008), [arXiv: 0807.0960].

[11] M. Burić and J. Madore, *On noncommutative spherically symmetric spaces*, arXiv:1401.3652.

[12] L. Tomassini, S. Viaggiu, *Building non-commutative spacetimes at the Planck length for Friedmann flat cosmologies*, Class. Quant. Grav. **31** 185001 (2014), [arXiv:1308.2767].

[13] Mir Faizal, *Noncommutative Quantum Gravity*, Mod.Phys.Lett. A **28**,1350034 (2013) [arXiv:1302.5156]

[14] A. Kobakhidze, C. Lagger and A. Manning, *Constraining noncommutative spacetime from GW150914*, Phys. Rev. D **94** 064033 (2016), [arXiv:1607.03776],

[15] D. Klammer and H. Steinacker, *Cosmological solutions of emergent noncommutative gravity*, Phys. Rev. Lett. **102** (2009) 221301, [arXiv:0903.0986].

[16] E. Harikumar and V. O. Rivelles, *Noncommutative Gravity*, Class. Quant. Grav. **23**, 7551-7560 (2006), [hep-th/0607115].

[17] M. Dobrski, *Background independent noncommutative gravity from Fedosov quantization of endomorphism bundle*, arXiv:1512.04504.

[18] M. Dobrski, *On some models of geometric noncommutative general relativity*, Phys. Rev. D **84**, 065005 (2011), [arXiv:1011.0165].

[19] M. Burić, T. Grammatikopoulos, J. Madore, G. Zoupanos, *Gravity and the Structure of Noncommutative Algebras*, JHEP **0604** 054, 2006, [hep-th/0603044].

[20] M. Burić, J. Madore, G. Zoupanos, *The Energy-momentum of a Poisson structure*, Eur. Phys. J. **C 55** 489-498, 2008, [arXiv:0709.3159].

[21] P. Aschieri and L. Castellani, *Noncommutative supergravity in $D = 3$ and $D = 4$*, JHEP **0906**, 087 (2009), [arXiv:0902.3823].

[22] L. Castellani, *Chern-Simons supergravities, with a twist*, JHEP **1307**, 133 (2013), [arXiv:1305.1566].

[23] M. Dimitrijević Ćirić, B. Nikolić and V. Radovanović, *Noncommutative gravity and the relevance of the θ-constant deformation*, Europhys. Lett. **118** no.2, 21002 (2017) [arXiv:1609.06469].

[24] M. Dimitrijević Ćirić, B. Nikolić and V. Radovanović, *NC $SO(2,3)_\star$ gravity: noncommutativity as a source of curvature and torsion*, Phys. Rev. D **96**, 064029 (2017) [arXiv:1612.00768].

[25] M. Dimitrijević, V. Radovanović and H. Štefančić, *AdS-inspired noncommutative gravity on the Moyal plane*, Phys. Rev. D **86**, 105041 (2012), [arXiv:1207.4675].

[26] M. Dimitrijević and V. Radovanović, *Noncommutative $SO(2,3)$ gauge theory and noncommutative gravity*, Phys. Rev. D **89**, 125021 (2014), [arXiv:1404.4213].

[27] P. Aschieri and L. Castellani, *Noncommutative $D = 4$ gravity coupled to fermions* JHEP, **0906**, 086 (2009), [arXiv:0902.3823].

[28] P. Aschieri and L. Castellani, *Noncommutative gravity coupled to fermions: second order expansion via Seiberg-Witten map*, JHEP **1207** (2012) [arXiv:1111.4822]

[29] P. Aschieri and L. Castellani, *Noncommutative gauge fields coupled to noncommutative gravity*, General Relativity and Gravitation **45** (2013), 3 [arXiv:1205.1911v1]

[30] P. Aschieri, *Extended gravity from noncommutativity*, Springer Proc. Phys. **145** (2014), 151 [arXiv:1207.5060]

[31] F. Bayen, M. Flato, C. Fronsdal, A. Lichnerowicz and D. Sternheimer, *Deformation theory and quantization*, Ann. Phys. **111**, 61 (1978).

[32] D. Sternheimer, *Deformation quantization: Twenty years after*, AIP Conf. Proc. **453**, 107 (1998), [math.qa/9809040].

[33] Maxim Kontsevich, *Deformation quantization of Poisson manifolds*, I, Lett. Math. Phys. **66**, 157-216 (2013), [q-alg/9709040].

[34] B. Jurčo, L. Möller, S. Schraml, P. Schupp and J. Wess, *Construction of non-Abelian gauge theories on noncommutative spaces*, Eur. Phys. J. C **21**, 383 (2001), [hep-th/0104153].

[35] N. Seiberg and E. Witten, *String theory and noncommutative geometry*, JHEP **09**, 032 (1999), [hep-th/9908142].

[36] K. S. Stelle and P. C. West, *Spontaneously broken de Sitter symmetry and the gravitational holonomy group*, Phys. Rev D **21**, 1466 (1980).

[37] F. Wilczek, *Riemann-Einstein structure from volume and gauge symmetry*, Phys. Rev. Lett. **80** (1998) 4851-4854, [hep-th/9801184].

[38] A. H. Chamseddine and V. Mukhanov, *Who Ordered the Anti-de Sitter Tangent Group?* JHEP **1311** 095 (2013), [arXiv:1308.3199]

[39] A. H. Chamseddine and V. Mukhanov, *Gravity with de Sitter and Unitary Tangent Groups* JHEP **1003** (2010) 033, [arXiv:1002.0541]

[40] P. Aschieri, L. Castellani and M. Dimitrijević, *Noncommutative gravity at second order via Seiberg-Witten map*, Phys.Rev. D **87**, 024017 (2013) [arXiv:1207.4346]

[41] F.K. Manasse and C.W. Misner, Fermi Normal Coordinates and Some Basic Concepts in Differential Geometry, J. Math. Phys. 4 (1963) 735-745.

[42] C. Chicone and B. Mashoon, Explicit Fermi coordinates and tidal dynamics in de Sitter and Godel spacetimes, Phys. Rev. D 74 (2006) 064019, [gr-qc/0511129].

[43] D. Klein and E. Randles, Fermi coordinates, simultaneity, and expanding space in Robertson-Walker cosmologies, Annales Henri Poincare 12 (2011) 303-328, [arXiv:1010.0588].

[44] D. Gočanin and V. Radovanović, *Dirac field and gravity in NC* $SO(2,3)_\star$ *model*, Eur. Phys. J. C, **78** (2018) 195

[45] M. Dimitrijević-Ćirić, D. Gočanin, N. Konjik and V. Radovanović, *Noncommutative Electrodynamics from* $SO(2,3)_\star$ *Model of Noncommutative Gravity* (in preparation)

[46] M. Burić and V. Radovanović, *The One loop effective action for quantum electrodynamics on noncommutative space* JHEP **0210** (2002) 074, [hep-th/0208204]

[47] R. Wulkenhaar, *Nonrenormalizability of theta expanded noncommutative QED*, JHEP **0203** (2002) 024,[hep-th/0112248]

[48] M. Burić, D. Latas, V. Radovanović, *Renormalizability of noncommutative SU(N) gauge theory*, JHEP **0602** (2006) 046, [hep-th/0510133]

Intrinsic non-commutativity in quantum gravity*

Laurent Freidel†
Perimeter Institute for Theoretical Physics, ON N2L 2Y5, Waterloo, Canada
Robert G. Leigh‡
Department of Physics, University of Illinois, Urbana, IL, 61801, U.S.A
Djordje Minic§
Department of Physics, Virginia Tech, Blacksburg, VA 24061, U.S.A.

Abstract

We summarize our recent work on the foundational aspects of string theory as a quantum theory of gravity. We emphasize the hidden quantum geometry (modular spacetime) behind the generic representation of quantum theory and then stress that the same geometric structure underlies a manifestly T-duality covariant formulation of string theory, that we call metastring theory. We also discuss the effective non-commutative description of closed strings as implied by an intrinsic non-commutativity of closed string theory. This fundamental non-commutativity is explicit in the metastring formulation of quantum gravity.

1. Introduction

In this talk we outline the essence of our recent work [1–8] on the foundational aspects of quantum gravity in the form of string theory. In the first section we describe the hidden quantum spacetime geometry underlying the generic representation of quantum theory (which renders it manifestly non-local) and then in second section we find that the same (and, in general, dynamical) geometric structure underlies metastring theory, a manifestly T-duality covariant formulation of string theory. Thus quantum gravity "gravitizes" the quantum spacetime geometry. Finally, in the last section we outline the effective description of closed strings at long distance that leads to a non-commutative effective field theory, as implied by an intrinsic non-commutativity of closed string theory, and we discuss some of its consequences.

* This presentation (by D.M) was supported by the Julian Schwinger Foundation.
† e-mail address: lfreidel@perimeterinstitute.ca
‡ e-mail address: rgleigh@illinois.edu
§ e-mail address: dminic@vt.edu

2. Quantum theory and quantum spacetime

We start by revealing the hidden quantum spacetime geometry of quantization [5], which will, surprisingly, take us all the way to quantum gravity. We focus on the Heisenberg (or Weyl-Heisenberg) group, which is generated, on the level of the corresponding algebra, by familiar position $\hat{q}^a$ and momentum $\hat{p}_b$ operators:

$$[\hat{q}^a, \hat{p}_b] = i\hbar\delta^a_b. \tag{1}$$

It will be convenient to introduce a length scale λ and a momentum scale ϵ, with $\lambda\epsilon = \hbar$. Then, let us introduce the following notation $\hat{x}^a \equiv \hat{q}^a/\lambda$, $\hat{\tilde{x}}_a \equiv \hat{p}_a/\epsilon$,, with $[\hat{x}^a, \hat{\tilde{x}}_b] = i\delta^a_b$. Even more compactly let us suggestively write

$$\mathbb{X}^A \equiv (x^a, \tilde{x}_a)^T, \quad [\hat{\mathbb{X}}^a, \hat{\mathbb{X}}^b] = i\omega^{AB}, \tag{2}$$

with $\frac{1}{2}\omega_{AB}dX^A dX^B = \frac{1}{\hbar}dp_a \wedge dq^a$, where $\omega_{AB} = -\omega_{BA}$ is the canonical symplectic form on phase space $\mathcal{P}$. The Heisenberg group $H_{\mathcal{P}}$ is generated by Weyl operators [9] ($\mathbb{K}$ stands for the pair $(\tilde{k}, k)$ and $\omega(\mathbb{K}, \mathbb{K}') = k \cdot \tilde{k}' - \tilde{k} \cdot k'$)

$$W_{\mathbb{K}} \equiv e^{2\pi i\omega(\mathbb{K},\mathbb{X})}. \tag{3}$$

These form a central extension of the translation algebra

$$W_{\mathbb{K}} W_{\mathbb{K}'} = e^{2\pi i\omega(\mathbb{K},\mathbb{K}')} W_{\mathbb{K}+\mathbb{K}'}. \tag{4}$$

The projection $\pi : H_{\mathcal{P}} \to \mathcal{P}$ (where $\pi : W_{\mathbb{K}} \to \mathbb{K}$) defines a line bundle over $\mathcal{P}$ (in principle a covariant phase space of quantum probes). In this formulation, states are sections of degree one

$$W_{\mathbb{K}'}\Phi(\mathbb{K}) = e^{2\pi i\omega(\mathbb{K},\mathbb{K}')}\Phi(\mathbb{K} + \mathbb{K}'). \tag{5}$$

In this language, geometric quantization means to take a Lagrangian $L \in \mathcal{P}$, so that states descend to square integrable functions on L.

A Lagrangian submanifold L is a maximally isotropic subspace L with $\omega|_L = 0$, and thus $\{\partial/\partial q^a\} \in T\mathcal{P}$ defines a Lagrangian submanifold, or "space". (Indeed, $\omega(\partial/\partial q^a, \partial/\partial q^b) = 0$.) This can be understood as a classical characterization of space (and in the covariant context, of spacetime), as a "slice" of phase space. How about a purely quantum characterization of space? We claim that quantum theory reveals a new notion of quantum space (and, more covariantly, a new notion of quantum spacetime).

Note that for space-like separations the operators of a local quantum field theory commute. Thus in order to understand the meaning of quantum spacetime (quantum Lagrangian), we need to look at a maximally commuting subalgebra of the Heisenberg algebra and the representation that diagonalizes it. Thus, borrowing from notions of non-commutative algebra and non-commutative geometry [10] (such as the theorem of Gelfand-Naimark [11]), we can say that a Lagrangian submanifold is a maximally commutative subgroup of the Heisenberg group. If we accept this notion of a Lagrangian, then the quantum regime is very different from the classical regime. In particular the vanishing Poisson bracket $\{f(q), g(p)\}$ requires either f or g to be constant. However, the vanishing commutator $[f(\hat{q}), g(\hat{p})] = 0$ requires only that the functions be commensurately periodic

$$e^{i\alpha\hat{p}}e^{i\beta\hat{q}} = e^{i\hbar\alpha\beta}e^{i\beta\hat{q}}e^{i\alpha\hat{p}}, \quad \alpha\beta = 2\pi/\hbar. \tag{6}$$

What is interesting here is that similar considerations led Aharonov to introduce *modular variables* to describe purely quantum phenomena, such as interference [12] (see also the prescient work of Schwinger [13]).

2.1. Modular variables

Modular variables are described in great detail in the very insightful book by Aharonov and Rohrlich [12], where one can find detailed bibliography on this subject[1]. The fundamental question posed there was as follows: how does one capture interference effects (due to the fundamental linearity of quantum theory) in terms of Heisenberg operators? For example, what are the quantum observables that can measure the relative phase responsible for interference in a double-slit experiment? No polynomial functions of the operators $\hat{q}$ and $\hat{p}$ can detect such phases, but operators that translate in space, such as $e^{iR\hat{p}/\hbar}$, do. Thus the modular variables denoted $[\hat{q}]$ and $[\hat{p}]$, which are defined modulo a length scale R (the slit spacing being a natural choice), play a central role, where

$$[p]_{h/R} = p \mod (h/R), \qquad [q]_R = q \mod (2\pi R), \tag{7}$$

and $h = 2\pi\hbar$. The shift operator $e^{iR\hat{p}/\hbar} = e^{iR[\hat{p}]/\hbar}$ shifts the position of a particle state (say an electron in the double-slit experiment) by a distance R and is a function of the modular momenta. These modular variables (the main examples being the Aharonov-Bohm and Aharonov-Casher phases [12]) satisfy non-local operator equations of motion. For example, given the Hamiltonian, $\hat{H} = \hat{p}^2/2m + V(\hat{q})$, the Heisenberg equation of motion for the shift operator is,

$$e^{-iR\hat{p}/\hbar}\frac{d}{dt}e^{iR\hat{p}/\hbar} = -\frac{iR}{\hbar}\left(\frac{V(\hat{q}+R) - V(\hat{q})}{R}\right). \tag{8}$$

Modular variables are fundamentally non-local in a non-classical sense, since we see here that their evolution depends on the value of the potential at distinct locations. Remarkably, thanks to the uncertainty principle, this dynamical non-locality does not lead to a violation of causality [12]. One of the characteristic features of these variables is that they do not have

[1] See also [13–15].

classical analogues; indeed, the limit $\hbar \to 0$ of $[p]_{h/R}$ is ill-defined. Also modular variables capture entanglement of continuous q, p variables.

Note that modular variables are, in general, covariant and, also, contextual[2]. In other words, they carry specific experimental information, such as the length R between the two-slits. However, in the context of quantum gravity such scales are automatically built in, and the contextuality is in principle removed. Also, the fundamental dynamical equations for modular variables are non-local in quantum gravity because of the presence of the fundamental length.

When exponentiated (i.e. when understood as particular Weyl operators), the modular variables naturally commute. In other words, given $[x^a, \tilde{x}_b] = \frac{i}{2\pi}\delta^a_b$, the following commutator of modular operators vanishes [5]

$$[e^{2\pi i x}, e^{2\pi i \tilde{x}}] = 0. \tag{9}$$

Thus a quantum algebra of modular variables possesses more commutative directions than a classical Poisson algebra, because the Poisson bracket of modular variables does not vanish, $\{e^{2\pi i x}, e^{2\pi i \tilde{x}}\} \neq 0$.

Here we make a historical note [16]: The above non-local equations of motion were essentially written by Max Born, in the very first paper which used the phrase "Quantum Mechanics" in its title, in 1924, one year before the Heisenberg breakthrough paper. Actually, Heisenberg crucially used Born's prescription of replacing classical equations by the corresponding difference equations, in order to derive what we now call the canonical commutation relations (properly written by Born and Jordan) from the Bohr-Sommerfeld quantization conditions.

2.2. Modular space and geometry of quantum theory

Returning to the subject of quantum Lagrangians, note that the quantum Lagrangian is analogous to a Brillouin cell in condensed matter physics. The volume and shape of the cell are given by λ and ϵ (i.e. $\hbar$ and G_N (α')) The uncertainty principle is implemented in a subtle way: we can specify a point in modular cell, but if so, we can't say *which* cell we are in.

This means that there is a more general notion of quantization [5], beyond that of geometric quantization. Instead of selecting a classical polarization L (the arguments of the wave function, or the arguments of a local quantum field) we can choose a *modular polarization.* In terms of the Heisenberg group all that is happening is that in order to have a commutative algebra, we need only

$$\omega(\mathbb{K}, \mathbb{K}') \in 2Z, \quad W_{\mathbb{K}} W_{\mathbb{K}'} = e^{2\pi i \omega(\mathbb{K}, \mathbb{K}')} W_{\mathbb{K}+\mathbb{K}'} = W_{\mathbb{K}'} W_{\mathbb{K}'}. \tag{10}$$

[2] Aharonov and collaborators have pushed the logic associated with modular variables to argue for a new kind of weak measurements of such non-local variables that capture the superposition principle of quantum theory. Similarly, Aharonov and collaborators argue for a time symmetric formulation of quantum theory [12].

This defines a lattice Λ in phase space $\mathcal{P}$. Finally, we specify a "lift" of the lattice from the phase space $\mathcal{P}$ to the Heisenberg group $H_{\mathcal{P}}$.

Maximally commuting subgroups $\hat{\Lambda}$ of the Heisenberg group correspond to lattices that are integral and self-dual with respect to ω [17, 18]. Given W_λ where $\lambda \in \Lambda$ there is a lift to $\hat{\Lambda}$ which defines "modular polarization"

$$U_\lambda = \alpha(\lambda) W_\lambda, \tag{11}$$

where $\alpha(\lambda)$ satisfies the co-cycle condition

$$\alpha(\lambda)\alpha(\mu)e^{\pi i \omega(\lambda,\mu)} = \alpha(\lambda+\mu), \quad \lambda, \alpha \in \Lambda. \tag{12}$$

One can parametrize a solution to the co-cycle condition by introducing a symmetric bilinear from η and setting (with $\eta(\mathbb{K}, \mathbb{K}') = k \cdot \tilde{k}' + \tilde{k} \cdot k',$)

$$\alpha_\eta(\lambda) \equiv e^{i\frac{\pi}{2}\eta(\lambda,\lambda)}. \tag{13}$$

Finally, when we choose a classical Lagrangian L, there is a special state that we associate with the vacuum: it is translation invariant (which in our context can be interpreted as "empty space"). In modular quantization there is no such translation invariant state (because of the lattice structure). The best we can do is to choose a state that minimizes an "energy", which requires the introduction of another symmetric bilinear form, that we call, again suggestively, H. This means, first, that we are looking for operators such that

$$[\hat{\mathbb{P}}_A, \Phi] = \frac{i}{2\pi}\partial_A \Phi, \qquad \Phi(\hat{\mathbb{X}} + \lambda) = \Phi(\hat{\mathbb{X}}), \tag{14}$$

where the modular observables $\Phi(\hat{\mathbb{X}} \mid \lambda) = \Phi(\mathbb{X})$ are generated by the lattice observables U_λ with $\lambda \in \Lambda$. Translation invariance would be the condition $\hat{\mathbb{P}}|0\rangle = 0$. Since this is not possible, the next natural choice is to minimize the translational energy. Therefore we pick a positive definite metric H_{AB} on $\mathcal{P}$, and we define [5]

$$\hat{E}_H \equiv H^{AB}\hat{\mathbb{P}}_A\hat{\mathbb{P}}_B, \tag{15}$$

and demand that $|0\rangle_H$ be the ground state of $\hat{E}_H$. This is indeed the most natural choice and it shows that we cannot fully disentangle the kinematics (i.e., the definition of translation generators) from the dynamics. In the Schrödinger case, since the translation generators commute, the vacuum state $\hat{E}|0\rangle = 0$ is also the translation invariant state and it carries no memory of the metric H needed to define the energy. In our context, due to the non-commutativity of translations, the operators $\hat{E}_H$ and $\hat{E}_{H'}$ do not commute. As a result the vacuum state depends on H, in other words $|0\rangle_H \neq |0\rangle_{H'}$, and it also possesses a non-vanishing zero point energy.

Thus, modular quantization involves the introduction of three quadratic forms (ω, η, H), i.e. what we call *Born geometry* [1,2], which underlies the geometry of modular variables. As we will see, in the context of metastring theory, a choice of polarization is a choice of a spacetime within $\mathcal{P}$ but the most general choice is a *modular polarization* that we have discussed above. From the foundational quantum viewpoint Born geometry (ω, η, H) arises as a parametrization of such quantizations, which results in a notion of quantum spacetime, that we call *modular spacetime*. Finally, large spacetimes of canonical string theory and general relativity result as a "many-body" phenomenon, through a process of tensoring of unit modular cells, that we refer to as "extensification" [5].

In particular, the symplectic structure ω $ds^2_\omega = \frac{1}{2}\omega_{AB} d\mathbb{X}^A d\mathbb{X}^B = \frac{1}{\hbar} dp_a \wedge dq^a$, is encoded in the canonical Heisenberg commutator between q^a and p_a. The generalized, quantum, metric H comes from the Born rule in quantum theory $ds^2_H = H_{AB} d\mathbb{X}^A d\mathbb{X}^B = \frac{1}{\hbar}(\frac{dq_a dq^a}{G_N} + G_N dp_a dp^a)$. For weak gravity, this metric reduces to the spacetime metric (where spacetime can be viewed as a slice of phase space). Due to gravity's extreme weakness, we only see spacetime metric at low energies. (The ratio ϵ/λ defines a tension; if this is identified with c^3/G_N, it is enormous, $\sim 10^{32} kg/sec.$) Therefore, in this formulation the usual dynamical spacetime metric is the low energy leftover of the quantum metric. Finally, the polarization (or locality metric) η encodes the distinction between spacetime-like and energy-momentum-like aspects of phase space (and in this sense it defines an analog of the "causal" structure in phase space) $ds^2_\eta = \eta_{AB} d\mathbb{X}^A d\mathbb{X}^B = \frac{2}{\hbar} dp_a dq^a$. This new metric captures the essence of relative locality - when η is constant we have absolute locality. Curving η also means "gravitizing the quantum". In general all three elements of Born geometry, ω, η and H are dynamical and curved in metastring theory, as we will discuss in what follows.

Also, we have that the Lorentz group (in D spacetime dimensions) lies at the intersection of the symplectic, neutral and doubly orthogonal groups [5],

$$\mathrm{O}(1, d-1) = \mathrm{Sp}(2d) \cap \mathrm{O}(d, d) \cap \mathrm{O}(2, 2(d-1)). \tag{16}$$

which sheds new light on the origin of quantum theory through compatibility of the causal (Lorentz) structure and the non-locality captured by the discreteness of the quantum spacetime. This also captures the role of relative (observer-dependent) locality [19] needed to resolve the apparent contradiction between discreteness of quantum spacetime and Lorentz symmetry.

Before doing so, let us end by a few comments regarding the Stone-von Neumann theorem [20–22] which asserts that all representations of the Heisenberg group are unitarily equivalent. Normally, we think of this as a choice of basis in phase space (a choice of polarization or classical Lagrangian), and all such choices are related by Fourier transform. Similarly, one can pass from a classical polarization (such as the Schrödinger representation) to a modular polarization via the Zak transform [23]. Note that, there is a connection on the line bundle over phase space that has unit flux through a modular cell. (This is very similar to integer quantum Hall effect systems.) A modular wave function is quasi-periodic

$$\Psi(x+a,\tilde{x}) = e^{2i\pi a\tilde{x}}\Psi(x,\tilde{x}), \quad \Psi(x,\tilde{x}+\tilde{a}) = \Psi(x,\tilde{x}). \tag{17}$$

The quasi-periods correspond to the tails of an Aharonov-Bohm [24] potential attached to a unit flux. In particular, vacuum states must have at least one zero in a cell, which leads to theta functions (the Zak transforms of Gaussians). Note that from the point of modular polarization, the familiar Schrödinger polarization is just a singular limit.

3. From modular spacetime to quantum gravity

The unexpected outcome of our research is that this fundamental quantum geometry of quantum theory can be realized in the context of metastring theory, where this quantum geometry is "gravitized" (i.e. dynamical). At the classical level, metastring theory [1–8] can be thought of as a formulation of string theory in which the target space is doubled in such a way that T-duality acts linearly on the coordinates. This doubling means that momentum and winding modes appear on an equal footing. We refer to the target space as a phase space since the metastring action requires the presence of a background symplectic form ω. The metastring formulation also requires the presence of geometrical structures that generalize to phase space the spacetime metric and the B-field (where the B-field originates from the symplectic structure ω). In fact, in the metastring we have not one but *two* notions of a metric. The first metric η is a neutral metric that defines a bi-Lagrangian structure and allows to define the classical spacetime as a Lagrangian sub-manifold[3] — more precisely, the classical spacetime is defined as a null subspace for η which is also Lagrangian for ω. The second metric H is a metric of signature $(2, 2(D-1))$ that encodes the geometry along the classical spacetime (of dimension D) as well as the transverse energy-momentum space geometry. In this formulation, T-duality exchanges the Lagrangian sub-manifold with its image under $J = \eta^{-1}H$. Classical metastring theory is defined by the following action [3] (see also the pioneering papers [25–27])

$$\hat{S} = \frac{1}{4\pi}\int_\Sigma d^2\sigma\Big(\partial_\tau\mathbb{X}^A(\eta_{AB}+\omega_{AB})(\mathbb{X})\partial_\sigma\mathbb{X}^B - \partial_\sigma\mathbb{X}^A H_{AB}(\mathbb{X})\partial_\sigma\mathbb{X}^B\Big), \tag{18}$$

[3]We remind the reader that in symplectic geometry, a Lagrangian subspace is a half-dimensional submanifold of phase space upon which the symplectic form pulls back to zero. In simple terms, a Lagrangian submanifold might be the subspace coordinatized by the q's within the phase space coordinatized by q's and p's.

where $\mathbb{X}^A$ are dimensionless coordinates on phase space and the fields η, H, ω are all dynamical (i.e., in general dependent on $\mathbb{X}$) phase space fields. In the context of a flat metastring we have constant η_{AB}, H_{AB} and ω_{AB}

$$\eta_{AB} \equiv \begin{pmatrix} 0 & \delta \\ \delta^T & 0 \end{pmatrix}, \quad H_{AB} \equiv \begin{pmatrix} h & 0 \\ 0 & h^{-1} \end{pmatrix}, \quad \omega_{AB} = \begin{pmatrix} 0 & \delta \\ -\delta^T & 0 \end{pmatrix}, \tag{19}$$

where δ^μ_ν is the d-dimensional identity matrix and $h_{\mu\nu}$ is the d-dimensional Lorentzian metric, T denoting transpose.

In this formulation [3] it is convenient, as suggested by the double field formalism [28,29], to introduce dimensionless coordinates $\mathbb{X}^A \equiv (X^\mu/\lambda, P_\mu/\varepsilon)^T$ on phase space, or equivalently, $\mathbb{X}^A \equiv (x^a, \tilde{x}_a)^T$, where λ and ϵ represent the fundamental spacetime and energy-momentum scales. As already stated, $\hbar = \lambda\epsilon$ and $\alpha' = \frac{\lambda}{\epsilon}$. Given a pair (H, η) it is natural to consider the operator $J \equiv \eta^{-1}H$. The consistency of string theory requires J to be a chiral structure, that is, a real structure ($J^2 = 1$) compatible with η, implying that J is an $O(D, D)$ transformation (realizing generalized T-duality in target space). These three structures, the symplectic $Sp(2D)$ ω, the $O(D, D)$ η and the $SO(2, 2(D-1))$ H, define the new concept of Born geometry [1–5] (see also [30]) which unifies the complex geometry of quantum theory with the metrical geometry of general relativity and the symplectic geometry of canonical Hamiltonian dynamics [31–35]. Note that in the phase space formulation the local phase space coordinates $\mathbb{X}$ are *quasiperiodic* $\mathbb{X}^A(\sigma+2\pi) = \mathbb{X}^A(\sigma) + \Delta^A$, where Δ^A is the corresponding quasiperiod (which either vanishes for the canonical Polyakov string or is given by the winding number in the usual treatment of T-duality on compact spaces.)

The worldsheet formulation of the metastring is chiral. Thus, even though the fields are doubled the central charges (left and right) are $c_L = c_R = D$ and we still have $D = 26$ for criticality. The metastring is not manifestly invariant under the worldsheet Lorentz transformations and it contains monodromies $\mathbb{X}^A(\sigma + 2\pi) = \mathbb{X}^A(\sigma) + \Delta^A$. The usual Polyakov string can be obtained by integrating out the dual $\tilde{X}$, for constant η and H backgrounds, and by supposing that the monodromies are in the kernel of $(\eta - \omega)$. T-duality is implemented in target space by the action of the chiral J operator ($J \equiv \eta^{-1}H$, $J^2 = 1$): $\mathbb{X} \to J(\mathbb{X})$.

The target space of the metastring is not spacetime, but, to first order, a chiral phase space $\mathcal{P}$ equipped by the symplectic structure ω, and the bilagrangian structure, and in particular, the polarization metric η which relates to the symplectic connection of the Fedosov deformation quantization [36] and thus leads to the star product of deformation quantization, and finally, the quantum H metric which relates to the complex structure in the context of geometric quantization [37], leading to the concept of Hilbert spaces. This classical Born geometry implements the ideas of Born duality in string theory [1, 2].

The classical equations of motion of the metastring $\partial_\tau \mathbb{X}^A - (J\partial_\sigma \mathbb{X})^A = 0$, implies the relation between momenta and monodromies $2\pi P = J(\Delta)$. There is *soldering* between worldsheet null coordinates $\sigma^\pm \equiv \sigma \pm \tau$ and the chiral target space structure $\partial_\pm \mathbb{X}^A - (P_\pm \mathbb{X})^A = 0$, where the chiral projector is defined as $2P_\pm = (1 \pm J)$. This allows us to liberate the left geometry from the right geometry (which is reminiscent of twistor theory). The careful analysis of the metastring action [3] shows that its symplectic form is $\Omega = \frac{1}{4\pi}\int \delta\mathbb{X}^A \eta_{AB} \nabla_\sigma \delta\mathbb{X}^B$, where ∇ is the generalized Fedosov connection found in the Fedosov deformation quantization approach [36].

Also, the operator product expansion of the metastring vertex operators $V_k = \epsilon_k e^{i\mathbb{K}\mathbb{X}}$, (i.e. modular variables) lead to the restriction of $\mathbb{K}$ on a double Lorentzian integral lattice Γ, that by modular invariance, must be self-dual. These exist in $D = 2mod(8)$, and are unique. Criticality gives a very unique lattice $\Gamma = \text{II}_{1,25} \times \text{II}_{1,25}$. This fact, in turn, leads to the large symmetry structure found by Borcherds in the study of the monstrous moonshine [38, 39][4].

As already noted, the metastring is chiral. This requires the introduction of a preferred worldsheet time coordinate which is fundamentally Lorentzian [3]. How can this be consistent with modular invariance? The answer is given by employing the Giddings-Wolpert-Krichever-Novikov construction [41, 42]: given a Riemann surface, provided a choice of local coordinates around punctures is labeled by one scalar, there exists a unique Abelian differentional e with imaginary periods. The real part of this Abelian differential is the modular invariant time $\tau = Re(e)$. The zeros of e represent interaction points where the worldsheet Lorentzian cones double. Cutting the Riemann surface along the real trajectory of e we obtain a string decomposition of the surface. The Nakamura graphs [43] encode this decomposition and give a very effective cell decomposition of moduli space. Thus Nakamura graphs are the natural Feynman diagrams for closed strings [44].

Finally, the metastring formulation points to an unexpected fundamental non-commutativity of closed string theory, that we address in what follows.

3.1. Intrinsic non-commutativity in metastring theory

It is well established that the structure of the zero mode algebra of the compactified closed string depends on a lattice of momenta $(\Lambda, 2\eta)$ which is integral and self-dual with respect to a neutral metric: a so-called Narain lattice [45]. In our recent work [1–8] we have refined this structure and we have shown that in fact the kinematical structure of the string zero modes depends on a *para-hermitian lattice*: a triple (Λ, η, ω), where Λ is a subgroup of $\mathbb{R}^{2d}$ that describes the lattice of wave-covectors $\lambda\mathbb{K}$, with λ the string length, η is a neutral metric, a symmetric bilinear form of signature

[4]For a string theory related discussion, see [40].

(d,d), and ω is an invertible two-form. This structure needs to satisfy two compatibility conditions: first, the lattice Λ is assumed to be integral with respect to the para-hermitian structure, i.e., $(\eta\pm\omega)(\lambda\mathbb{K},\lambda\mathbb{K}')\in\mathbb{Z}$, for $\lambda\mathbb{K},\lambda\mathbb{K}'\in\Lambda$. Second, the metric η and the 2-form ω must be compatible, in the sense that $\eta^{-1}\omega := K$ is a product structure, that satisfies the condition $K^2=1$.

These two conditions are a consequence of mutual locality on the worldsheet (i.e. worldsheet causality). It is clear that if (Λ,η,ω) is a para-hermitian lattice, then $(\Lambda,2\eta)$ is a Narain lattice, so the kinematical structure that we highlight is a refinement of the usual one. The extra information is contained in the 2-form ω. This form does not enter expressions for the spectrum or the partition function and this why it is usually ignored. It does enter however crucially in the definition of the vertex operator algebra and parameterizes what is usually referred to as a cocycle. The role of ω is to promote the zero mode double space $\mathcal{P}\simeq\mathbb{R}^{2d}$ dual to $\mathbb{R}[\Lambda]$ to the status of phase space: $\mathcal{P}$ should be viewed as a symplectic manifold. At the quantum level, both geometrical structures η and ω enter in the commutation relations of string operators. ω controls the non-commutativity of the zero-modes while η controls the non-commutativity of the string oscillator modes. This can be seen if one introduces a double notation for the string coordinate $\mathbb{X}(\sigma)$ that includes the string map X and its dual $\tilde{X}$. The string commutation relations, were derived in [7, 8]

$$[\mathbb{X}^A(\sigma),\mathbb{X}^B(\sigma')] = 2i\lambda^2\left[\pi\omega^{AB}-\eta^{AB}\theta(\sigma-\sigma')\right], \tag{20}$$

where $\theta(\sigma)$ is the staircase distribution, i.e., a solution of $\theta'(\sigma)=2\pi\delta(\sigma)$; it is odd and quasi-periodic with period 2π.

Following standard practice, all indices are raised and lowered using η and η^{-1}. The momentum density operator is given by $\mathbb{P}_A(\sigma)=\frac{1}{2\pi\alpha'}\eta_{AB}\partial_\sigma\mathbb{X}^B(\sigma)$ and the previous commutation relation implies that it is conjugate to $\mathbb{X}^A(\sigma)$. The two-form ω appears when one integrates this canonical commutation relation to include the zero-modes, the integration constant being uniquely determined by worldsheet causality. Denoting by $(\hat{\mathbb{X}},\hat{\mathbb{P}})$ the zero mode components of the string operators $\mathbb{X}(\sigma)$ and $\mathbb{P}(\sigma)$ we simply have that

$$[\hat{\mathbb{P}}_A,\hat{\mathbb{P}}_B]=0, [\hat{\mathbb{X}}^A,\hat{\mathbb{P}}_B]=i\hbar\delta^A{}_B, [\hat{\mathbb{X}}^A,\hat{\mathbb{X}}^B]=2\pi i\lambda^2\omega^{AB}. \tag{21}$$

This is a deformation of the doubled Heisenberg algebra involving the string length λ as a deformation parameter.

So far we have assumed that the background is trivial, with the fields (η,ω) *constant* and given by $\eta(\mathbb{K},\mathbb{K}')=k\cdot\tilde{k}'+\tilde{k}\cdot k'$, and $\omega(\mathbb{K},\mathbb{K}')=k\cdot\tilde{k}'-\tilde{k}\cdot k'$. As shown in [7], we can turn on non-trivial backgrounds encoded into ω by changing the $\mathrm{O}(d,d)$ frame $\mathbb{X}\to O\mathbb{X}$. This change of frame preserves η but transforms ω. Any constant ω can be obtained this way. Since ω has an interpretation as the symplectic form on the space of

$\mathbb{X}$'s, modifying ω affects the commutation relations[5] $[\hat{\mathbb{X}}^A, \hat{\mathbb{X}}^B] = 2\pi i\lambda^2 \Pi^{AB}$,, with $\Pi^{AB}\omega_{BC} = \delta^A{}_C$, where we have introduced the Poisson tensor $\Pi = \omega^{-1}$.

For instance, under a constant B-field transformation $\mathbb{X} = (x^a, \tilde{x}_a) \mapsto (x^a, \tilde{x}_a + B_{ab}x^b)$, the trivial symplectic form $\omega(\mathbb{K}, \mathbb{K}') = k \cdot \tilde{k}' - \tilde{k} \cdot k'$ is mapped onto $\omega(\mathbb{K}, \mathbb{K}') = k_a \tilde{k}'^a - k'_a \tilde{k}^a - 2B_{ab}\tilde{k}^a \tilde{k}'^b$, and the commutators read

$$[\hat{x}^a, \hat{x}^b] = 0, \qquad [\hat{x}^a, \hat{\tilde{x}}_b] = 2\pi i\lambda^2 \delta^a{}_b, \qquad [\hat{\tilde{x}}_a, \hat{\tilde{x}}_b] = -4\pi i\lambda^2 B_{ab}. \tag{22}$$

We see that the effect of the B-field is to render the dual coordinates non-commutative (and that the B-field originates from the symplectic structure ω). More generally, we can parameterize an arbitrary $O(d,d)$ transformation as $g = e^{\hat{B}}\hat{A}e^{\hat{\beta}}$, where $\hat{A} \in GL(d)$ and $e^{\hat{B}} = \left(\begin{smallmatrix} 1 & 0 \\ B & 1 \end{smallmatrix}\right)$ and $e^{\hat{\beta}} = \left(\begin{smallmatrix} 1 & \beta \\ 0 & 1 \end{smallmatrix}\right)$ are nilpotent. $e^{\hat{B}}$ is the B-field transformation discussed above, and is associated with the usual B-field deformation in string theory. We note that the transformation of $(x^a, \tilde{x}_a)$ given above does not modify x^a, and thus fields that depend only on x^a are unmodified. The β-transformation on the other hand corresponds to the map $(x^a, \tilde{x}_a) \mapsto (x^a + \beta^{ab}\tilde{x}_b, \tilde{x}_a)$. Equivalently, it has the effect of mapping the symplectic structure to $\omega(\mathbb{K}, \mathbb{K}') = k_a \tilde{k}'^a - k'_a \tilde{k}^a + 2\beta^{ab}k_a k'_b$, and yields commutation relations

$$[\hat{x}^a, \hat{x}^b] = 4\pi i\lambda^2 \beta^{ab}, \qquad [\hat{x}^a, \hat{\tilde{x}}_b] = 2\pi i\lambda^2 \delta^a{}_b, \qquad [\hat{\tilde{x}}_a, \hat{\tilde{x}}_b] = 0. \tag{23}$$

Dramatically, the coordinates that are usually thought of as the spacetime coordinates have become themselves non-commutative. Since this is the result of an $O(d,d)$ transformation, we know that it can be thought of in similar terms as the B-field; these are related by T-duality. We are familiar with the B-field background because we have, in the non-compact case, a fixed notion of locality in the target space theory. However, in the non-geometric β-field background, we do not have such a notion of locality but we can access it through T-duality.

4. Effective quantum fields and manifest non-locality

What is the effective description of closed strings that incorporates the above intrinsic non-commutativity? For a closed sting on a circle of radius R (where the dual radius $\tilde{R}$, is defined as $R\tilde{R} = 2\lambda^2$ and the respective winding integers are n and w) this effective description is captured by the generalized field [7, 8]

$$\Phi(x, \tilde{x}) \equiv \sum_w \Phi_w(x) e^{iw\tilde{x}/\tilde{R}}. \tag{24}$$

[5]The algebraic structure that we are working with here has an analogy in electromagnetism in the presence of monopoles. In that analogy, the string length becomes the magnetic length, and the form ω becomes the magnetic field. Another analogy occurs in quantum Hall liquids, the algebra being the magnetic algebra of the lowest Landau level.

This meshes well with the observation [7, 8] that the string product is essentially a representation of the Heisenberg group, which suggests that one should consider the "quantization" map

$$\Phi(x,\tilde{x}) \to \hat{\Phi} = \sum_{w} \Phi_w(\hat{x}) e^{iw\hat{\tilde{x}}/\tilde{R}}, \tag{25}$$

from generalized fields to non-commutative fields.[6] Under this map the T-duality transformation becomes "localized" and is expressed as the exchange of $\hat{x}$ with $\hat{\tilde{x}}$. The T-dual expression is given by [7, 8]

$$\hat{\Phi} = \sum_{n} e^{in\hat{x}/R} \Phi_n(\hat{\tilde{x}} - \pi n\tilde{R}) = \sum_{n} \Phi_n(\hat{\tilde{x}}) e^{in\hat{x}/R}, \tag{26}$$

which has a similar form to (25). We see that the non-commutativity of $\hat{x}$ with $\hat{\tilde{x}}$ allows one to reabsorb all the shifts in terms of a simple reordering that exchanges $\hat{x}$ with $\hat{\tilde{x}}$ and is the expression of T-duality. The "quantized" field is simply expanded in terms of modes as

$$\hat{\Phi} \equiv \sum_{w,n} e^{in\hat{x}/R} \Phi(n,w) e^{iw\hat{\tilde{x}}/\tilde{R}}. \tag{27}$$

It is useful at this point to generalize the construction to higher dimensional tori. This can be done in a straightforward manner by introducing the modes $\mathbb{K}^A = (\tilde{k}^a, k_a)$, generalizing $(w/\tilde{R}, n/R)$. The integrality condition for the lattice Λ of admissible modes $\mathbb{K}, \mathbb{K}' \in \Lambda$ reads in this notation as[7] $(\eta \pm \omega)(\lambda\mathbb{K}, \lambda\mathbb{K}') \in \mathbb{Z}$. We now write $\Phi(\mathbb{K}) = \langle \mathbb{K}|\Phi\rangle$ with the ordering chosen as $\langle \mathbb{K}| = \langle 0|\hat{U}_{-\mathbb{K}}$, where $\hat{U}_{\mathbb{K}} \equiv e^{ik\cdot\hat{x}} e^{i\tilde{k}\cdot\hat{\tilde{x}}}$. This ordering can be seen to be related to the choice of an $O(d,d)$ frame, where we place the operator associated with x on the left and the operator associated with the dual space $\tilde{x}$ on the right. The key point is that this choice of frame is entirely encoded into the choice of symplectic potential ω and the vertex operator can be covariantly written in terms of $\mathbb{K} = (\tilde{k}, k)$ and $\mathbb{X} = (x, \tilde{x})$ as

$$\hat{U}_{\mathbb{K}} = e^{\frac{i}{2}(\eta+\omega)(\mathbb{K},\hat{\mathbb{X}})} e^{\frac{i}{2}(\eta-\omega)(\mathbb{K},\hat{\mathbb{X}})}. \tag{28}$$

Given this notation we can write the string product covariantly as [7, 8]

$$(\Phi \circ \Psi)(\mathbb{K}) = \sum_{\mathbb{K}'+\mathbb{K}''=\mathbb{K}} \Phi(\mathbb{K}') e^{i\pi(\eta-\omega)(\lambda\mathbb{K}',\lambda\mathbb{K}'')} \Psi(\mathbb{K}''). \tag{29}$$

The non-commutativity of the string product is encoded in terms of a π-flux due to ω. As it turns out the phase factor is exactly the same as the cocycle factor $\epsilon(\mathbb{K},\mathbb{K}') = e^{i\pi(\eta-\omega)(\lambda\mathbb{K},\lambda\mathbb{K}')}$ that appears in the definition of the vertex operator product [7, 8].

[6] Here, we have chosen a specific operator ordering. Given this ordering, the mapping is well-defined and consistent with the string product.

[7] In the one dimensional case where $\mathbb{K} = (w/\tilde{R}, n/R)$ this follows directly from $(\eta + \omega)(\lambda\mathbb{K}, \lambda\mathbb{K}') = nw'$ and similarly $(\eta - \omega)(\lambda\mathbb{K}, \lambda\mathbb{K}') = wn'$, given that $n, n', w, w' \in \mathbb{Z}$.

4.1. Non-local excitations

To summarize, the above manifestly T-duality covariant formulation of closed strings (i.e. the metastring) implies intrinsic non-commutativity of zero-modes. It is thus instructive to formulate a particle-like limit of the metastring that we call the *metaparticle*. Given the form for the symplectic structure of the zero modes derived in section 4. of [7] (equation (67) of that paper, without the contribution coming from string oscillators), the action $S \equiv \int d\tau L$ of the metaparticle is governed by the following Lagrangian L

$$L = p_\mu \dot{x}^\mu + \tilde{p}^\mu \dot{\tilde{x}}_\mu - \pi\alpha' p_\mu \dot{\tilde{p}}^\mu - \frac{e}{2}\left(p_\mu p^\mu + \tilde{p}_\mu \tilde{p}^\mu - m^2\right) + \tilde{e}\left(p_\mu \tilde{p}^\mu - \mu^2\right). \quad (30)$$

where e and $\tilde{e}$ are the Lagrange multipliers for the two constraints that follow from the Hamiltonian ($H \equiv \partial_\sigma \mathbb{X}^A H_{AB} \partial_\sigma \mathbb{X}^B = 0$) and diffeomorphism constraints ($D \equiv \partial_\sigma \mathbb{X}^A \eta_{AB} \partial_\sigma \mathbb{X}^B = 0$) of the metastring [3, 6]. Note that the usual particle limit is obtained, at least classically, by taking $\mu \to 0$ and $\tilde{p} \to 0$. The theory of metaparticles can be viewed as the theory of the zero modes of the closed string, which fully takes into account its intrinsic non-commutativity. Given the form of the above Lagrangian, the metaparticle looks like two particles that are entangled through a Berry phase-like $p_\mu \dot{\tilde{p}}^\mu$ factor. The metaparticle is fundamentally non-local, and thus it should not be associated with effective local field theory. In particular, by looking at the metaparticle constraints $p^2 + \tilde{p}^2 = m^2$ and $p\tilde{p} = \mu^2$, we note that the momenta p and $\tilde{p}$ can be, in principle, widely separated. For example, if m is of the order of the Planck energy, and μ of the order of one TeV (which could be understood as a characteristic particle physics scale), then the momentum p can be of the order of the Planck energy, and the momentum $\tilde{p}$ of the vacuum energy scale! Thus the metaparticle theory is able to naturally relate widely separated scales, which transcends the usual reasoning based on Wilsonian effective field theory (and should be relevant for the naturalness and hierarchy problems).

We expect that the correct field theoretic description of the metaparticle is in terms of the above non-commutative (modular) field theory $\Phi(x, \tilde{x})$ limit of the metastring. Like their particle cousins, metaparticles should be detectable, and they might even present good candidates for dark matter quanta [46–51]. Such effective non-commutative field theory (similar in spirit to [52–54]) can be also useful in illuminating the vacuum energy problem, via the sequester mechanism [55] in which the dimensional parameters in the effective action of the matter sector depend on the four-volume element of the universe. Finally, we note that the concept of metaparticles might be argued from the compatibility of the quantum spacetime that underlies the generic representations of quantum theory, as discussed in [5], and thus the metaparticle might be as ubiquitous as the concept of antiparticles which is demanded by the compatibility of relativity and quantum theory.

5. Conclusions

In this talk we have discussed intrinsic non-commutativity in quantum gravity [7, 8] related to a new concept of quantum spacetime, called modular spacetime [4,5] that also appears as a habitat for metastring theory [3,6] and that is deeply rooted in the foundations of quantum theory. Note that this concept stems from a quantization of spacetime, and not from quantization of gravitational field/metric. Even the flat space is quantized according to our approach to quantum gravity. This allows for superposition and entanglement of spacetimes. Also, this formulation provides for an explicit construction of spacetime quanta or qubits, and a new non-perturbative definition of quantum gravity as "gravitization of the quantum" [1, 2].

References

[1] L. Freidel, R. G. Leigh and D. Minic, Phys. Lett. B **730**, 302 (2014), arXiv:1307.7080 [hep-th].

[2] L. Freidel, R. G. Leigh and D. Minic, Int. J. Mod. Phys. D **23**, no. 12, 1442006 (2014), arXiv:1405.3949 [hep-th].

[3] L. Freidel, R. G. Leigh and D. Minic, JHEP **1506**, 006 (2015) arXiv:1502.08005 [hep-th].

[4] L. Freidel, R. G. Leigh and D. Minic, Int. J. Mod. Phys. D **24**, no. 12, 1544028 (2015).

[5] L. Freidel, R. G. Leigh and D. Minic, Phys. Rev. **D 94**, 104052 (2016), arXiv:1606.01829 [hep-th].

[6] L. Freidel, R. G. Leigh and D. Minic, J. Phys. Conf. Ser. **804**, no. 1, 012032 (2017). doi:10.1088/1742-6596/804/1/012032

[7] L. Freidel, R. G. Leigh and D. Minic, JHEP **1709**, 060 (2017) [arXiv:1706.03305 [hep-th]].

[8] L. Freidel, R. G. Leigh and D. Minic, Phys. Rev. D **96**, no. 6, 066003 (2017) [arXiv:1707.00312 [hep-th]].

[9] H. Weyl, *The theory of groups and quantum mechanics* (New York: Dover) (1931).

[10] A. Connes, *Noncommutative Geometry* (Cambridge: Academic Press) (1994).

[11] I. M. Gelfand, and M. A. Naimark, *Mat.Sbornik* **12** 197 (1943).

[12] Y. Aharonov, and D. Rohrlich, *Quantum Paradoxes: Quantum Theory for the Perplexed* (New York: Wiley) (2005).

[13] J. Schwinger, *Quantum kinematics and dynamics* (Boulder: Westview Press) (2000).

[14] Y. Aharonov, A. Petersen and H. Pendleton, Int. J. Theor. Phys. **2**, 213 (1969).

[15] G. 't Hooft, arXiv:1405.1548 [quant-ph]

[16] B. L. van der Waerden, *Sources of Quantum Mechanics* (New York: Dover) (1968).

[17] G. Mackey, *Acta Math.* **99** 265 (1958).

[18] G. Mackey, *Annals Math.* **55** 101 (1952).

[19] G. Amelino-Camelia, L. Freidel, J. Kowalski-Glikman and L. Smolin, Phys. Rev. D **84**, 084010 (2011), arXiv:1101.0931 [hep-th].

[20] M. Stone, *Proc.Nat.Acad.Sci.* **16** 172 (1930).

[21] J. von Neuman, *Math.Ann.* **104** 570 (1931).

[22] G. Mackey, *Proc.Nat.Acad.Sci.* **35** 537 (1949).

[23] J. Zak, *Phys.Rev.Lett.* **19** 1385 (1967).

[24] Y. Aharonov and D. Bohm, Phys. Rev. **115**, 485 (1959).

[25] A. A. Tseytlin, *Nucl. Phys.* B **350** 395 (1991).

[26] A. A. Tseytlin, *Phys. Lett.* B **242** 163 (1990).

[27] R. Floreanini, and R. Jackiw, *Phys. Rev. Lett.* **59** 1873 (1987).

[28] W. Siegel, it Phys. Rev. D **47** 5453 (1993).

[29] C. Hull, and B. Zwiebach, *JHEP* **0909** 099 (2009).

[30] L. Freidel, F. J. Rudolph and D. Svoboda, JHEP **1711**, 175 (2017) [arXiv:1706.07089 [hep-th]].

[31] G. W. Gibbons, *J. Geom. Phys.* **8** 147 (1992).

[32] A. Ashtekar, and t. Schilling, gr-qc/9706069.

[33] D. Minic, and H. C. Tze, *Phys. Rev.* D **68** 061501 (2003).

[34] D. Minic, and H. C. Tze, *Phys. Lett.* B **581** 111 (2004).

[35] V. Jejjala, M. Kavic, and D. Minic, *Int. J. Mod. Phys.* A **22** 3317 (2007).

[36] B. Fedosov, *Deformation quantization and index theory* (Berlin: Akademie-Verlag) (1996).

[37] N. Woodhouse, *Geometric Quantization* (Oxford: Clarendon Press) (1991).

[38] R. E. Borcherds, *Proc. Nat. Acad. Sci.* **83** 3068 (1986).

[39] R. E. Borcherds, *Inventiones Mathematicae* **109** 405 (1992).

[40] G. W. Moore, *Preprint* [hep-th/9305139]

[41] S. B. Giddings, and S. A. Wolpert, *Commun. Math. Phys.* **109** 177 (1987).

[42] I. M. Krichever, and S. P. Novikov, *Funct.Anal.Appl.* **21** 294 (1987).

[43] S. Nakamura, *Tokyo J. Math.* **23** 87 (2000).

[44] L. Freidel, D. Garner, and S. Ramgoolam, *Phys. Rev.* D **91** 126001 (2015).

[45] K. S. Narain, Phys. Lett. **169B**, 41 (1986). doi:10.1016/0370-2693(86)90682-9

[46] C. M. Ho, D. Minic and Y. J. Ng, Phys. Lett. B **693**, 567 (2010) [arXiv:1005.3537 [hep-th]].

[47] C. M. Ho, D. Minic and Y. J. Ng, Gen. Rel. Grav. **43**, 2567 (2011) [Int. J. Mod. Phys. D **20**, 2887 (2011)] [arXiv:1105.2916 [gr-qc]].

[48] C. M. Ho, D. Minic and Y. J. Ng, Phys. Rev. D **85**, 104033 (2012) [arXiv:1201.2365 [hep-th]].

[49] D. Edmonds, D. Farrah, C. M. Ho, D. Minic, Y. J. Ng and T. Takeuchi, Astrophys. J. **793**, 41 (2014) [arXiv:1308.3252 [astro-ph.CO]].

[50] D. Edmonds, D. Farrah, C. m. Ho, D. Minic, Y. J. Ng and T. Takeuchi, Int. J. Mod. Phys. A **32**, no. 18, 1750108 (2017) [arXiv:1601.00662 [astro-ph.CO]].

[51] D. Edmonds, D. Farrah, D. Minic, Y. J. Ng and T. Takeuchi, arXiv:1709.04388 [astro-ph.CO].

[52] M. R. Douglas and N. A. Nekrasov, Rev. Mod. Phys. **73**, 977 (2001) arXiv:hep-th/0106048.

[53] R. J. Szabo, Phys. Rept. **378**, 207 (2003) arXiv:hep-th/0109162.

[54] H. Grosse and R. Wulkenhaar, Commun. Math. Phys. **256**, 305 (2005) arXiv:hep-th/0401128.

[55] N. Kaloper and A. Padilla, Phys. Rev. Lett. **112**, no. 9, 091304 (2014) [arXiv:1309.6562 [hep-th]].

Introduction to teleparallel gravities

Alexey Golovnev*
Asia Pacific Center for Theoretical Physics,
67 Cheongam-ro, Nam-gu, Pohang 37673, Republic of Korea;
Department of High Energy Physics, Saint Petersburg State University,
Ulyanovskaya ulitsa 1, Stary Petergof, Saint Petersburg 198506, Russia.

Abstract

We give a pedagogical introduction into the field of (modified) teleparallel theories of gravity. Our presentation is fairly self-contained. In particular, we carefully explain the basic principles of metric-affine approaches to gravity. This contribution is based on our talk "Teleparallel gravity, its modifications, and the local Lorentz invariance" at the 9th Mathematical Physics Meeting: School and Conference on Modern Mathematical Physics in Belgrade, September 2017.

PACS: 04.20.Cv; 04.20.Fy; 04.50.Kd

1. Introduction

The great success of general relativity in describing experimental data does clearly show that we are on the right track, and our geometric theory of gravity works extremely well. From experimental side the only reservations come from cosmology.

Even though we have a perfectly good model also there (ΛCDM), it is very unpleasant that we must confess we don't have any reasonable idea as to the nature of some 95% of the energy budget of the Universe [1]. Numerous searches for new physics beyond the Standard Model of elementary particles are yet futile.

Can we do better via some (infrared) modifications of gravity? Very probably, the answer could be in positive. It seems to be not an easy endeavor, but nevertheless the one which is worth trying.

In order to have an interesting model, we need to start from a deep understanding of gravity. Simply changing the Newtonian limit for incorporation of, say, flat rotation curves in galaxies is way too easy. MOND is an archetypal example [2]. Embedding it into a wider picture is very arbitrary and hardly was ever successfull per se [3].

Of course, modifications to the full general relativity can also be done in a large number of different ways. However even the most straightforward

* e-mail address: agolovnev@yandex.ru

consistency tests prove to restrict the freedom a lot. A very good example is massive gravity for which even the most basic task of avoiding the Boulware-Deser ghost turned out a highly non-trivial problem [4] with interesting mathematical structures behind [5, 6].

Even very simple extensions such as (metric) $f(R)$ models lead to a new degree of freedom due to the higher derivative nature of any available curvature invariants; and the problem of screening the extra degrees of freedom from low energy phenomenology emerges [7]. One notable exception from this rule is the Palatini $f(R)$ gravity which is however subject to its own problems [8].

We must conclude that the search for a viable modification to the theory of gravitational interactions is anything but a simple task, and very often the models acquire their own deep academic interest even irrespective of possible phenomenological applications. And despite a large number of reasonable ideas on the market, we cannot claim a definitive success of any particular approach to cosmology-motivated modification of general relativity.

From this vantage point, it would be natural to consider models of teleparallel type. Indeed, teleparallel gravity is just an equivalent (modulo global topological issues of parallelisability) description of general relativity in terms of different geometry. However, if one attempts at modifying the teleparallel Lagrangian then the resulting modified gravity models from simple modifications would generically be different from those which can be obtained by analogous procedures applied to general relativity in the standard formulation. Therefore it is an independent direction of research which is definitely worth trying.

It is precisely this viewpoint which we will take regarding modified teleparallel gravities in this contribution. In Section 2 we explain the essence of metric-affine approaches to gravity. In Section 3 we describe the teleparallel equivalent of general relativity in its classical standing with vanishing spin connection, and also present some possible extensions. In Section 4 we discuss covariantised versions and related problems. Finally, in Section 5 we conclude.

2. Metric-affine preliminaries

In principle, a metric and an affine connection are two different geometrical structures on a manifold. The former defines lenghts and distances, and also gives a natural measure $\sqrt{-g} \cdot d^4x$, while the latter provides us with the rule of parallel transport. Apparently, we don't need to think of the metric structure in order to have some choice of the horizontal section in the tangent bundle.

2.1. Affine connection

Leaving an invariant discussion for textbooks on differential geometry, we will use the plain language of components for the purposes of this elementary introduction. The connection coefficients Γ are used to tell us how the components of a vector are changed under an infinitesimal parallel transport, δx^μ in terms of coordinates: $\delta A^\nu = -\Gamma^\nu_{\mu\alpha} A^\alpha \delta x^\mu$, or analogously for 1-forms $\delta A_\nu = \Gamma^\alpha_{\mu\nu} A_\alpha \delta x^\mu$. The same coefficients with different signs are used in the two rules in order to ensure that scalar quantities such as $A_\mu B^\mu$ are not changed under the parallel transport.

One can treat the parallelly transported vector as being the same vector at a different point. Covariant derivatives are then simply defined by subtracting this trivial change above from the actual change of vector field components:

$$\nabla_\mu A^\nu \equiv \partial_\mu A^\nu + \Gamma^\nu_{\mu\alpha} A^\alpha, \tag{1}$$

and

$$\nabla_\mu A_\nu \equiv \partial_\mu A_\nu - \Gamma^\alpha_{\mu\nu} A_\alpha \tag{2}$$

for differential 1-forms. Note that different conventions can be encountered at this point. Ours is that the index which corresponds to the derivative is the left one among the lower indices of the connection coefficients.

The rules (1) and (2) were given in one particular coordinate system. One can however deduce the transformation law for Γ demanding that the covariant derivatives be tensors. It does not depend on any particular choice of the affine connection, and a simple calculation from any textbook on general relativity gives

$$\Gamma^\kappa_{\mu\nu} = \frac{\partial x^\kappa}{\partial x'^\rho} \left(\frac{\partial x'^\alpha}{\partial x^\mu} \frac{\partial x'^\beta}{\partial x^\nu} \Gamma'^\rho{}_{\alpha\beta} + \frac{\partial^2 x'^\rho}{\partial x^\mu \partial x^\nu} \right). \tag{3}$$

Now we see that symmetric connections, $\Gamma^\kappa_{\mu\nu} = \Gamma^\kappa_{\nu\mu}$, are somewhat distinguished from the viewpoint of the equivalence principle since they can be set to zero at any single point by a mere coordinate transformation. On the other hand, the antisymmetric part is a tensor

$$T^\alpha{}_{\mu\nu} = \Gamma^\alpha_{\mu\nu} - \Gamma^\alpha_{\nu\mu} \tag{4}$$

which is known under the name of torsion. It will be very important for us in what follows.

2.1.1. Geodesic lines

Given an affine connection, one can generalise the notion of a straight line. Indeed, let us call a line $x^\mu(\tau)$ geodesic if an only if its tangent vector

$$e^\mu(\tau) \equiv \frac{dx^\mu(\tau)}{d\tau}$$

remains tangent when parallelly transported along the line. In other words, the tangent vector $e^\mu(\tau)$ is covariantly constant along the line.

More precisely, under an infinitesimal change of the parameter τ we demand that $\delta \frac{dx^\mu}{d\tau} = -\Gamma^\mu_{\nu\alpha} \frac{dx^\alpha}{d\tau} \delta x^\nu = -\Gamma^\mu_{\nu\alpha} \frac{dx^\alpha}{d\tau} \frac{dx^\nu}{d\tau} \delta\tau$. It gives the geodesic equation

$$\frac{d^2 x^\mu}{d\tau^2} + \Gamma^\mu_{\nu\alpha} \frac{dx^\alpha}{d\tau} \frac{dx^\nu}{d\tau} = 0. \tag{5}$$

This equation (5) is invariant under affine changes of τ. Under a general non-linear reparametrisation, this equation becomes more complicated. There is a preferred class of choices for the parameterisation of the line, the class of affine parameters. It stems from demanding that $e^\mu(\tau)$ is strictly preserved when being parallelly transported along a geodesic line, not only up to a scalar factor, and allows one to define the null infinity in general relativity such that a null geodesic is infinitely long if its affine parameter reaches an infinite value.

2.2. Curvature tensor

Of course, if one simply introduces curvilinear coordinates in a Euclidean space, connection coefficients already become non-trivial. One may ask how to distinguish between a real curvature and a (locally) flat space in curvilinear coordinates. In Riemannian geometry the answer is very simple (in general we will only need to add torsion and non-metricity to this discussion, see below).

It is difficult to compare two vectors at a distance from each other, but it is fairly easy to do so if they are at the same point in the space. If one parallelly transports a vector along a closed contour, then the resulting vector always coincides with the initial one if the space is flat. Therefore, if after such a procedure a vector has changed, it is a clear indication of non-trivial geometry (the converse is also true but not so elementary, even for rigorous understanding of what it means).

Let us perform a parallel transport of a vector ξ^μ along a closed infinitesimal contour $\mathcal{C}$. After one revolution we have

$$\delta \xi^\mu = \oint_{\mathcal{C}} d\tau \dot{\xi}^\mu$$

where the vector function $\xi^\mu(x(\tau))$ is determined by the parallel transport equation $\dot{\xi}^\mu = -\Gamma^\mu_{\nu\alpha} \xi^\alpha \frac{dx^\nu}{d\tau}$. Moreover, we can parallelly transport the vector to everywhere in a small neighbourhood of the contour.

Let us assume for simplicity that the origin of the coordinate system, i.e. $x^\mu = 0$, is chosen inside the contour and Taylor expand the vector field components and the connection coefficients around the origin. As at the lowest order the integral is proportional to $\oint dx^\mu = 0$, we look at the first order correction:

$$\delta \xi^\mu = -\oint_{\mathcal{C}} d\tau \left(\Gamma^\mu_{\nu\alpha} \cdot (\partial_\rho \xi^\alpha) x^\rho + (\partial_\rho \Gamma^\mu_{\nu\alpha}) x^\rho \cdot \xi^\alpha + \mathcal{O}(x^2) \right) \frac{dx^\nu}{d\tau}$$

where the only x-dependence is the explicit one with all functions being taken at $x^\mu = 0$.

In the first term we use the parallel transport equation for the vector field $\partial_\rho \xi^\alpha = -\Gamma^\alpha_{\rho\sigma}\xi^\sigma$ once more to obtain

$$\delta\xi^\mu = -\oint_{\mathcal{C}} d\tau \left(-\Gamma^\mu_{\nu\alpha}\Gamma^\alpha_{\rho\sigma}\xi^\sigma x^\rho + (\partial_\rho \Gamma^\mu_{\nu\sigma})\xi^\sigma x^\rho + \mathcal{O}(x^2)\right) \frac{dx^\nu}{d\tau}.$$

Omitting the $\mathcal{O}(x^2)$-corrections, the integral is proportional to

$$\oint x^\rho \frac{dx^\nu}{d\tau} d\tau = -\oint x^\nu \frac{dx^\rho}{d\tau} d\tau = \int\int dx^\rho \wedge dx^\nu,$$

the antisymmetric area element (encircled by the contour $\mathcal{C}$) which we will denote by $S^{\rho\nu} = -S^{\nu\rho}$.

The antisymmetrised coefficient must also be a tensor, which can of course be checked directly by using equation (3), and finally we get

$$\delta\xi^\mu = \oint_{\mathcal{C}} d\tau \dot{\xi}^\mu = -\frac{1}{2} R^\mu{}_{\sigma\rho\nu}\xi^\sigma S^{\rho\nu}$$

where the curvature tensor is defined as

$$R^\mu{}_{\sigma\rho\nu} = \partial_\rho \Gamma^\mu_{\nu\sigma} - \partial_\nu \Gamma^\mu_{\rho\sigma} + \Gamma^\mu_{\rho\alpha}\Gamma^\alpha_{\nu\sigma} - \Gamma^\mu_{\nu\alpha}\Gamma^\alpha_{\rho\sigma}. \tag{6}$$

By definition, it is always antisymmetric

$$R^\mu{}_{\sigma\rho\nu} = -R^\mu{}_{\sigma\nu\rho}, \tag{7}$$

and this is the only symmetry property of the curvature tensor for a generic connection.

Of course, we could also transport a one-form instead of a vector around the contour. It yields the same tensor:

$$\delta\zeta_\mu = \oint_{\mathcal{C}} d\tau \dot{\zeta}_\mu = \frac{1}{2} R^\sigma{}_{\mu\rho\nu}\zeta_\sigma S^{\rho\nu},$$

where the change of sign ensures again that the scalar quantities remain unchanged.

2.2.1. Commutator of covariant derivatives

There is also another way to introduce the curvature. Normally, the partial derivatives commute. However, it is not true of the covariant ones. Let us compute the commutator:

$$[\nabla_\mu\,,\ \nabla_\nu]\,\xi^\alpha = \nabla_\mu\left(\partial_\nu \xi^\alpha + \Gamma^\alpha_{\nu\beta}\xi^\beta\right) - \nabla_\nu\left(\partial_\mu \xi^\alpha + \Gamma^\alpha_{\mu\beta}\xi^\beta\right) =$$
$$= \partial_\mu\left(\partial_\nu \xi^\alpha + \Gamma^\alpha_{\nu\beta}\xi^\beta\right) + \Gamma^\alpha_{\mu\rho}\left(\partial_\nu \xi^\rho + \Gamma^\rho_{\nu\beta}\xi^\beta\right) - \Gamma^\rho_{\mu\nu}\left(\partial_\rho \xi^\alpha + \Gamma^\alpha_{\rho\beta}\xi^\beta\right) -$$
$$- \partial_\nu\left(\partial_\mu \xi^\alpha + \Gamma^\alpha_{\mu\beta}\xi^\beta\right) - \Gamma^\alpha_{\nu\rho}\left(\partial_\mu \xi^\rho + \Gamma^\rho_{\mu\beta}\xi^\beta\right) + \Gamma^\rho_{\nu\mu}\left(\partial_\rho \xi^\alpha + \Gamma^\alpha_{\rho\beta}\xi^\beta\right) =$$
$$= R^\alpha{}_{\beta\mu\nu}\xi^\beta - T^\rho{}_{\mu\nu}\,\nabla_\rho\,\xi^\alpha$$

where $T^\rho{}_{\mu\nu} = \Gamma^\rho_{\mu\nu} - \Gamma^\rho_{\nu\mu}$ is the torsion tensor (4).

Therefore, we have

$$[\nabla_\mu\,,\ \nabla_\nu]\,\xi^\alpha = R^\alpha{}_{\beta\mu\nu}\xi^\beta - T^\rho{}_{\mu\nu}\,\nabla_\rho\,\xi^\alpha, \tag{8}$$

or analogously for a 1-form:

$$[\nabla_\mu\,,\ \nabla_\nu]\,\zeta_\alpha = -R^\beta{}_{\alpha\mu\nu}\zeta_\beta - T^\rho{}_{\mu\nu}\,\nabla_\rho\,\zeta_\alpha. \tag{9}$$

2.3. Torsion and non-metricity

In Riemannian geometry, the curvature tensor is a precise diagnostic tool to see if the geometry is (locally) trivial. For general metric-affine models it is no longer the case, and we need more information. Of course, one obvious indicator of non-trivial geometry is the torsion tensor (4)

$$T^\alpha_{\ \mu\nu} = \Gamma^\alpha_{\mu\nu} - \Gamma^\alpha_{\nu\mu}.$$

One more tool is easy to discover if we recall that in Riemannian geometry the metric tensor is covariantly constant, $\nabla_\alpha g_{\mu\nu} = 0$. Let us therefore introduce another tensorial quantity, the non-metricity tensor

$$Q_{\mu\alpha\beta} = \nabla_\mu g_{\alpha\beta} \tag{10}$$

which also signals non-trivial metric-affine geometry.

Now we finally have all necessary ingredients in the sense that the postulated equation (10), $\partial_\mu g_{\alpha\beta} = \Gamma^\rho_{\mu\alpha} g_{\rho\beta} + \Gamma^\rho_{\mu\beta} g_{\alpha\rho} + Q_{\mu\alpha\beta}$, together with a prescribed torsion fixes the affine connection uniquely.

Indeed, following the standard derivation of the Levi-Civita connection, we compute the combination $\partial_\mu g_{\nu\beta} + \partial_\nu g_{\mu\beta} - \partial_\beta g_{\mu\nu}$ and use it to determine the components of Γ, this time keeping track of its antisymmetric part which is T and of $Q \neq 0$. The result is

$$\Gamma^\alpha_{\mu\nu} = \frac{1}{2} g^{\alpha\beta}\left(\partial_\mu g_{\nu\beta} + \partial_\nu g_{\mu\beta} - \partial_\beta g_{\mu\nu}\right)$$
$$+ \frac{1}{2}\left(T^\alpha{}_{\mu\nu} + T_\nu{}^\alpha{}_\mu + T_\mu{}^\alpha{}_\nu\right) - \frac{1}{2}\left(Q_{\mu\nu}{}^\alpha + Q_{\nu\mu}{}^\alpha - Q^\alpha{}_{\mu\nu}\right). \tag{11}$$

We see that possible connections differ from each other by contributions of torsion and non-metricity. As it should have been, difference of two connections is always a tensor.

2.3.1. Symmetric connections

In case of symmetric connections we put torsion to zero, and the curvature tensor has further symmetry properties. A simple inspection of the definition (6) readily shows that for symmetric Γ we have

$$R^{\mu}{}_{\alpha\beta\gamma} + R^{\mu}{}_{\beta\gamma\alpha} + R^{\mu}{}_{\gamma\alpha\beta} = 0$$

and also a differential (Bianchi) identity

$$\nabla_{\alpha} R^{\mu}{}_{\nu\beta\gamma} + \nabla_{\beta} R^{\mu}{}_{\nu\gamma\alpha} + \nabla_{\gamma} R^{\mu}{}_{\nu\alpha\beta} = 0.$$

Note that for non-symmetric connections, we would have got torsion terms in the right hand sides.

Any model which deals with different metric fields in the (otherwise Riemannian) geometric part and in the matter coupling prescription can be viewed as a model with non-metricity and without torsion. Indeed, being Levi-Civitian for some metric, the connection is definitely symmetric, however it is not metric compatible from the viewpoint of the physical metric which governs the motion of matter. Concrete examples include scalar-tensor models such as $f(R)$ gravity in the Einstein frame.

2.3.2. Metric compatible connections

If we have a metric, we can raise and lower the indices. Moreover, if the connection is metric compatible $Q = 0$, we can commute the metric with the covariant derivatives. It implies new symmetry properties of the curvature tensor. We can rewrite the commutators of covariant derivatives (8), (9) as

$$[\nabla_{\mu}\,,\ \nabla_{\nu}]\,\xi_{\alpha} = R_{\alpha\beta\mu\nu}\xi^{\beta} - T^{\beta}{}_{\mu\nu}\,\nabla_{\beta}\,\xi_{\alpha},$$

$$[\nabla_{\mu}\,,\ \nabla_{\nu}]\,\zeta_{\alpha} = -R_{\beta\alpha\mu\nu}\zeta^{\beta} - T^{\beta}{}_{\mu\nu}\,\nabla_{\beta}\,\zeta_{\alpha}.$$

and deduce:

$$R_{\alpha\beta\mu\nu} = -R_{\beta\alpha\mu\nu}. \tag{12}$$

Note that the antisymmetry (12) is valid also in presence of torsion. Teleparallel gravity belongs to this class of models.

2.3.3. The Levi-Civita connection

Finally, if we set both torsion and non-metricity to zero, then our expression for affine connection (11) reduces to the Levi-Civita one

$$\Gamma^{\alpha}_{\mu\nu} = \frac{1}{2} g^{\alpha\beta} \left(\partial_{\mu} g_{\nu\beta} + \partial_{\nu} g_{\mu\beta} - \partial_{\beta} g_{\mu\nu}\right).$$

In this case we are back to Riemannian geometry, and one can check that our definition of the geodesic line (5) coincides with that of the path of the shortest distance between two points (provided they are not too far from each other). Also one can prove another symmetry of the curvature tensor

$$R_{\mu\nu\alpha\beta} = R_{\alpha\beta\mu\nu}$$

for the Levi-Civita connection.

As a side note, let us mention that, evidently, the curvature tensor has the maximal amount of symmetry properties when in Riemannian geometry, and therefore it is often much simpler to calculate with the Levi-Civita connection. On the other hand, it is anyway second order (6) in the connection coefficients, and substituting $\Gamma^{\alpha}_{\mu\nu} = \overset{(0)}{\Gamma}{}^{\alpha}_{\mu\nu}(g) + \delta\Gamma^{\alpha}{}_{\mu\nu}$ where $\overset{(0)}{\Gamma}{}^{\alpha}_{\mu\nu}(g)$ is the Levi-Civita connection for a given metric $g_{\mu\nu}$ and $\delta\Gamma$ is always a tensor, we have a relation

$$R^{\alpha}{}_{\beta\mu\nu}(\Gamma) = R^{\alpha}{}_{\beta\mu\nu}(\overset{(0)}{\Gamma}) + \overset{(0)}{\nabla}{}_{\mu}\delta\Gamma^{\alpha}{}_{\nu\beta} - \overset{(0)}{\nabla}{}_{\nu}\delta\Gamma^{\alpha}{}_{\mu\beta} + \delta\Gamma^{\alpha}{}_{\mu\rho} \cdot \delta\Gamma^{\rho}{}_{\nu\beta} - \delta\Gamma^{\alpha}{}_{\nu\rho} \cdot \delta\Gamma^{\rho}{}_{\mu\beta} \qquad (13)$$

which often gives a nicer way of approaching a modified gravity model.

2.4. Tetrad formulation

If only for introducing fermions, one needs the tetrad description of general relativity. And so is for teleparallel gravity. We don't need much more from this than is available in any reasonable exposition of standard general relativity in terms of the frame field.

The main essence is that we have a non-degenerate matrix e^{μ}_{a} composed of components of four vector fields e_a, $a = 0, 1, 2, 3$ which form an orthonormal ($g_{\mu\nu}e^{\mu}_{a}e^{\nu}_{b} = \eta_{ab}$) basis in the tangent space at each point. The inverse matrix is denoted by e^{a}_{μ}, and the metric components can be calculated as

$$g_{\mu\nu} = e^{a}_{\mu}e^{b}_{\nu}\eta_{ab}, \qquad g^{\mu\nu} = e^{\mu}_{a}e^{\nu}_{b}\eta^{ab}.$$

Very importantly, if we are given a metric, the tetrad fields e^{a}_{μ} are only defined up to an arbitrary local Lorentz rotation $e^{a}_{\mu}(x) \to \Lambda^{a}_{b}(x) \cdot e^{b}_{\mu}(x)$ since this is the natural freedom of choosing an orthonormal basis.

Now, that we have a non-degenerate matrix with indices of two kinds, we can define components of tensor fields with Latin indices by

$$\mathcal{T}^{a_1,\ldots,a_n}{}_{b_1,\ldots,b_m} \equiv e^{a_1}_{\alpha_1} \cdots e^{a_n}_{\alpha_n} \mathcal{T}^{\alpha_1,\ldots,\alpha_n}{}_{\beta_1,\ldots,\beta_m} e^{\beta_1}_{b_1} \cdots e^{\beta_m}_{b_m}$$

which amounts to relating tensorial quantities to the chosen frame.

We ought to extend our notion of parallel transport to quantities with the new type of indices. Therefore we introduce some new connection coefficients $\omega^{a}{}_{\mu b}$ which are usually called spin connection. And now the adopted recipe is that we use Γ-terms for Greek indices, and ω-terms for Latin indices. For example, $\nabla_\mu T^{a\nu} = \partial_\mu T^{a\nu} + \Gamma^\nu_{\mu\rho} T^{a\rho} + \omega^a{}_{\mu c} T^{c\nu}$.

Of course, the spin connection must indeed transform as a connection under the local Lorentz transformations in the space of tetrads:

$$e^a_\mu \longrightarrow \Lambda^a_c e^c_\mu\,, \quad \omega^a{}_{\mu b} \longrightarrow \Lambda^a_c \omega^c{}_{\mu d} (\Lambda^{-1})^d_b - (\Lambda^{-1})^a_c \partial_\mu \Lambda^c_b\,, \tag{14}$$

and with respect to its Latin indices it is assumed to belong to the Lie algebra of the Lorentz group.

If we treat objects with different types of indices as different representations of the same invariant entities, then it is natural to demand that the two notions of parallel transport coincide with each other. Formally it means that the tetrad field (which changes the type of an index) must commute with taking a covariant derivative of the field. This is ensured by vanishing of the "full covariant derivative" of the tetrad:

$$\partial_\mu e^a_\nu + \omega^a{}_{\mu b} e^b_\nu - \Gamma^\alpha_{\mu\nu} e^a_\alpha = 0 \tag{15}$$

which is usually imposed when working with tetrads, and which we will also adopt in what follows.

Condition (15) can be solved straightforwardly to obtain

$$\Gamma^\alpha_{\mu\nu} = e^\alpha_a \left(\partial_\mu e^a_\nu + \omega^a{}_{\mu b} e^b_\nu \right) \equiv e^\alpha_a\, \mathfrak{D}_\mu e^a_\nu \tag{16}$$

with $\mathfrak{D}_\mu$ being the Lorentz-covariant (with respect to the Latin index only) derivative, or another way around

$$\omega^a{}_{\mu b} = e^a_\alpha \Gamma^\alpha_{\mu\nu} e^\nu_b - e^\nu_b \partial_\mu e^a_\nu. \tag{17}$$

In particular, one can find the spin connection $\overset{(0)}{\omega}$ which corresponds to the Levi-Civita connection $\overset{(0)}{\Gamma}(g)$ of a given metric g and obtain the tetrad description of the standard general relativity.

Note that as we have already mentioned, under the condition (15) of vanishing full covariant derivative of the tetrad, both $\Gamma^\alpha_{\mu\beta}$ and $\omega^a{}_{\mu b}$ represent one and the same connection in different disguises. And indeed, a simple calculation shows that the two curvatures,

$$R^a{}_{b\mu\nu}(\omega) = \partial_\mu \omega^a{}_{\nu b} - \partial_\nu \omega^a{}_{\mu b} + \omega^a{}_{\mu c}\omega^c{}_{\nu b} - \omega^a{}_{\nu c}\omega^c{}_{\mu b}$$

and

$$R^\alpha{}_{\beta\mu\nu}(\Gamma) = \partial_\mu \Gamma^\alpha_{\nu\beta} - \partial_\nu \Gamma^\alpha_{\mu\beta} + \Gamma^\alpha_{\mu\rho}\Gamma^\rho_{\nu\beta} - \Gamma^\alpha_{\nu\rho}\Gamma^\rho_{\mu\beta}\,,$$

are related by a mere change of the type of indices:

$$R^\alpha{}_{\beta\mu\nu}(\Gamma) = e^\alpha_a R^a{}_{b\mu\nu}(\omega) e^b_\beta\,.$$

3. Teleparallel gravity

Let us now see how we can use different geometries in the tetrad approach. First, we have to note that the non-metricity is automatically equal to zero due to the adopted condition (15):

$$\begin{aligned}\nabla_\alpha g_{\mu\nu} \equiv \eta_{ab}\left(\partial_\alpha\left(e^a_\mu e^b_\nu\right) - \Gamma^\beta_{\alpha\mu} e^a_\beta e^b_\nu - \Gamma^\beta_{\alpha\nu} e^a_\mu e^b_\beta\right)&\\ = -e^b_\mu e^c_\nu\left(\eta_{ab}\omega^a{}_{\alpha c} + \eta_{ac}\omega^a{}_{\alpha b}\right) = 0&\end{aligned}$$

where in the last step we have taken into account that the matrices $\omega^{\cdot}{}_{\alpha\cdot}$ belong to the Lie algebra of the Lorentz group $SO(1,3)$. Therefore we have the antisymmetry (12) property $R_{\alpha\beta\mu\nu} = -R_{\beta\alpha\mu\nu}$ for the curvature tensor.

Given the absence of non-metricity, our connection (11) is of the form

$$\Gamma^\alpha_{\mu\nu} = \overset{(0)}{\Gamma}{}^\alpha_{\mu\nu}(g) + K^\alpha{}_{\mu\nu}$$

where $\overset{(0)}{\Gamma}{}^\alpha_{\mu\nu}(g)$ is the Levi-Civita connection of the metric g, while the tensor

$$K_{\alpha\mu\nu} = \frac{1}{2}\left(T_{\alpha\mu\nu} + T_{\nu\alpha\mu} + T_{\mu\alpha\nu}\right) = \frac{1}{2}\left(T_{\mu\alpha\nu} + T_{\nu\alpha\mu} - T_{\alpha\nu\mu}\right) \tag{18}$$

is known under the name of contortion. It is obviously antisymmetric with respect to the two lateral indices:

$$K_{\alpha\mu\nu} = -K_{\nu\mu\alpha}.$$

In the curvature variation (13) we can simply substitute $\delta\Gamma$ by K to get

$$R^\alpha{}_{\beta\mu\nu}(\Gamma) = R^\alpha{}_{\beta\mu\nu}(\overset{(0)}{\Gamma}) + \overset{(0)}{\nabla}_\mu K^\alpha{}_{\nu\beta} - \overset{(0)}{\nabla}_\nu K^\alpha{}_{\mu\beta} + K^\alpha{}_{\mu\rho}K^\rho{}_{\nu\beta} - K^\alpha{}_{\nu\rho}K^\rho{}_{\mu\beta}.$$

After necessary contractions we obtain the fundamental relation for the scalar curvatures:

$$R(\Gamma) = R(\overset{(0)}{\Gamma}) + 2\overset{(0)}{\nabla}_\mu T^\mu + \mathbb{T} \tag{19}$$

where the torsion vector is

$$T_\mu \equiv T^\alpha{}_{\mu\alpha} = -T^\alpha{}_{\alpha\mu},$$

and the torsion scalar can be written in several equivalent ways:

$$\begin{aligned}\mathbb{T} &= \frac{1}{2}K_{\alpha\beta\mu}T^{\beta\alpha\mu} - T_\mu T^\mu\\ &= \frac{1}{2}T_{\alpha\beta\mu}S^{\alpha\beta\mu}\\ &= \frac{1}{4}T_{\alpha\beta\mu}T^{\alpha\beta\mu} + \frac{1}{2}T_{\alpha\beta\mu}T^{\beta\alpha\mu} - T_\mu T^\mu\end{aligned}$$

with the "superpotential"

$$S^{\alpha\mu\nu} \equiv K^{\mu\alpha\nu} + g^{\alpha\mu}T^{\nu} - g^{\alpha\nu}T^{\mu} \tag{20}$$

which satisfies the same antisymmetry condition $S^{\alpha\mu\nu} = -S^{\alpha\nu\mu}$ as the torsion tensor itself.

The basic idea of teleparallel gravity would be to set the curvature to zero: $R^{\alpha}{}_{\beta\mu\nu}(\Gamma) = 0$ and in particular $R(\Gamma) = 0$.

3.1. Teleparallel equivalent of general relativity

In the classical formulation of teleparallel gravity, one uses the Weitzenböck connection given by

$$\omega^{a}{}_{\mu b} = 0 \tag{21}$$

or equivalently $\Gamma^{\alpha}_{\mu\nu} = e^{\alpha}_{a}\partial_{\mu}e^{a}_{\nu}$ which is obviously curvature-free, $R^{\alpha}{}_{\beta\mu\nu} = 0$, and has the following torsion:

$$T^{\alpha}{}_{\mu\nu} = e^{\alpha}_{a}\left(\partial_{\mu}e^{a}_{\nu} - \partial_{\mu}e^{a}_{\nu}\right).$$

We can denote the determinant of e^{a}_{μ} by $\|e\| = \sqrt{-g}$ and see from the fundamental relation (19) that, under the Weitzenböck assumption (21), the action

$$S = -\int d^4x \|e\| \cdot \mathbb{T} \tag{22}$$

is equivalent to the action $\int d^4x\sqrt{-g} \cdot R(\overset{(0)}{\Gamma})$ of general relativity modulo a surface term with $2\overset{(0)}{\nabla}_{\mu}T^{\mu}$.

Note that we discuss pure gravity ignoring the matter fields. There is no problem with minimally coupling bosonic fields since they simply have to interact with the metric $g_{\mu\nu} = e^{a}_{\mu}e^{b}_{\nu}\eta_{ab}$ in the usual way. Fermionic fields would require the spin connection $\overset{(0)}{\omega}$ which is admittedly not so natural in this framework, but it won't be our concern in this paper.

Let us also mention straitaway that the choice of $\omega = 0$ is locally Lorentz breaking as is evident, if nothing else, from the transformation law (14). This choice allowed us to formulate the model without any explicit use of the spin connection. Sometimes the action (22) with $\omega = 0$ is referred to as a pure tetrad gravity, as opposed to genuine teleparallel which would have used an arbitrary flat spin connection. Note though that this distinction is not common in the literature and very often it may cause a confusion as to what was precisely assumed in a given paper.

For the teleparallel equivalent of general relativity this issue is not particularly important since the local Lorentz violation is totally contained in the surface term without affecting the equations of motion. However, it will play a role for modified models. We will come to this point soon.

3.1.1. Equations of motion

Equations of motion are easily derived for the action (22). By using the standard method of (first order) varying the inverse matrices $\delta e_a^\mu = -e_b^\mu e_a^\nu \delta e_\nu^b$ and determinants $\delta\|e\| = \|e\| \cdot e_a^\mu \delta e_\mu^a$ we easily get

$$\delta S = -\int d^4x \|e\| \cdot \left(-2S^{\alpha\mu\nu}T_{\alpha\beta\nu}e_a^\beta \delta e_\mu^a + \mathbb{T}e_a^\mu \delta e_\mu^a - 2S_\beta{}^{\mu\alpha}e_a^\beta \mathfrak{D}_\alpha \delta e_\mu^a\right)$$

where we use the Lorentz-covariant derivative to account for a non-trivial spin connection if there was one. In the current case $\mathfrak{D}$ is just a partial derivative since $\omega = 0$.

We need to perform integration by parts in the last term which gives $2\delta e_\mu^a \cdot \left(\partial_\alpha \left(\|e\| \cdot S_\beta{}^{\mu\alpha}e_a^\beta\right) - \|e\| \cdot \omega^b{}_{\alpha a} S_\beta{}^{\mu\alpha} e_b^\beta\right)$. In teleparallel gravity it is often used as it is, in this very non-covariant form. However one can easily do better. Indeed, it is very tempting to make the derivative in the first term into a Levi-Civita-covariant one, $\overset{(0)}{\nabla}_\nu S_a{}^{\mu\nu} = \frac{1}{\|e\|}\partial_\nu \left(\|e\| S_a^{\mu\nu}\right) - \overset{(0)}{\omega}{}^b{}_{\nu a} S_b{}^{\mu\nu}$. Correcting for the mismatch of connections $\omega^b{}_{\nu a} - \overset{(0)}{\omega}{}^b{}_{\nu a} = K^b{}_{\nu a}$, one obtains a nice contribution to the equation of motion: $\overset{(0)}{\nabla}_\nu S_a{}^{\mu\nu} - K^c{}_{\nu a} S_c{}^{\mu\nu}$.

Finally, using the non-degeneracy of tetrads, we get the equation of motion in the form

$$\overset{(0)}{\nabla}_\alpha S_\beta{}^{\mu\alpha} - S^{\alpha\mu\nu}\left(T_{\alpha\beta\nu} + K_{\alpha\nu\beta}\right) + \frac{1}{2}\mathbb{T}\delta_\beta^\mu = 0. \tag{23}$$

Is it equivalent to general relativity? Yes! Directly substituting

$$R^\alpha{}_{\beta\mu\nu}(\overset{(0)}{\Gamma}) = -\left(\overset{(0)}{\nabla}_\mu K^\alpha{}_{\nu\beta} - \overset{(0)}{\nabla}_\nu K^\alpha{}_{\mu\beta} + K^\alpha{}_{\mu\rho}K^\rho{}_{\nu\beta} - K^\alpha{}_{\nu\rho}K^\rho{}_{\mu\beta}\right)$$

into the Einstein equation $G_\beta^\mu(\overset{(0)}{\Gamma}, g(e)) = 0$, one can prove that it is.

3.2. Possible extensions

Having found the action (22), one is naturally driven to consider possible modifications of gravity in the teleparallel framework. One of the first ideas in this direction was the so-called new GR of Ref. [9] which introduces the Lagrangian density $c_1 T_{\alpha\beta\mu}T^{\alpha\beta\mu} + c_2 T_{\alpha\beta\mu}T^{\beta\alpha\mu} + c_3 T_\mu T^\mu$ with modified coefficients c_i. On can also add parity violating terms and explore non-linear functions of torsion scalars. And once non-linear functions are introduced, one can also put the divergence $2\overset{(0)}{\nabla}_\mu T^\mu$ inside as an argument [10].

Rather obviously, if the Weitzenböck connection $\omega = 0$ is assumed, generically all these modified gravities would violate local Lorentz invariance in the tetrad space, even at the level of equations of motion. This fact used to be a cause of much confusion about probably the most simple modification, the $f(\mathbb{T})$ gravity [11]. Its Lagrangian density is a non-linear function of the torsion scalar from relation (19) which is used in the teleparallel equivalent of general relativity. As such it is in a way very similar to $f(R)$ gravity which has one extra degree of freedom due to higher derivatives of the metric in the curvature scalar. The false expectation [12] was that $f(\mathbb{T})$ should not increase the number of degrees of freedom since $\mathbb{T}$ contains only first derivatives of the tetrad. Of course the catch is that the local Lorentz invariance is broken and this simple count of degrees of freedom is not justified. It has been realised a bit later [13, 14].

Another related issue is that Lorentz violating models do care about a particular choice of the tetrad for a given metric. In particular, an ansatz for a spherically symmetric solution in spherical coordinates would normally fail for the most natural (diagonal) choice of the tetrad unless a non-trivial flat spin connection is inserted to account for a transition from a Cartesian to a spherical frame.

It even led to introducing the notion of good and bad tetrads [15]. The former represent such ansätze which do go through smoothly, while the latter do fail typically requiring $\frac{d^2 f(\mathbb{T})}{d\mathbb{T}^2} = 0$ which brings us back to the teleparallel equivalent of general relativity with local Lorentz invariance being restored in equations of motion. This is probably a very unique instance in science when failing with a bad ansatz gave rise to a whole new concept.

4. Review of covariantisation

As we have already discussed, teleparallel gravity can be naturally covariantised if a spin connection with the proper transformation law (14) is introduced as an additional physical field. Note that it is very important that the spin connection be flat.

First, otherwise it is not a teleparallel model any longer.

Second, with an arbitrary spin connection the (exact) variation of the torsion tensor $\delta_\omega T^\alpha{}_{\mu\nu} = \delta\omega^\alpha{}_{\mu\nu} - \delta\omega^\alpha{}_{\nu\mu}$ can be used to show that even the simplest action (22) yields a trivial model instead of general relativity. Indeed, the equation of motion is $T^\mu{}_{\alpha\nu} + T_\nu \delta^\mu_\alpha - T_\alpha \delta^\mu_\nu = 0$ which (in spacetime dimension $d \neq 2$) entails $T_\mu = 0$ upon tracing, with the totally trivial final result of $T^\mu{}_{\alpha\nu} = 0$.

Sometimes the spin connection is treated as not a dynamical field, so that one does not have to make a variation with respect to it. In this case the action with any given choice of the flat spin connection is not invariant, however there is the freedom of making this choice which renders the whole thing invariant again. We don't subscribe to this viewpoint. It seems kind of awkward to have a spacetime-dependent quantity in the action of a fundamental theory which is not subject to variation. Moreover, we feel that such covariantisation is a very superficial one.

This can be compared to making a model with preferred direction isotropic (i.e. SO(3)-invariant) by explicitly introducing a vector field in the preferred direction with the correct transformation law for the components under rotations. It is only our description which gets covariantised in this manner unless the vector, or the spin connection, is subject to an independent dynamics. A concrete example would be about massive bodies thrown near the surface of Earth. Does it really put all three directions on equal footing if we covariantly introduce the vector of free fall acceleration with arbitrary components instead of pointing it along the usual z-direction? The real physics is objectively anisotropic at this level, and one would have to underpin the emergence of the preferred direction in a more fundamental theory.

4.1. Teleparallel equivalent of general relativity

Clearly, the proper procedure would be to vary the spin connection constraining it to be flat (curvature-free) or, as some like to call it, purely inertial. Applying a gauge transformation (14) to the zero Weitzenböck connection (21), we see that the spin connection we need is of the following form:

$$\omega^{a}{}_{\mu b} = -(\Lambda^{-1})^{a}_{c}\partial_{\mu}\Lambda^{c}_{b} \tag{24}$$

where $\Lambda(x) \in SO(1,3)$ is an arbitrary Lorentz matrix. Literally it means that there exists a frame in which $\omega = 0$ (Weitzenböck), but a local Lorentz rotation has been done to get away from this frame.

Guided by this idea, we replace the teleparallel (pure tetrad) action (22) with a slightly more refined version:

$$S = -\int d^4x \|e\| \cdot \mathbb{T}(e, \omega(\Lambda)) \tag{25}$$

in which e and Λ should be thought of as independent variables, and the spin connection is given by equation (24). Of course, keeping Λ fixed but arbitrary would precisely correspond to the "superficial covariantisation" mentioned above.

Varying with respect to Λ in the action (25) does not have the disastrous consequences of varying with respect to an arbitrary spin connection. Actually, it does not have any consequences at all. Indeed, from the fundamental relation (19) we see that the combination $2\overset{(0)}{\nabla}_{\mu}T^{\mu} + \mathbb{T}$ does not depend on Λ at all since $R(\omega(\Lambda)) = 0$, and therefore the dependence of the action (25) on the spin connection is only a surface term effect [16, 17, 18].

In other words, we have $\delta_{\Lambda}\mathbb{T} = \delta_{\Lambda}R(\omega) - 2\overset{(0)}{\nabla}_{\mu}(\delta_{\Lambda}T^{\mu})$ with the usual chain rule $\delta_{\Lambda}(...) = \delta_{\omega}(...) \cdot \delta_{\Lambda}\omega$, and since $R(\omega(\Lambda)) \equiv 0$ the variation is a surface term and does not produce any new equation of motion. For those who are not satisfied with these simple (and exhaustive) arguments, explicit calculations are given in the Ref. [18].

4.1.1. A few remarks

Note that one could also impose the flatness condition $R^a{}_{b\mu\nu}(\omega) = 0$ with a Lagrange multiplier instead of our representation (24) having the Lagrangian density of the form $\mathbb{T}(e,\omega) + \lambda_a{}^{b\mu\nu} R^a{}_{b\mu\nu}(\omega)$ where $\lambda_a{}^{b\mu\nu}$ is a Lagrange multiplier with postulated symmetry properties $\lambda^{ab\mu\nu} = -\lambda^{ab\nu\mu}$ and $\lambda^{ab\mu\nu} = -\lambda^{ba\mu\nu}$.

This is a possible approach [19, 20] in its own right. Moreover, it can be used in the metric language instead of tetrads, and can also be applied to models of gravity in terms of non-metricity with neither curvature nor torsion [21]. We think however that for (modified) teleparallel models it might be more instructive to keep the fields Λ as the new legitimate players in the game.

Another remark concerns some fundamental considerations about choosing the spin connection. Namely, one might be interested in imposing proper asymptotic conditions for rendering the action finite [16, 22] which could be an important point for quantisation of teleparallel gravity. Moreover, sometimes it is claimed that actually there is a spin connection which is naturally associated to a given tetrad [23].

Such claims refer to a much more elaborated framework, that is to gauging the translations group [9, 23]. In this approach, due to whatever a reason, one assumes that there exists a global translational symmetry $x^a \to x^a + \epsilon^a$ which can be gauged. Then any tetrad can be separated into two parts, one of which is obtained by a local Lorentz transformation of the "trivial" tetrad $\partial_\mu x^a$ thereby prescribing the preferred (flat) spin connection, and another being a non-trivial piece which serves as a gauge potential for the torsion field [23]. This is certainly a more conventional gauge theory compared to our covariantised teleparallel gravity which, having gauged the Lorentz group, keeps it in purely unphysical sector with vanishing field strength.

It is very unfortunate that there is no accepted terminological distinction to set pure torsion geometry and gauged translations apart. Once we understand that the latter approach starts from the global translations group, it perhaps should no longer come as an unexplainable miracle that people are able to define covariant conservation laws and separate gravity from inertia in this framework [23], something which is not to get on well with the basic ideas of general relativity.

Probably, the best hopes associated with these constructions hinge upon the idea that the gauge theory approach, which certainly proved so well in doing with other fundamental interactions, might also guide us towards better understanding of gravity [21, 23], not without a reference to problems of quantisation. However, personally we do not feel that it is time to abandon the unique geometrical flavour of general relativity which makes it so exciting. After all, it's awesome to compute amplitudes of creation from nothing; and all the singularities, Cauchy horizons, causality violations, and time machines might finally be inherent parts of a big beautiful picture.

4.2. Covariantisation of modified models

Covariantisation procedures cannot work the same way in modified teleparallel gravities with generic dependence on the tetrad e^a_μ and the torsion $T^a{}_{\mu\nu}(e, \omega(\Lambda))$. Since the dependence on the spin connection in generalised models cannot be reduced to a surface term, the variation with respect to the (flat) spin connection does produce non-trivial equation of motion, though a very benign one.

One can easily prove that in any covariant teleparallel model this equation is redundant. More precisely, it coincides with the antisymmetric part of the equation of motion for the tetrad field [18]. Indeed, by the very definition, the action of a covariant model is identically invariant under combined transformations (14) which for the flat spin connection are obviously in a one-to-one correspondence with variations of the Lorentz matrix Λ in its representation (24). Therefore, any variation of the action with respect to the flat spin connection (24) can be precisely compensated by an appropriate Lorentz rotation of the tetrad. However, the latter is a legitimate variation of the tetrad field which by itself must keep the action stationary, as is independently ensured by the antisymmetric part of the tetrad equation of motion.

In our opinion, this simple argument constitutes a fairly rigorous proof. However, explicit calculations are given in Ref. [18] concerning mostly the $f(\mathbb{T})$ gravity, and also in Refs. [24, 25] for more general models.

4.2.1. Further on $f(\mathbb{T})$ gravity

For the simplest example, in covariant $f(\mathbb{T})$ gravity

$$S = -\int d^4x \|e\| \cdot f\left(\mathbb{T}(e, \omega(\Lambda)\right) \tag{26}$$

one can easily perform all variations and obtain the tetrad equation of motion in the following form [18]:

$$f'(\mathbb{T}) \cdot \overset{(0)}{R}{}^{\mu\nu} + K^{\mu\nu\alpha}\partial_\alpha f'(\mathbb{T}) - T^\mu \partial^\nu f'(\mathbb{T}) + \frac{1}{2} f(\mathbb{T}) \cdot g^{\mu\nu} = 0 \tag{27}$$

which, due to the local Lorentz breaking in its pure tetrad ($\omega = 0$) restriction, has a non-trivial antisymmetric part

$$T^{\alpha\mu\nu}\partial_\alpha f'(\mathbb{T}) + T^\nu \partial^\mu f'(\mathbb{T}) - T^\mu \partial^\nu f'(\mathbb{T}) = 0 \tag{28}$$

which in turn coincides [18] with equation of motion coming from variation with respect to Λ.

Note that it is this antisymmetric part (28) which normally turns a "bad tetrad" down. It contains only first derivatives of ω for which reason the flat spin connection of covariant teleparallel gravities is sometimes considered as a non-dynamical variable [25]. We must disagree on this point. For example, Hamiltonian equations are always first order in time derivatives which does not entail that Hamiltonian systems are not truely dynamical.

This analogy might even go deeper than it seems to, with components of the spin connection being canonically conjugate to each other which would be nicely in line with the $3 = \frac{6}{2}$ new degrees of freedom compared to general relativity. It is nothing but a mere speculation for now. However, it is evident that we desperately need better understanding of dynamics even in such a simple model as $f(\mathbb{T})$. It is not inconceivable that it can be achieved with an accurate treatment of $\Lambda^a_b(x)$ as new fields in the model.

It would be very good to have a new count of the number of degrees of freedom, more palatable than the truely heroic brute force approach of the Ref. [26] and simultaneously more detailed and reliable than the nice general arguments of the Ref. [27]. The role of the remnant symmetry (those local Lorentz transformations which remain unbroken [28]) should also be clarified. Last but not least on this very concise list is the conundrum of cosmological perturbations [29]. What is the fundamental reason for the extra modes to be absent from linear cosmological perturbations around any spatially flat Friedman universe? It is relatively simple to spot an accidental restoration of the local Lorentz invariance in the quadratic (pure tetrad) action for perturbations around Minkowski space, however the general cosmological hide-and-seek looks much more mysterious.

5. Conclusions

Teleparallel gravity provides us with an unprecedented opportunity for constructing new and interesting modified gravity models. And indeed, it is very actively used for cosmological model building [30]. However, as we have seen, the very foundations of the modified teleparallel gravity models do urgently call for better consideration. We hope to see major new progress in this direction quite soon.

The Author is greatly indepted to Tomi Koivisto, José Pereira, Martin Krššák, Yen Chin Ong, and many other people for interesting and fruitful discussions on various topics of teleparallel gravity. Needless to say, this acknowledgement cannot be used to entail that they would agree with any of those views on the subject which have been expressed in this paper.

References

[1] Planck Collaboration, *Astronomy and Astrophys.* **594** (2016) A13.

[2] M. Milgrom, *Astrophys. J.* **270** (1983) 371.

[3] A. Golovnev and N. Masalaeva, *Gen. Relativ. Gravit.* **46** (2014) 1754.

[4] C. de Rham, *Living Rev. Relativity* **17** (2014) 7.

[5] A. Golovnev and F. Smirnov, *Phys. Lett. B* **770** (2017) 209.
[6] A. Golovnev and F. Smirnov, *J. High Ener. Phys.* (2017) JHEP06(2017)130.
[7] Ph. Brax, *Class. Quant. Grav.* **30** (2013) 214005.
[8] Th. P. Sotiriou and V. Faraoni, *Rev. Mod. Phys.* **82** (2010) 451.
[9] K. Hayashi and T. Shirafuji, *Phys. Rev. D* **19** (1979) 3524.
[10] S. Bahamonde, Ch. G. Böhmer and M. Krššák, *Phys. Lett. B* **775** (2017) 37.
[11] G. R. Bengochea and R. Ferraro, *Phys. Rev. D* **79** (2009) 124019 .
[12] E. V. Linder, *Phys. Rev. D* **81** (2010) 127301.
[13] B. Li, Th. P. Sotiriou and J. D. Barrow, *Phys. Rev. D* **83** (2011) 064035.
[14] Th. P. Sotiriou, B. Li and J. D. Barrow, *Phys. Rev. D* **83** (2011) 104030.
[15] N. Tamanini and Ch. G. Böhmer, *Phys. Rev. D* **86** (2012) 044009.
[16] M. Krššák, *Eur. Phys. J. C* **77** (2017) 44.
[17] M. Krššák and E. N. Saridakis, *Class. Quant. Grav.* **33** (2016) 115009.
[18] A. Golovnev, T. Koivisto and M. Sandstad, *Class. Quant. Grav.* **34** (2017) 145013.
[19] M. Blagojević, *Gravitation and gauge symmetries*, IoP Publishing, 2002.
[20] J. M. Nester and Y. Ch. Ong, *arXiv:* 1709.00068 (2017).
[21] J. Beltrán Jiménez, L. Heisenberg and T. Koivisto, *arXiv:* 1710.03116 (2017).
[22] M. Krššák, *arXiv:* 1705.01072 (2017).
[23] R. Aldrovandi and J. G. Pereira, *Teleparallel Gravity*, Springer, 2013.
[24] M. Hohmann, L. Järv, M. Krššák and Ch. Pfeifer, *arXiv:* 1711.09930 (2017).
[25] M. Hohmann, L. Järv and U. Ualikhanova, *arXiv:* 1801.05786 (2018).
[26] M. Li, R.-X. Miao and Y.-G. Miao, *J. High Ener. Phys.* (2011) JHEP07(2011)108.
[27] P. Chen, K. Izumi, J. M. Nester and Y. Ch. Ong, *Phys. Rev. D* **91** (2015) 064003.
[28] R. Ferraro and F. Fiorini, *Phys. Rev. D* **91** (2015) 064019.
[29] K. Izumi and Y. Ch. Ong, *J. Cosm. Astropart. Phys.* (2013) JCAP06(2013)029.
[30] Y.-F. Cai, S. Capozziello, M. De Laurentis and E. N. Saridakis, *Rept. Prog. Phys.* **79** (2016) 106901.

Quintessence, unified dark energy and dark matter, and confinement/deconfinement mechanism

Eduardo Guendelman*
Physics Department, Ben Gurion University of the Negev
Beer Sheva, Israel

Emil Nissimov and Svetlana Pacheva†
Institute for Nuclear Research and Nuclear Energy
Bulgarian Academy of Sciences, Sofia, Bulgaria

ABSTRACT

We describe a new type of generalized gravity-matter models where $f(R) = R + R^2$ gravity couples in a non-conventional way to a scalar "inflaton" field, to a second scalar "darkon" field responsible for dark energy/dark matter unification, as well as to a non-standard nonlinear gauge field system containing a square-root of the ordinary Maxwell Lagrangian, which is responsible for a charge confining/deconfinfing mechanism. The essential non-conventional feature of our models is employing the formalism of non-Riemannian volume forms, *i.e.* metric-independent non-Riemannian volume elements on the spacetime manifold, defined in terms of auxiliary antisymmetric tensor gauge fields. Although being (almost) pure-gauge degrees of freedom, the non-Riemannian volume-forms trigger a series of important features unavailable in ordinary gravity-matter models. Upon passing to the physical Einstein frame we obtain an effective matter-gauge-field Lagrangian of quadratic "k-essence" type both w.r.t. the "inflaton" and the "darkon", with the following properties: (i) Remarkable effective "inflaton" potential possessing two infinitely large flat regions with vastly different heights ("vacuum" energy densities) describing the "early" and "late" Universe; (ii) Nontrivial effective gauge coupling constants running with the "inflaton", in particular, effective "inflaton"-running coupling constant of the square-root Maxwell term, which determines the strength of the charge confienement; (iii) The confinement-strength gauge coupling constant is non-zero in the "late" Universe, *i.e.*, charge confinement is operating, whereas it vanishes in the "early" Universe, *i.e.*, confinement-free epoch; (iv) The unification of dark energy and dark matter is explicitly seen within the FLRW reduction, where they appear as dynamically generated effective vacuum energy density and dynamically induced dust-like matter, correspondingly.

* e-mail address: guendel@bgu.ac.il
† e-mail address: nissimov@inrne.bas.bg, svetlana@inrne.bas.bg

1. Introduction

Our main task, which pertains to the interface of particle physics and cosmology [1], is to present a non-canonical model of extended ($f(R) = R+R^2$) [2] gravity interacting in a non-standard way with two scalar fields and a strongly nonlinear gauge field, which is capable to provide a systematic description of:

(a) Unified treatment of dark energy and dark matter (for a background, see [3, 4]) revealing them as manifestations of a single material entity – the first non-canonical scalar "darkon" field. A number of proposals already exist for an adequate description of dark energy's and dark matter's dynamics within the framework of standard general relativity or its modern extensions, among them: "Chaplygin gas" models [5], "purely kinetic k-essence" models [6], "mimetic" dark matter models [7]. In Section 2 we briefly review our own approach [8].

(b) Quintessential scenario driven by the remarkable dynamically generated effective potential of the second "inflaton" scalar field with a consistent explanation of the vast difference between the energy scales of the "early" and the "late" Universe;

(c) Charge confinement/deconfinement mechanism triggered via special interplay between the "inflaton" and the nonlinear gauge field dynamics explaining absence of charge confinement in the "early" Universe and exhibiting confinement in the "late" Universe.

The principal ingredient of our approach is the method of non-Riemannian volume-forms (metric-independent volume elements) on the pertinent spacetime manifold (see refs.[9, 10, 11] for a consistent geometrical formulation, which is an extension of the originally proposed method [12]). Non-Riemannian volume-forms are constructed in terms of auxiliary maximal-rank antisymmetric tensor gauge fields, which are shown to be essentially pure-gauge degrees of freedom, *i.e.*, they do not introduce additional propagating field-theoretic degrees of freedom. Yet, they leave a trace in a form of several dynamically induced integration constants, which trigger a series of important features unavailable in ordinary gravity-matter models.

Another important ingredient of our approach is the inclusion of interaction of the extended gravity with a nonlinear gauge field system containing alongside the ordinary Maxwell Lagrangian also a square-root of the latter. In flat spacetime such gauge field system is known [13, 14] to yield a simple implementation of 't Hooft's idea [15] about confinement being produced due to the presence in the energy density of electrostatic field configurations of a term *linear* w.r.t. electric displacement field in the infrared region (presumably as an appropriate infrared counterterm). It has been shown in [13] (for flat spacetime) and in appendix B of [11] (in curved spacetime) that the strength of confinement in this model is measured by the corresponding coupling constant of the square-root Maxwell term. Let us also note that one could start as well with the non-Abelian version of the above nonlinear gauge theory containing square-root of the Yang-Mills

Lagrangian. For static spherically symmetric solutions, the non-Abelian theory effectively reduces to an Abelian one as pointed out in [13].

In Section 3 we obtain upon passing to the physical Einstein frame an effective matter-gauge-field Lagrangian of quadratic "k-essence" type [16] w.r.t. both the "inflaton" and the "darkon" scalar fields with several remarkable properties:

(i) It contains an effective "inflaton" potential possessing two infinitely large flat regions with vastly different heights – dynamically generated "vacuum" energy densities thanks to the appearance of the above mentioned integration constants – which describe the "early" and "late" Universe, accordingly;

(ii) The Einstein-frame Lagrangian contains nontrivial effective gauge coupling constants running with the "inflaton". Particularly important is the effective "inflaton"-running coupling constant of the square-root Maxwell term, which determines the "inflaton"-dependent strength of the charge confienement;

(iii) We show the confinement-strength gauge coupling constant to be non-zero in the "late" Universe, *i.e.*, charge confinement is operating there. On the other hand, the confinement-strength gauge coupling constant vanishes in the "early" Universe, *i.e.*, the latter being confinement-free epoch;

In Section 4 we consider a cosmological FLRW reduction of the Einstein-frame effective matter-gauge-field Lagrangian where the unification of dark energy and dark matter is explicitly seen upon identifying them as dynamically generated effective vacuum energy density and dynamically induced dust-like matter, correspondingly.

2. A Simple Model of Dark Energy and Dark Matter Unification

We start with a simple particular case of a non-conventional gravity-scalar-field action – a member of the general class of the non-Riemannian-volume-element-based gravity-matter theories [17, 11] (we use for simplicity units where the Newton constant $G_N = 1/16\pi$):

$$S = \int d^4x\sqrt{-g}\,R + \int d^4x(\sqrt{-g} + \Phi(C))L(u,Y)\,. \tag{1}$$

R denotes the standard Riemannian scalar curvature for the spacetime Riemannian metric $g_{\mu\nu}$. $L(u,Y)$ is general-coordinate invariant Lagrangian of a single scalar field $u(x)$, the simplest example being:

$$L(u,Y) = Y - V(u) \quad , \quad Y \equiv -\frac{1}{2}g^{\mu\nu}\partial_\mu u \partial_\nu u\,, \tag{2}$$

It is coupled symmetrically to two mutually independent spacetime volume-elements (integration measure densities) – the standard Riemannian $\sqrt{-g}$

and to an alternative non-Riemannian (metric-independent) one:

$$\Phi(C) = \frac{1}{3!}\varepsilon^{\mu\nu\kappa\lambda}\partial_\mu C_{\nu\kappa\lambda} \, . \tag{3}$$

As a result of the equations of motion w.r.t. $C_{\mu\nu\lambda}$ we obtain a crucial new property – *dynamical constraint* on $L(u,Y)$:

$$\partial_\mu L(u,Y) = 0 \longrightarrow L(u,Y) = -2M_0 = \text{const} \; , \; \text{i.e.} \; Y = V(u) - 2M_0 \, . \tag{4}$$

The integration constant M_0 will play below (Eq.(5)) the role of dynamically generated cosmological constant.

The pertinent energy-momentum tensor $T_{\mu\nu}$ can be cast into a relativistic hydrodynamical form (taking into account (4)):

$$T_{\mu\nu} = -2M_0 g_{\mu\nu} + \rho_0 u_\mu u_\nu \; , \tag{5}$$

$$\rho_0 \equiv \Bigl(1 + \frac{\Phi(C)}{\sqrt{-g}}\Bigr) 2Y \quad , \quad u_\mu \equiv -\frac{\partial_\mu u}{\sqrt{2Y}} \quad , \quad u^\mu u_\mu = -1 \; , \tag{6}$$

with the corresponding pressure p and energy density ρ:

$$p = -2M_0 = \text{const} \quad , \quad \rho = \rho_0 - p = 2M_0 + \Bigl(1 + \frac{\Phi(C)}{\sqrt{-g}}\Bigr) 2Y \, . \tag{7}$$

Because of the constant pressure ($p = -2M_0$) the covariant energy-momentum conservation $\nabla^\nu T_{\mu\nu} = 0$ implies *both* conservation of a hidden Noether symmetry current $J^\mu = \rho_0 u^\mu$, as well as *geodesic fluid motion*:

$$\nabla_\mu J^\mu \equiv \nabla_\mu (\rho_0 u^\mu) = 0 \quad , \quad u_\nu \nabla^\nu u_\mu = 0 \, . \tag{8}$$

The *hidden strongly nonlinear Noether symmetry* giving rise to the current $J^\mu = \rho_0 u^\mu$ results from the invariance (up to a total derivative) of the scalar field action in (1), exclusively due to the presence of the non-Riemannian volume element $\Phi(C)$, under the following nonlinear symmetry transformations:

$$\delta_\epsilon u = \epsilon\sqrt{Y} \quad , \quad \delta_\epsilon g_{\mu\nu} = 0 \quad , \quad \delta_\epsilon \mathcal{C}^\mu = -\epsilon \frac{1}{2\sqrt{Y}} g^{\mu\nu}\partial_\nu u (\Phi(C) + \sqrt{-g}) \; , \tag{9}$$

where $\mathcal{C}^\mu \equiv \frac{1}{3!}\varepsilon^{\mu\nu\kappa\lambda} C_{\nu\kappa\lambda}$.

Thus, $T_{\mu\nu}$ (5)-(7) represents an exact sum of two contributions of the two "dark" material species with corresponding pressures and energy densities:

$$p = p_{\text{DE}} + p_{\text{DM}} \quad , \quad \rho = \rho_{\text{DE}} + \rho_{\text{DM}} \tag{10}$$

$$p_{\text{DE}} = -2M_0 \; , \; \rho_{\text{DE}} = 2M_0 \quad ; \quad p_{\text{DM}} = 0 \; , \; \rho_{\text{DM}} = \rho_0 \; , \tag{11}$$

i.e., according to (8) the dark matter component is a dust fluid flowing along geodesics. This is explicit unification of dark energy and dark matter originating from the non-canonical dynamics of a single scalar field – the "darkon" $u(x)$ coupled symmetrically to a standard Riemannian and another non-Riemanian (metric-independent) volume element. Let us also note that the physical result $T_{\mu\nu}$ (5) does *not* depend on the explicit form of the "darkon" potential $V(u)$.

Upon reduction to the cosmological class of Friedmann-Lemaitre-Robertson-Walker (FLRW) metrics:

$$ds^2 = -dt^2 + a^2(t)\Big[\frac{dr^2}{1-Kr^2} + r^2(d\theta^2 + \sin^2\theta d\phi^2)\Big] , \qquad (12)$$

the J^μ-current conservation (8) now reads:

$$\nabla^\mu(\rho_0 u_\mu) = 0 \;\to\; \frac{d}{dt}(a^3\rho_0) = 0 \;\to\; \rho_0 = \frac{c_0}{a^3} \;,\; c_0 = \mathrm{const}, \qquad (13)$$

$$p = -2M_0 \;,\; \rho = 2M_0 + \frac{c_0}{a^3} \;, \qquad (14)$$

with the dark matter contribution ρ_0 being the typical cosmological dust solution (*e.g.* [18]).

3. Noncanonical Gravity-Matter System Coupled to Charge-Confining Nonlinear Gauge Field

Let us now extend the simple gravity-"darkon" model (1) to $f(R) = R + R^2$ gravity coupled non-canonically to both "inflaton" $\varphi(x)$ and "darkon" $u(x)$ scalar fields within the non-Riemannian volume-form formalism, as well as we will also add coupling to a non-standard non-linear gauge field subsystem:

$$S = \int d^4x\, \Phi(A)\Big[g^{\mu\nu}R_{\mu\nu}(\Gamma) + X - V_1(\varphi) - \frac{1}{2}f_0\sqrt{-F^2}\Big] + \int d^4x\, \Phi(B)\Big[\epsilon R^2 - \frac{1}{4e^2}F^2 + \frac{\Phi(H)}{\sqrt{-g}}\Big] + \int d^4x(\sqrt{-g} + \Phi(C))L(u,Y) \,. \qquad (15)$$

Here the following notations are used:

- $\Phi(A)$ and $\Phi(B)$ are two new metric-independent non-Riemannian volume-elements:

$$\Phi(A) = \frac{1}{3!}\varepsilon^{\mu\nu\kappa\lambda}\partial_\mu A_{\nu\kappa\lambda} \;,\; \Phi(B) = \frac{1}{3!}\varepsilon^{\mu\nu\kappa\lambda}\partial_\mu B_{\nu\kappa\lambda} \;, \qquad (16)$$

  defined in terms of field-strengths of two auxiliary 3-index antisymmetric tensor gauge fields, apart from $\Phi(C)$;

- $\Phi(H) = \frac{1}{3!}\varepsilon^{\mu\nu\kappa\lambda}\partial_\mu H_{\nu\kappa\lambda}$ is the dual field-strength of an additional auxiliary tensor gauge field $H_{\nu\kappa\lambda}$ whose presence is crucial for the consistency of (15).
- Let us specifically emphasize the importance to use here the Palatini formalism: $R = g^{\mu\nu}R_{\mu\nu}(\Gamma)$; $g_{\mu\nu}$, $\Gamma^\lambda_{\mu\nu}$ – metric and affine connection are *apriori* independent.
- The "inflaton" part reads:

$$X \equiv -\frac{1}{2}g^{\mu\nu}\partial_\mu\varphi\partial_\nu\varphi \quad , \quad V_1(\varphi) = f_1\exp\{-\alpha\varphi\} \,. \tag{17}$$

- $F_{\mu\nu}$ is the field-strength of an (Abelian) gauge field $\mathcal{A}_\mu$:

$$F_{\mu\nu} = \partial_\mu\mathcal{A}_\nu - \partial_\nu\mathcal{A}_\mu \quad , \quad F^2 = F_{\mu\nu}F_{\kappa\lambda}g^{\mu\kappa}g^{\nu\lambda} \tag{18}$$

- The coupling parameters f_1, f_0 and α are dimensionful positive constants (mass^4, mass^2, mass^{-1}).

The specific form of the action (15) apart from the "darkon" field part (1) has been fixed by the requirement for invariance under global Weyl-scale transformations:

$$g_{\mu\nu} \to \lambda g_{\mu\nu} \,,\; \Gamma^\mu_{\nu\lambda} \to \Gamma^\mu_{\nu\lambda} \,,\; \varphi \to \varphi + \frac{1}{\alpha}\ln\lambda \,,\; \mathcal{A}_\mu \to \mathcal{A}_\mu \,,$$
$$A_{\mu\nu\kappa} \to \lambda A_{\mu\nu\kappa} \quad , \quad B_{\mu\nu\kappa} \to \lambda^2 B_{\mu\nu\kappa} \quad , \quad H_{\mu\nu\kappa} \to H_{\mu\nu\kappa} \,. \tag{19}$$

As shown in [11], the solution to the equations of motion w.r.t. independent affine connection $\Gamma^\mu_{\nu\lambda}$ yield the latter as a Levi-Civitta connection:

$$\Gamma^\mu_{\nu\lambda} = \Gamma^\mu_{\nu\lambda}(\bar{g}) = \frac{1}{2}\bar{g}^{\mu\kappa}\left(\partial_\nu\bar{g}_{\lambda\kappa} + \partial_\lambda\bar{g}_{\nu\kappa} - \partial_\kappa\bar{g}_{\nu\lambda}\right) \,, \tag{20}$$

w.r.t. the Weyl-rescaled metric:

$$\bar{g}_{\mu\nu} = (\chi_1 + 2\epsilon\chi_2 R)g_{\mu\nu} \quad , \quad \chi_1 \equiv \frac{\Phi_1(A)}{\sqrt{-g}} \quad , \quad \chi_2 \equiv \frac{\Phi_2(B)}{\sqrt{-g}} \,. \tag{21}$$

Varying the action (15) w.r.t. auxiliary tensor gauge fields $A_{\mu\nu\lambda}$, $B_{\mu\nu\lambda}$ and $H_{\mu\nu\lambda}$ yields the equations:

$$\partial_\mu\Big[R + X - V_1(\varphi) - \frac{1}{2}f_0\sqrt{-F^2}\Big] = 0$$
$$\partial_\mu\Big[\epsilon R^2 - \frac{1}{4e^2}F^2 + \frac{\Phi(H)}{\sqrt{-g}}\Big] = 0 \quad , \quad \partial_\mu\Big(\frac{\Phi_2(B)}{\sqrt{-g}}\Big) = 0 \,, \tag{22}$$

whose solutions are:

$$R + X - V_1(\varphi) - \frac{1}{2} f_0 \sqrt{-F^2} = M_1 = \text{const} \;,$$

$$\epsilon R^2 - \frac{1}{4e^2} F^2 + \frac{\Phi(H)}{\sqrt{-g}} = -M_2 = \text{const} \;\;,\;\; \frac{\Phi_2(B)}{\sqrt{-g}} \equiv \chi_2 = \text{const} \,. \tag{23}$$

Here M_1 and M_2 are arbitrary dimensionful (mass^4) and χ_2 arbitrary dimensionless integration constants. The appearance of M_1, M_2 signifies *dynamical spontaneous breakdown* of global Weyl-scale invariance under (19) due to the scale non-invariant solutions (second and third ones) in (23).

The physical meaning of the integration constants M_0, $M_{1,2}$, χ_2, has been elucidated in refs.[10, 19, 11] from the point of view of the canonical Hamiltonian formalism. Namely, it was shown that M_0, $M_{1,2}$, χ_2, which remain the only traces of the auxiliary gauge fields $C_{\mu\nu\lambda}$, $A_{\mu\nu\lambda}$, $B_{\mu\nu\lambda}$, $H_{\mu\nu\lambda}$, are identified as conserved Dirac-constrained canonical momenta conjugated to (certain components of) the latter.

Using the equations of motion of the starting action (15) w.r.t. $g_{\mu\nu}$ as well as Eqs.(23) we obtain the following relation between the original metric $g_{\mu\nu}$ and the Weyl-rescaled one (21) (with $\chi_{1,2}$ the same as in (21), Y - same as in (2)):

$$\bar{g}_{\mu\nu} = \chi_1 \Omega g_{\mu\nu} \;\;,\;\; \chi_1 \Omega = \frac{\chi_1 + 2\epsilon\chi_2 (V_1(\varphi) + M_1)}{1 + 2\epsilon\chi_2 (\bar{X} - \frac{f_0}{2}\sqrt{-\bar{F}^2})} \,, \tag{24}$$

$$\chi_1 = (V_1(\varphi) + M_1)^{-1} \Big[2\chi_2 M_2 - 4M_0 - (1+\chi_4) Y \Big] \,, \tag{25}$$

where:

$$\bar{X} \equiv -\frac{1}{2} g^{\mu\nu} \partial_\mu \varphi \partial_\nu \varphi \;\;,\;\; \bar{F}^2 \equiv F_{\mu\nu} F_{\kappa\lambda} \bar{g}^{\mu\kappa} \bar{g}^{\nu\lambda} \;\;,\;\; \chi_4 \equiv \frac{\Phi(C)}{\sqrt{-g}} \tag{26}$$

Now, following the same steps as in refs.[11] we derive from (15) the physical *Einstein-frame* theory w.r.t. Weyl-rescaled Einstein-frame metric $\bar{g}_{\mu\nu}$ (24) and perform an additional "darkon" field redefinition $u \to \widetilde{u}$:

$$\frac{\partial \widetilde{u}}{\partial u} = (V_1(u) - 2M_0)^{-\frac{1}{2}} \;\;;\;\; Y \to \widetilde{Y} = -\frac{1}{2} \bar{g}^{\mu\nu} \partial_\mu \widetilde{u} \partial_\nu \widetilde{u} \,. \tag{27}$$

The explicit form of the Einstein-frame matter action reads:

$$\begin{aligned} L_{\text{eff}} = \bar{X} - \frac{f_0}{2}\sqrt{-\bar{F}^2} - \frac{\chi_2}{e^2}\bar{F}^2 + \epsilon\chi_2 \Big(\bar{X} - \frac{f_0}{2}\sqrt{-\bar{F}^2}\Big)^2 \\ -\widetilde{Y}\Big(V(\varphi) + M_1\Big)\Big[1 + 2\epsilon\chi_2\Big(\bar{X} - \frac{f_0}{2}\sqrt{-\bar{F}^2}\Big)\Big] \\ +\widetilde{Y}^2 \Big\{\chi_2\Big[M_2 + \epsilon(V(\varphi) + M_1)^2\Big] - 2M_0\Big\} \,. \end{aligned} \tag{28}$$

To reveal the total matter effective potential $U_{\rm total}$, *i.e.*, the *minus* part of $L_{\rm eff}$ (28) for static (spacetime independent) matter fields φ, F^2 and u (*not* $\widetilde{u}$ (27)), we note that relations (4), (24) and (27) imply $\widetilde{Y} = \frac{1}{\chi_1 \Omega}$, meaning that according to (24)-(25) for static matter fields:

$$\widetilde{Y}\Big|_{\rm static} = \frac{(V(\varphi)+M_1)(1-\epsilon\chi_2 f_0\sqrt{-\bar{F}^2})}{2\Big\{\chi_2\Big[M_2+\epsilon(V(\varphi)+M_1)^2\Big]-2M_0\Big\}} \,. \tag{29}$$

Upon inserting (29) in (28) we find:

$$U_{\rm total}(\varphi,\bar{F}^2) = -L_{\rm eff}\Big|_{\rm static} = \mathcal{U}(\varphi) + \frac{1}{2} f_{\rm eff}(\varphi)\sqrt{-\bar{F}^2} + \frac{1}{4e^2_{\rm eff}(\varphi)}\bar{F}^2 \,, \tag{30}$$

$$\mathcal{U}(\varphi) \equiv \frac{(V_1(\varphi)+M_1)^2}{4\Big\{\chi_2\Big[M_2+\epsilon(V_1(\varphi)+M_1)^2\Big]-2M_0\Big\}} \,, \tag{31}$$

$$\text{(recall (17) } V_1(\varphi) = f_1\exp\{-\alpha\varphi\}) \,,$$

with running "inflaton"-dependent gauge coupling constants:

$$f_{\rm eff}(\varphi) = f_0(1-4\epsilon\chi_2\mathcal{U}(\varphi)) \;,\quad \frac{1}{e^2_{\rm eff}(\varphi)} = \frac{\chi_2}{e^2}\Big[1+\epsilon e^2 f_0^2(1-4\epsilon\chi_2\mathcal{U}(\varphi))\Big] \tag{32}$$

Concluding this Section let us note that $f_{\rm eff}(\varphi)$ measures the strength of charge confinement. Indeed, as shown in Appendix B of [11], for static spherically symmetric fields in a static spherically symmetric metric the presence of the term $-\frac{1}{2}f_{\rm eff}(\varphi)\sqrt{-F^2}$ will produce an effective "Cornell"-type [20] potential $V_{\rm eff}(L)$ between charged quantized fermions, L being the distance between the latter:

$$V_{\rm eff}(L) = -\frac{e^2_{\rm eff}(\varphi)}{2\pi}\frac{1}{L} + e_{\rm eff}(\varphi) f_{\rm eff}(\varphi)\sqrt{2}\,L + (L\text{–independent const}) \,, \tag{33}$$

i.e., a linear confining plus a Coulomb piece. Thus, we will consider below the dynamics of the Abelian nonlinear gauge system as mimicking quantum chromodynamics.

4. Cosmological Implications

4.1. Flat Regions of the Total "Inflaton" Effective Potential

The total matter effective potential (30) possesses non-trivial "inflaton"-dependent gauge field vacuum ($\frac{\partial}{\partial \bar{F}^2} U_{\rm total}\,|_{\bar{F}^2_{\rm vac}} = 0$):

$$\sqrt{-\bar{F}^2_{\rm vac}} = e^2_{\rm eff}(\varphi) f_{\rm eff}(\varphi) \,, \tag{34}$$

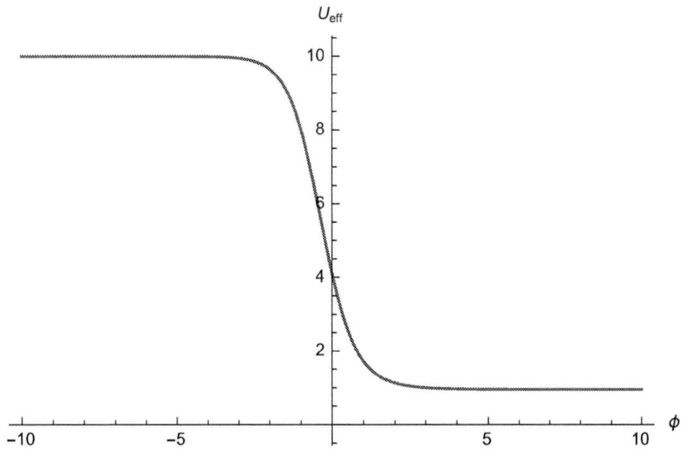

Figure 1: Qualitative shape of the total effective "inflaton" potential $U_{\rm eff}(\varphi)$ (35).

Upon inserting (34) in (30) we get the following total effective "inflaton" potential $U_{\rm eff}(\varphi)$ (taking into account expressions (31)-(32)):

$$U_{\rm eff}(\varphi) = \mathcal{U}(\varphi) + \frac{1}{4}e^2_{\rm eff}(\varphi)f^2_{\rm eff}(\varphi) = \mathcal{U}(\varphi) + \frac{e^2 f_0^2 \left(1 - 4\epsilon\chi_2\mathcal{U}(\varphi)\right)^2}{4\chi_2[1 + e^2\epsilon f_0^2 \left(1 - 4\epsilon\chi_2\mathcal{U}(\varphi)\right)]} \tag{35}$$

As clearly seen from the graphic representation of (35) on Fig.1, the total effective "inflaton" potential possesses the following reparkable property – two infinitely large flat regions corresponding to large negative and large positive "inflaton" values (the sub-/super-scripts $(\pm)$ label belonging to the respective flat region):

- $(-)$ flat region – for large negative φ-values:

$$\mathcal{U}(\varphi) \simeq \mathcal{U}_{(-)} = \frac{1}{4\epsilon\chi_2} \quad , \quad U_{\rm eff}(\varphi) \simeq U^{(-)}_{\rm eff} = \mathcal{U}_{(-)} = \frac{1}{4\epsilon\chi_2} \, , \tag{36}$$

$$f_{\rm eff}(\varphi) \simeq f_0^{(-)} = 0 \quad , \quad e^2_{\rm eff}(\varphi) \simeq e^2_{(-)} = \frac{e^2}{\chi_2} \, . \tag{37}$$

  The first relation in (37) implies according to (33) that there is ***no*** **charge confinement in the $(-)$ flat region**.

- $(+)$ flat region – for large positive φ-values:

$$\mathcal{U}(\varphi) \simeq \mathcal{U}_{(+)} = \frac{M_1^2}{4\left[\chi_2(M_2 + \epsilon M_1^2) - 2M_0\right]} \, , \tag{38}$$

$$U_{\rm eff}(\varphi) \simeq U^{(+)}_{\rm eff} = \frac{\epsilon\chi_2 M_1^2 + \epsilon e^2 f_0^2(\chi_2 M_2 - 2M_0)}{4\epsilon\chi_2\left[(\chi_2 M_2 - 2M_0)(1+\epsilon e^2 f_0^2) + \epsilon\chi_2 M_1^2\right]} , \quad (39)$$

$$f_{\rm eff}(\varphi) \simeq f_0^{(+)} = f_0 \frac{(\chi_2 M_2 - 2M_0)}{\chi_2 M_2 - 2M_0 + \epsilon\chi_2 M_1^2} , \quad (40)$$

$$\frac{1}{e^2_{\rm eff}(\varphi)} \simeq \frac{1}{e^2_{(+)}} = \frac{\chi_2}{e^2}\Big[1 + \frac{\epsilon e^2 f_0^2(\chi_2 M_2 - 2M_0)}{\chi_2 M_2 - 2M_0 + \epsilon\chi_2 M_1^2}\Big] . \quad (41)$$

Relation (40) implies according to (33) that **charge confinement *does take place* in the** $(+)$ **flat region**. Also, there is a non-trivial rescaling of the fundamental electric charge unit when passing from the $(-)$ flat “inflaton” region (second relation in (37)) to the $(+)$ flat “inflaton” region (41).

Let us specifically point out that both the vacuum energy densities $U^{(\pm)}_{\rm eff}$ as well as the confinement/deconfinement phenomena in both flat “inflaton” regions are entirely *dynamically* induced.

Associating appropriately the integration constants $M_{1,2}$, M_0 and the coupling constants f_0, ϵ with the fundamental physical constants M_{EW} (electroweak scale), M_{Pl} (Planck mass) and M_{infl} (inflationary energy scale):

$$M_1 \sim M^4_{EW} \;\; , \;\; M_2 \sim M^4_{Pl} \;\; , \;\; e^2 f_0^2 \sim \frac{M_1}{M_2} \;\; , \;\; \chi_2\epsilon \sim M^{-4}_{infl} \; , \quad (42)$$

allow us to identify – through their respective vacuum energy densities $U^{(-)}_{\rm eff} \sim 10^{-8} M^4_{Pl} \sim M^4_{infl}$ (39) and $U^{(+)}_{\rm eff} \sim 10^{-122} M^4_{Pl}$ (36) – the $(-)$ flat “inflaton” region as describing the “early” Universe, whereas the $(+)$ flat “inflaton” region will be describing the “late” Universe in accordance with the PLANCK collaboration data [21].

4.2. FLRW Reduction and Dark Energy/Dark Matter Unification

Let us now consider a reduction of the Einstein-frame gauge-matter action (28) to the class of cosmological FLRW class of metrics (12) (here for simplicity we will take $K = 0$ – flat spacial sections):

$$ds^2 = \bar{g}_{\mu\nu} dx^\mu dx^\nu = -N^2(t)dt^2 + a^2(t) d\vec{x}.d\vec{x} \; , \quad (43)$$

where now $\bar{X} = \frac{1}{2}\,\dot{\varphi}^2(t)$, $\widetilde{Y} = \frac{1}{2}\,\dot{\tilde{u}}^2(t)$, and $\sqrt{-F^2} = \dot{A}(t)$. The resulting full FLRW Einstein-frame action reads (in the gauge $N(t) = 1$):

$$S_{\rm FLRW} = \int dt \Big[-6a\,\dot{a}^2 + a^3 L_{\rm FLRW}\Big] \quad (44)$$

with the FLRW matter-gauge Lagrangian, or the pressure p:

$$p \equiv L_{\rm FLRW} = -\frac{f_0}{2}\Big[1+\epsilon\chi_2\Big(\dot{\varphi}^2 - \dot{\tilde{u}}^2\,(V_1(\varphi)+M_1)\Big)\Big]\,\dot{A}$$
$$+\frac{\chi_2}{4e^2}(1+\epsilon e^2 f_0^2)\,\dot{A}^2 + \frac{\dot{\varphi}^2}{2} + \epsilon\chi_2\frac{\dot{\varphi}^4}{4} - \frac{\dot{\tilde{u}}^2}{2}(V_1(\varphi)+M_1)(1+\epsilon\chi_2\,\dot{\varphi}^2)$$
$$+\frac{\dot{\tilde{u}}^4}{4}\Big[\chi_2 M_2 - 2M_0 + \epsilon\chi_2(V_1(\varphi)+M_1)^2\Big]\,. \qquad (45)$$

The canonical FLRW Hamiltonian corresonding to $L_{\rm FLRW} \equiv p$ (45), or the energy density ρ is given by:

$$\rho = \frac{\pi_A}{a^3} + \frac{p_\varphi}{a^3} + \frac{p_{\tilde{u}}}{a^3} - L_{\rm FLRW}\,, \qquad (46)$$

where the canonical momenta are defined via:

$$\frac{\pi_A}{a^3} = \frac{\partial L_{\rm FLRW}}{\partial \dot{A}}\,,\quad \frac{p_\varphi}{a^3} = \frac{\partial L_{\rm FLRW}}{\partial \dot{\varphi}}\,,\quad \frac{p_{\tilde{u}}}{a^3} = \frac{\partial L_{\rm FLRW}}{\partial \dot{\tilde{u}}}\,. \qquad (47)$$

From the first Eq.(47) we find:

$$\dot{A} = \frac{2e^2}{\chi_2(1+\epsilon e^2 f_0^2)}\Big\{\frac{\pi_A}{a^3} + \frac{f_0}{2}\Big[1+\epsilon\chi_2\Big(\dot{\varphi}^2 - \dot{\tilde{u}}^2\,(V_1(\varphi)+M_1)\Big)\Big]\Big\}\,, \qquad (48)$$

whereas the second and the third Eqs.(47) are a system of two mixed cubic algebraic equations w.r.t. the velocities $\dot{\varphi}$ and $\dot{\tilde{u}}$:

$$\frac{p_\varphi}{a^3} = \frac{\epsilon\chi_2\,\dot{\varphi}^3}{1+\epsilon e^2 f_0^2} + \frac{\dot{\varphi}}{1+\epsilon e^2 f_0^2}\Big[1 - 2\epsilon e^2 f_0 \frac{\pi_A}{a^3} - \epsilon\chi_2(V_1(\varphi)+M_1)\,\dot{\tilde{u}}^2\Big]\,, \qquad (49)$$
$$\frac{p_{\tilde{u}}}{a^3} = \dot{\tilde{u}}^3\Big[\chi_2 M_2 - 2M_0 + \frac{\epsilon\chi_2}{1+\epsilon e^2 f_0^2}(V_1(\varphi)+M_1)^2\Big]$$
$$-\,\dot{\tilde{u}}\,\frac{(V_1(\varphi)+M_1)}{1+\epsilon e^2 f_0^2}\Big(1 - 2\epsilon e^2 f_0 \frac{\pi_A}{a^3} - \epsilon\chi_2\,\dot{\varphi}^2\Big)\,. \qquad (50)$$

(i) In the (+) flat region of the total effective "inflaton" potential, i.e. in the "late" Universe where φ is large positive, negleting $V_1(\varphi) = f_1 e^{-\alpha\varphi}$ in (49)-(50), we obtain approximate (up to higher powers of $\frac{1}{a^3}$) solutions for the velocities $\dot{\varphi}$ and $\dot{\tilde{u}}$ as functions of the canonical momenta:

$$\dot{\varphi} = \frac{p_\varphi}{a^3}(1+\epsilon e^2 f_0^2)\,\frac{(\chi_2 M_2 - 2M_0)(1+\epsilon e^2 f_0^2) + \epsilon\chi_2 M_1^2}{(\chi_2 M_2 - 2M_0)(1+\epsilon e^2 f_0^2)} + \mathrm{O}(a^{-6})\,, \qquad (51)$$

$$\dot{\tilde{u}}= \Big[\frac{M_1}{(\chi_2 M_2 - 2M_0)(1+\epsilon e^2 f_0^2) + \epsilon\chi_2 M_1^2}\Big]^{\frac{1}{2}} + \frac{1}{2}(1+\epsilon e^2 f_0^2)\times$$
$$\times\Big[M_1\Big((\chi_2 M_2 - 2M_0)(1+\epsilon e^2 f_0^2) + \epsilon\chi_2 M_1^2\Big)\Big]^{-\frac{1}{2}} \frac{p_{\tilde{u}}}{a^3} + \mathrm{O}(a^{-6}) \,. \quad (52)$$

Inserting (51)-(52) into (45)-(46) we obtain for the pressure and energy density in the "late" Universe:

$$p_{(+)} = -U_{\rm eff}^{(+)} + \mathrm{O}(a^{-6})\,, \quad (53)$$

$$\rho_{(+)} = U_{\rm eff}^{(+)} + \Big\{p_{\tilde{u}}\Big[\frac{M_1}{(\chi_2 M_2 - 2M_0)(1+\epsilon e^2 f_0^2) + \epsilon\chi_2 M_1^2}\Big]^{\frac{1}{2}}$$
$$+\pi_A \frac{e^2 f_0}{\chi_2(1+\epsilon e^2 f_0^2)} \frac{(\chi_2 M_2 - 2M_0)(1+\epsilon e^2 f_0^2) - \epsilon\chi_2 M_1^2}{(\chi_2 M_2 - 2M_0)(1+\epsilon e^2 f_0^2) + \epsilon\chi_2 M_1^2}\Big\} \frac{1}{a^3} + \mathrm{O}(a^{-6})\,, \quad (54)$$

where $U_{\rm eff}^{(+)}$ is the dynamically induced vacuum energy density in the "late" Universe (39). Note that there is *no* $\mathrm{O}(a^{-3})$ in the expression for the pressure (53), while the leading terms in (53)-(54) are with opposite signs. Thus, comparing with (14) we can identify the expressions (53)-(54) as describing a unification in the "late" Universe of the dark energy ($\rho^{\rm DE}_{(+)} = -p^{\rm DE}_{(+)} = U_{\rm eff}^{(+)}$) and dust dark matter ($p^{\rm DM}_{(+)} = 0$, $\rho^{\rm DM}_{(+)} = \mathrm{O}(a^{-3})$ term in (54)) contributions..

(ii) In the $(-)$ flat region of the total effective "inflaton" potential, i.e. in the "early" Universe where φ is large negative, the approximate solutions for $\dot{\varphi}$ and $\dot{\tilde{u}}$ as functions of the canonical momenta – counterparts of Eqs.(51)-(52) – read:

$$\dot{\varphi}= \Big[\frac{(1+\epsilon e^2 f_0^2)p_\varphi}{2\epsilon\chi_2}\Big]^{1/3} a^{-1} + \mathrm{O}(a^{-2})\,, \quad (55)$$

$$\dot{\tilde{u}}= \frac{1}{\sqrt{\epsilon\chi_2 f_1}} - \frac{1}{2}\sqrt{\frac{\epsilon\chi_2}{f_1}}\Big[\frac{(1+\epsilon e^2 f_0^2)p_\varphi}{2\epsilon\chi_2}\Big]^{2/3} a^{-2} + \mathrm{O}(a^{-3})\,. \quad (56)$$

Inserting (55)-(56) into (45)-(46) we obtain for the pressure and energy density in the "early" Universe:

$$p_{(-)} = -U_{\rm eff}^{(-)} + \mathrm{O}(a^{-4})\,, \quad (57)$$

$$\rho_{(+)} = U_{\rm eff}^{(-)} + \frac{p_{\tilde{u}} e^{\frac{1}{2}\varphi}}{\sqrt{\epsilon\chi_2 f_1}} a^{-3} + \mathrm{O}(a^{-4})\,, \quad (58)$$

where $U_{\rm eff}^{(-)}$ is the dynamically induced vacuum energy density in the "early" Universe (36). Once again we see that there is *no* $\mathrm{O}(a^{-3})$ in the expression for the pressure (57) as in (53), while the leading terms in (57)-(58) are with opposite signs. Thus, comparing with (14) we can again identify the expressions (57)-(58) as describing a unification in the "early" Universe of the dark energy ($\rho^{\rm DE}_{(-)} = -p^{\rm DE}_{(-)} = U^{(-)}_{\rm eff}$) and dust dark matter ($p^{\rm DM}_{(-)} = 0$, $\rho^{\rm DM}_{(-)} = \mathrm{O}(a^{-3})$ term in (58)) contributions.

5. Conclusions

We have demonstrated above that the method of non-Riemanninan spacetime volume-forms (metric-independent volume elements), combined with other ingredients such as non-canonical gravity interactions with "inflaton" and "darkon" scalar fields and special type of non-linear gauge fields, provides a plausible description of various basic features of cosmological evolution:

(a) Unified description of dark energy and dark matter – through the impact of the dynamics of the non-canonical scalar "darkon" field;

(b) Natural explanation of the huge difference between the vacuum energy scales in the "early" and "late" Universe – thanks to the generation of dimensionful integration constants as an impact of the dynamics of the non-Riemannian volume-form fields;;

(c) Natural explanation of the absence of charge confinement in the "early" Universe epoch, and the appearance of QCD-like confinement in the "late" Universe - through an "inflaton"-running coupling constants in the physical Einstein-frame theory.

Further physically relevant phenomena adequately described using the method of non-Riemanninan spacetime volume-forms are:

(d) A novel mechanism for the supersymmetric Brout-Englert-Higgs effect, namely, the appearance of a dynamically generated cosmological constant triggering spontaneous supersymmetry breaking and mass generation for the gravitino [9, 19].

(e) Gravity-assisted spontaneous symmetry breaking of electroweak gauge symmetry – when adding interactions with the bosonic fields of the electroweak sector of the standard model of elementary particles gravity triggers the generation of a Higgs-like spontaneous symmetry breaking effective potential in the "late" Universe, whereas in the "early" Universe gravity keeps the Higgs-like scalar isodoublet massless, *i.e.*, no spontaneous electroweak breaking in the "early" Universe [22].

Finally, let us mention some other modifications of the method of non-Riemannian volume elements – the so called gravity models with dynamical spacetime [23], which were further developed into models of interacting diffusive unified dark energy and dark matter ([24] and references therein).

Acknowledgments

E.G. and E.N. are sincerely grateful to Prof. Branko Dragovich and the organizers of the *Ninth Meeting in Modern Mathematical Physics* in Belgrade for cordial hospitality. E.G., E.N. and S.P. gratefully acknowledge support of our collaboration through the academic exchange agreement between the Ben-Gurion University in Beer-Sheva, Israel, and the Bulgarian Academy of Sciences. E.N. and E.G. have received partial support from European COST actions MP-1405 and CA-16104, and from CA-15117 and CA-16104, respectively. E.N. and S.P. are also thankful to Bulgarian National Science Fund for support via research grant DN-18/1.

References

[1] E. Kolb and M. Turner, *"The Early Universe* (Addison Wesley, 1990);
A. Linde, *"Particle Physics and Inflationary Cosmology"*, (Harwood, Chur, Switzerland, 1990);
A. Guth, *"The Inflationary Universe"* (Addison-Wesley, 1997);
A. Liddle and D. Lyth, *"Cosmological Inflation and Large-Scale Structure"* (Cambridge Univ. Press, 2000);
S. Dodelson, *"Modern Cosmology"* (Acad. Press, 2003);
V. Mukhanov, *"Physical Foudations of Cosmology"* (Cambridge Univ. Press, 2005);
S. Weinberg, *"Cosmology"* (Oxford Univ. Press, 2008).

[2] A. De Felice and S. Tsujikawa, *Liv. Rev. Rel.* **13** (2010) 3 (*arxiv:1002.4928*);
S.Capozziello and V. Faraoni, *"Beyond Einstein Gravity"* (Springer, 2011).

[3] A.G. Riess, *et.al.*, *Astron. J.* **116** (1998) 1009-1038 (*arXiv:astro-ph/9805201*);
S. Perlmutter, *et.al.*, *Astrophys. J.* **517** (1999) 565-586 (*arXiv:astro-ph/9812133*);
A.G. Riess, *et.al.*, *Astrophys. J.* **607** (2004) 665-687 (*arXiv:astro-ph/0402512*);
M.S. Turner, in *Third Stromle Symposium The Galactic Halo*, ASP Conference Series Vol.666, ed. by B.K. Gibson, T.S. Axelrod and M.E. Putman, (1999);
N. Bahcall, J. Ostriker, S. Perlmutter and P. Steinhardt, *Science* **284** (1999) 1481;
V. Sahni and A.A. Starobinsky, *Int. J. Mod. Phys.* **D9** (2000) 373 (*arXiv:astro-ph/9904398*);
P.J.E. Peebles and B. Ratra, *Rev. Mod. Phys.* **75** (2003) 559 (*arXiv:astro-ph/0207347]*).

[4] M.J. Rees, *Phil. Trans. R. Soc. Lond.* **A361** (2003) 2427 (*arXiv:astro-ph/0402045*);
K. Garrett and G. Duda, *Adv. Astron.* **2011** (2011) 968283 (*arXiv:1006.2483*);
V. Sahni, *Lect. Notes Phys.* **653** (2004) 141 (*arXiv:astro-ph/0403324*).

[5] A.Yu. Kamenshchik, U. Moschella and V. Pasquier, *Phys. Lett.* **B511** (2001) 265 (*arXiv:gr-qc/0103004*);
N. Bilic, G. Tupper and R. Viollier, *Phys. Lett.* **535** (2002) 17 (*arXiv:astro-ph/0111325*);
N. Bilic, G. Tupper and R. Viollier, *J. Phys.* **A40** (2007) 6877 (*arXiv:gr-qc/0610104*).

[6] R.J. Scherrer, *Phys. Rev. Lett.* **93** (2004) 011301 (*arXiv:astro-ph/0402316*);
D. Bertacca, M. Bruni, O. Piattella and D. Pietrobon, *JCAP* **02** (2011)018 (*arXiv:1011.6669*).

[7] A. Chamseddine and V. Mukhanov, *JHEP* **1311** (2013 135 (*arXiv:1308.5410*);
A. Chamseddine, V. Mukhanov and A. Vikman, *JCAP* **1406** (2014) 017 (*arXiv:1403.3961*);
M. Chaichian, J. Kluson, M. Oksanen and A. Tureanu, *JHEP* **1412** (2014)102

(*arXiv:1404.4008*);
R. Myrzakulov, L. Sebastiani and S. Vagnozzi, *Eur. Phys. J.* **C75** (2015) 444 (*arXiv:1504.07984*).

[8] E. Guendelman, E. Nissimov and S. Pacheva, *Euro. Phys. J.* **C75** (2015) 472-479 (*arXiv:1508.02008*);
E. Guendelman, E. Nissimov and S. Pacheva, *Euro. Phys. J.* **C76** (2016) 90 (*arXiv:1511.07071*).

[9] E. Guendelman, E. Nissimov and S. Pacheva, *Bulg. J. Phys.* **41** (2014) 123 (*arXiv:1404.4733*).

[10] E. Guendelman, E. Nissimov and S. Pacheva, in *Eight Mathematical Physics Meeting*, ed. by B. Dragovich and I. Salom (Belgrade Inst. Phys. Press, 2015), pp. 93-103 (*arXiv:1407.6281*).

[11] E. Guendelman, E. Nissimov and S. Pacheva, *Int. J. Mod. Phys.* **A30** (2015) 1550133 (*arXiv:1504.01031*).

[12] E. Guendelman, *Mod. Phys. Lett.* **A14** (1999) 1043-1052 (*arXiv:gr-qc/9901017*);
E. Guendelman and A. Kaganovich, *Phys. Rev.* **D60** (1999) 065004 (*arXiv:gr-qc/9905029*);
E. Guendelman and O. Katz, *Class. Quant. Grav.* **20** (2003) 1715-1728 (*arXiv:gr-qc/0211095*).

[13] P. Gaete and E. Guendelman, *Phys. Lett.* **640B** (2006) 201-204 (*arxiv:0607113*).

[14] P. Gaete, E. Guendelman and E. Spalluci, *Phys. Lett.* **649B** (2007) 217 (*arxiv:0702067*);
E. Guendelman, *Int. J. Mod. Phys.* **A19** (2004) 3255 (*arxiv:0306162*);
E. Guendelman, *Mod. Phys. Lett.* **A22** (2007) 1209-1215 (*arxiv:0703139*);
I. Korover and E. Guendelman, *Int. J. Mod. Phys.* **A24** (2009) 1443-1456;
E. Guendelman, *Int. J. Mod. Phys.* **A25** (2010) 4195-4220 (*arxiv:1005.1421*).

[15] G. 't Hooft, *Nucl. Phys. B (Proc. Suppl.)* **121** (2003) 333-340 (*arxiv:0208054*[hep-th]).

[16] T. Chiba, T.Okabe and M. Yamaguchi, *Phys. Rev.* **D62** (2000) 023511 (*arXiv:astro-ph/9912463*);
C. Armendariz-Picon, V. Mukhanov and P. Steinhardt, *Phys. Rev. Lett.* **85** (2000) 4438 (*arXiv:astro-ph/0004134*);
C. Armendariz-Picon, V. Mukhanov and P. Steinhardt, *Phys. Rev.* **D63** (2001) 103510 (*arXiv:astro-ph/0006373*);
T. Chiba, *Phys. Rev.* **D66** (2002) 063514 (*arXiv:astro-ph/0206298*).

[17] E. Guendelman, R. Herrera, P. Labrana, E. Nissimov and S. Pacheva, *Gen. Rel. Grav.* **47** (2015) art.10 (*arXiv:1408.5344v4*).

[18] A. Liddle, *"Introduction to Modern Cosmology"*, 2nd ed., (John Wiley & Sons, Chichester, West Sussex, 2003).

[19] E. Guendelman, E. Nissimov and S. Pacheva, in *Eight Mathematical Physics Meeting*, ed. by B. Dragovich and I. Salom (Belgrade Inst. Phys. Press, 2015), pp. 105-115 (*arXiv:1501.05518*).

[20] E. Eichten, K. Gottfried, T. Kinoshita, J. Kogut, K. Lane and T.-M. Yan, *Phys. Rev. Lett.* **34** (1975) 369-372;
W. Buchmüller (ed.), *"Quarkonia"*, *Current Physics Sources and Comments*, vol.9 (North Holland, 1992);
M. Karliner, B. Keren-Zur, H. Lipkin and J. Rosner, *Ann. of Phys.* **324** (2009) 2-15 (*0804.1575*).

[21] R. Adam *et al.* (Planck Collaboration), *Astron. Astrophys.* **571** (2014) A22 (*arxiv:1303.5082* [astro-ph.CO]);
R. Adam *et al.* (Planck Collaboration), *Astron. Astrophys.* **586** (2016) A133 (*arxiv:1409.5738* [astro-ph.CO]).

[22] E. Guendelman, E. Nissimov and S. Pacheva, *Int. J. Mod. Phys.* **D25** (2016) 1644008 (*1603.06231*);
E. Guendelman, E. Nissimov and S. Pacheva, *Bulg. J. Phys.* **44** (2017) 15-30 (*1609.06915*).

[23] E. Guendelman, *Int. J. Mod. Phys.* **A25** (2010) 4081 (*arXiv:0911.0178*).

[24] D. Benisty and E. Guendelman, *Eur. Phys. J.* **C77** (2017) 396 (*arXiv:1701.08667*).

Hyperspherical three-body variables applied to lattice QCD Data*

James Leech†
University of St. Andrews, The Old Burgh School,
Abbey Walk, St Andrews, Fife KY16 9LB, Scotland

Milovan Šuvakov‡
Institute of Physics Belgrade, University of Belgrade,
Pregrevica 118, Zemun, P.O.Box 57, 11080 Beograd, Serbia

Veljko Dmitrašinović§
Institute of Physics Belgrade, University of Belgrade,
Pregrevica 118, Zemun, P.O.Box 57, 11080 Beograd, Serbia

Abstract

Is the three-quark confinement potential better described by the Y-string or the Δ-string form? In order to answer that question we have re-analysed the recent lattice QCD calculations by Koma & Koma [1] and the (much) older results of Takahashi et al. [2] using hyperspherical three-body variables. The (presently) extant lattice data do not allow a conclusive answer to the above question, but we show that hyperspherical coordinates lead to: 1) a definitive criterion for the discrimination between the Y and Δ strings; and 2) a more discriminating choice of lattice points to be made; both of these points are based on the dynamical O(2) symmetry of the Y-string.

1. Introduction

Lattice QCD offers a method to calculate three-quark potentials *ab initio*. The calculations are limited by the lattice size as well as by the computing- and man-power available. The results are therefore subject to systematic and statistical errors which must be estimated and interpreted as in any experimental result.

---

* This work was done while J.L. was visiting IPB in an IAESTE student exchange program. We thank Dr Igor Salom for serving as liasion with the Belgrade office of IAESTE. The work of M. Š and V.D. was supported by the Serbian Ministry of Science and Technological Development under Grants No. OI 171037 and III 41011.

† e-mail address: jl200@st-andrews.ac.uk

‡ e-mail address: suki@ipb.ac.rs

§ e-mail address: dmitrasin@ipb.ac.rs

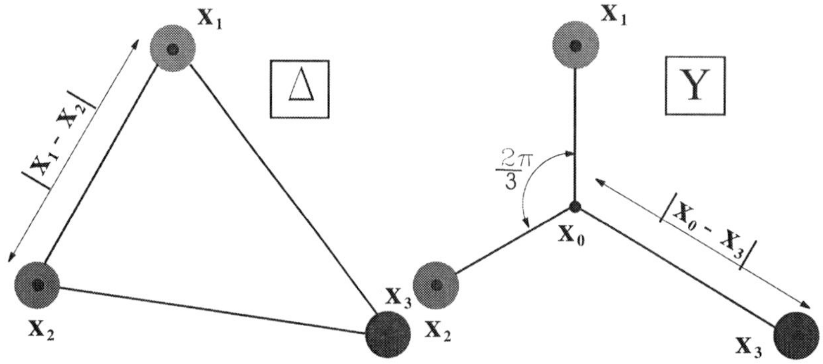

Figure 1: The Δ and Y- strings. The Δ-string potential goes as the perimeter (sum of sides) of the triangle subtended by the three quarks, whereas the Y-string potential goes as the sum of lengths from each quark to the Fermat-Torricelli point of the three quark system.

The form of the three-heavy-quark potential in lattice QCD is not well known. Two older calculations of the effective static three-quark potential on the lattice are the one by Alexandrou et al [3], who proposed the Δ-string form, Fig. 1, and the one by Takahashi et al. [2], who proposed the Y-string, Fig. 1. In the Y-string description, the potential depends on the sum of distances from the three quarks to the Fermat-Torricelli point (at the centre) of the system. In the Δ-string description, the potential goes as the perimeter (the sum of sides) of the triangle subtended by the three quarks.

Explicit formulae for the functional forms of the Y and Δ string potentials are given in Refs. [4, 5, 6]. Both potentials depend linearly on the "overall size" variable, the hyper-radius R (see Eq. (4) below) of the three-quark system. One way to distinguish the Y-string from the Δ-string is to note that the Y-string has an O(2) dynamical symmetry [4, 5, 6]. This dynamical O(2) symmetry is visible to the naked eye when using permutation-adapted hyper-spherical coordinates, see Fig. 2 in [5]. The dynamical O(2) symmetry ought to be used in any conclusive attempt to distinguish between the Y-string and the Δ-string potential.

The recent calculation of the three-quark potential by Koma & Koma [1], on a larger (24^4) lattice than before, has re-ignited interest in this question. These authors did not test the potential for the O(2) dynamical symmetry, however, so their work cannot be conclusive with regard to the Y vs. Δ dilemma.

The aim of this work is to use hyper-spherical coordinates to re-analyse both the Koma and the Takahashi data so as to see if it is possible on the basis of presently available lattice data to decide on the Y vs. Δ dilemma. We shall show that a clear resolution is not possible, due to an unfortunate choice of triangle shapes.

2. Hyper-spherical coordinates

The three bodies' Cartesian coordinates can be expressed in terms of the centre-of-mass variable $\mathbf{R}_{CM}$, and the two relative Jacobi vectors $\boldsymbol{\rho}$ and $\boldsymbol{\lambda}$.

$$\boldsymbol{\rho} = \frac{1}{\sqrt{2}}(\mathbf{x_1} - \mathbf{x_2}), \tag{1}$$

$$\boldsymbol{\lambda} = \frac{1}{\sqrt{6}}(\mathbf{x_1} + \mathbf{x_2} - 2\mathbf{x_3}), \tag{2}$$

$$\mathbf{R}_{\mathrm{CM}} = \frac{1}{3}(\mathbf{x_1} + \mathbf{x_2} + \mathbf{x_3}), \tag{3}$$

which obscures the permutation symmetry, however. The confining potential must be invariant under: 1) translations - the potential must not depend on $\mathbf{R}_{CM}$, but only on the relative Jacobi vectors; 2) rotations - the potential can only depend on the scalar products of $\boldsymbol{\rho}$ and $\boldsymbol{\lambda}$. 3) Permutations of particle labels. There is one preferred set of hyper-angles that makes the permutation symmetry manifest. The hyper-radius R

$$R^2 = \boldsymbol{\rho}^2 + \boldsymbol{\lambda}^2 \tag{4}$$

determines, loosely speaking, the size of the triangle subtended by the three particles. The hyper-angles α and ϕ, which describe the shape of the triangle, are defined as follows

$$\phi = \arctan\left(\frac{2\boldsymbol{\rho}\cdot\boldsymbol{\lambda}}{\boldsymbol{\rho}^2 - \boldsymbol{\lambda}^2}\right) \tag{5}$$

$$\alpha = \arccos\left(\frac{2(\boldsymbol{\rho}\times\boldsymbol{\lambda})}{\boldsymbol{\rho}^2 + \boldsymbol{\lambda}^2}\right) \tag{6}$$

The two hyper-angles determine a point on a unit-radius shape sphere. Figure 1 in Ref. [6] shows a view of the shape sphere from infinity above the North Pole.

3. Lattice QCD data in terms of hyper-spherical coordinates

The data from the papers was converted to our symmetric hyper-spherical coordinates by the equations described in section 2.. All of the data points are taken from the surface of a single hemisphere of the shape sphere. We plotted the data points calculated by Takahashi and Koma & Koma in the x-y projection of the shape sphere. Figure 2 includes all permutations of the three-body system. There are three lines representing isosceles triangles - that cross the origin. There are also three lines representing the right-angled triangles - one such line is at $y = -0.5$. Equilateral triangles are located at the center of the circle (origin of the coordinate system). Collinear configurations all lie on a circle of radius 1 about the origin.

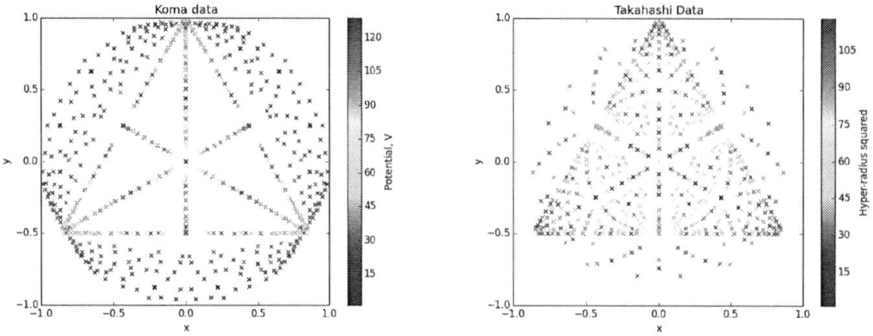

Figure 2: (a): Hyper-radius as a function of hyper-angles for Koma & Koma data points. (b): Hyper-radius as a function of hyper-angles for Takahashi data points.

Note that the two data sets contain strikingly different shapes: whereas most of the Koma points (triangle shapes) lie outside of the straight lines (triangle) defined by the right-triangle shapes, most of the Takahashi points lie inside of this boundary. Note moreover, that most of the Takahashi points that lie inside the right-triangle lines are low-hyper-radii ones (blue to greenish color - see the colour code), which makes this data set unsuitable to address the Y vs. Δ dilemma, as one needs high(er) values of the hyper-radius, so as to suppress the Coulomb term which QCD generates in addition to the confining part.

4. Analysis of lattice data

Following standard lattice QCD treatises, Refs. [7], [8], we assume that the total three-quark potential V_{3q} has the form

$$V_{3q}(\alpha, \phi, R) = -\frac{A(\alpha, \phi)}{R} + B(\alpha, \phi)R + C, \tag{7}$$

henceforth referred to as the Coulomb + linear potential Ansatz. The first term represents the sum of QCD Coulomb pairwise interactions, which is dominant at small values of the hyper-radius R. The second term represents the confinement potential, which is linear in R and dominant at large values of hyper-radius R, and the third term - C - is a constant. Here $A(\phi, \alpha)$ is assumed to be the (standard) sum of of pair-wise Coulomb terms, and $B(\phi, \alpha)$ is the unknown hyper-angular dependence of the linearly rising confining potential.

Our initial goal was to determine the unknown hyper-angular dependence $B(\phi, \alpha)$ of the linearly rising confining part of the three-quark potential using the lattice data and the well-known hyper-angular and hyper-radial dependences of the two-body Coulomb term.

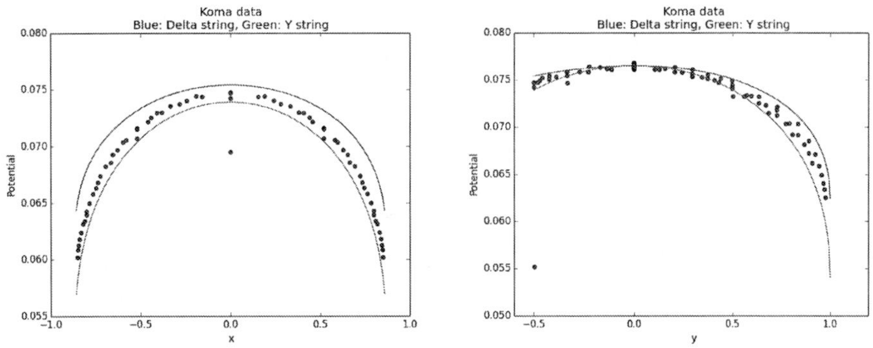

Figure 3: Plot of the hyperangular part of confining potential as a function of x or y, for isosceles and right-angled triangle configurations in the Koma & Koma data set - blue line is the Δ string ad the green line is the Y string prediction. (l.h.s. panel): Right-angled triangles; (r.h.s. panel): Isosceles triangles.

Surprisingly, however, Koma & Koma [1] have noticed that the first term - the sum of pairwise Coulomb terms - does not adequately describe the lattice data as the hyper-radius R decreases: the discrepancy amounts up to 26 % of the total Coulomb potential, depending on the shape of the triangle (see Eq. (34) in Sect. III.E "The functional form of the three-quark potential"). This indicates existence of a non-negligible three-body dependence of the QCD Coulomb potential that is not predicted by perturbative QCD calculations. This fact/circumstance additionally complicates our analysis and attempts to extract $B(\phi, \alpha)$.

In order to minimize the influence of the Coulomb term, and of the constant C[1], we have used only lattice configurations that are either (a) far away from the two-body collision points, where the Coulomb term rises to infinity, on the shape sphere; or (b) have large values of the hyper-radius R, where the Coulomb term is suppressed as compared with the confining term. There are only a few such sets of such configurations in the Koma and the Takahashi data sets: (i) the isosceles; (ii) the right-angled triangles. We shall use them both.

(i) It can be seen in Fig. 3.b that for the isosceles triangle configurations $\phi = const.$ in the Koma & Koma data set, the $B(\alpha)$ values form a (more or less continuous) line in between the Δ and Y-string potentials' functional forms. This result supports the conclusions of the Koma & Koma [1] that the three-quark potential is neither pure Y nor pure Δ string. The Takahashi data set does not fit well with either functional form and is strongly scattered.

[1]Of course, there is the possibility that the "constant" C is actually a function of the shape-sphere angles $C(\phi, \alpha)$.

(ii) The right-angled triangle configurations provide little insight: The functional forms for the Y and Δ string potentials are almost identical and do not allow clear separation. The Koma & Koma data points follow a similar shape in Fig. 3.a, but could be attributed to either functional form. In the Takahashi data set the points are too scattered to draw any conclusion.

5. Conclusions

We have analysed the lattice QCD data, Refs. [2, 1] in terms of permutation-adapted hyperspherical variables with a view towards resolving the Δ vs. Y string dilemma.

The analysis along the lines of isosceles triangles gives no conclusive answer to the question of the confinement potential form, in agreement with Ref. [1]. The Takahashi data, Ref. [2], is too scattered, while the Koma & Koma data, Ref. [1], suggests some mixture of the Y and Δ string. Similarly, the values of the potential for the right-angled triangle configurations provide little insight into the Δ vs. Y string dilemma.

This calls for new lattice calculations, where: 1) there is a constant hyper-radius; and 2) one of the hyper-angles is held constant while varying the other.

References

[1] Y. Koma and M. Koma, Phys. Rev. D **95**, no. 9, 094513 (2017).

[2] T. T. Takahashi, H. Suganuma, Y. Nemoto and H. Matsufuru, Phys. Rev. D **65**, 114509 (2002)

[3] C. Alexandrou, P. de Forcrand and O. Jahn, Nucl. Phys. Proc. Suppl. **119**, 667 (2003).

[4] V. Dmitrašinović, T. Sato and M. Šuvakov , Phys. Rev. D **80**, 054501 (2009).

[5] M. Šuvakov, and V. Dmitrašinović, Phys. Rev. E **83**, 056603 (2011).

[6] V. Dmitrašinović, and Milovan Šuvakov, p. 201 - 206, Proceedings of the "Sixth Mathematical Physics Meeting", eds. B. Dragovich, and Z. Rakić, Institute of Physics, Belgrade, Serbia, (2011).

[7] Heinz J. Rothe, "Lattice Gauge Theories: An Introduction", (3rd edition) World Scientific Lecture Notes in Physics vol. 74 (2005).

[8] Christof Gattringer, and Christian B. Lang, "Quantum Chromodynamics on the Lattice - An Introductory Presentation", Lecture Notes in Physics 788; Springer, Heidelberg: (2010).

Quasi-classical limit of the open Jordanian XXX spin chain*

Nenad Manojlović†
Grupo de Física Matemática da Universidade de Lisboa,
Av. Prof. Gama Pinto 2, PT-1649-003 Lisboa, Portugal
and
Departamento de Matemática, F. C. T., Universidade do Algarve,
Campus de Gambelas, PT-8005-139 Faro, Portugal

Nuno Cirilo António‡
Centro de Análise Funcional e Aplicações
Instituto Superior Técnico, Universidade de Lisboa
Av. Rovisco Pais, 1049-001 Lisboa, Portugal

Igor Salom§
Institute of Physics, University of Belgrade
P.O. Box 57, 11080 Belgrade, Serbia

Abstract

We study the open deformed XXX spin chain. In particular we obtain the explicit expression of the Sklyanin monodromy matrix in terms of the entries of the local Lax operator of the Jordanian chain. These results are essential in the study of the so-called quasi-classical limit of the system.

1. Introduction

A particularly interesting feature of quantum groups is the so-called twist transformation [1]. It yields new quantum groups form already known ones. More precisely, a twist of a quantum group, or more generally, of a Hopf algebra $\mathcal{A}$ is a similarity transformation of the co-product $\Delta : \mathcal{A} \to \mathcal{A} \otimes \mathcal{A}$ by an invertible twist element

$$\Delta(a) \mapsto \Delta_t(a) = \mathcal{F}\Delta(a)\mathcal{F}^{-1}, \qquad \forall a \in \mathcal{A}. \tag{1}$$

*N. M. and I. S. acknowledge partial financial support by the FCT project PTDC/MAT-GEO/3319/2014. I.S. was supported in part by the Serbian Ministry of Science and Technological Development under grant number ON 171031.

† e-mail address: nmanoj@ualg.pt
‡ e-mail address: nantonio@math.ist.utl.pt
§ e-mail address: isalom@ipb.ac.rs

In order to guarantee the co-associativity property of the twisted co-product the element $\mathcal{F}$ must satisfy the so-called twist equation [1]

$$\mathcal{F}_{12}\,(\Delta \otimes \mathrm{id})\,(\mathcal{F}) = \mathcal{F}_{23}\,(\mathrm{id} \otimes \Delta)\,(\mathcal{F})\,, \tag{2}$$

where

$$(\Delta \otimes \mathrm{id})\,\mathcal{F} = (\Delta \otimes \mathrm{id}) \sum_j f_j^{(1)} \otimes f_j^{(2)} = \sum_j \Delta \left(f_j^{(1)} \right) \otimes f_j^{(2)} \in \mathcal{A} \otimes \mathcal{A} \otimes \mathcal{A}. \tag{3}$$

Although the twist transformation generates an equivalence relation between quantum groups they produce different R-matrices. Namely, the transformation law of the co-product also determines how the corresponding universal R-matrix changes,

$$\mathcal{R} \mapsto \mathcal{R}^{(t)} = \mathcal{F}_{21} \mathcal{R} \mathcal{F}^{-1}, \quad \text{here} \quad \mathcal{F}_{21} = \sum_j f_j^{(2)} \otimes f_j^{(1)}. \tag{4}$$

This new R-matrix allows building and studying new integrable models [2, 3].

2. Deformed Yang R-matrix and the corresponding K-matrix

As our initial step, we briefly review the Jordanian twist element, as a particular solution of the twist equation. We consider $s\ell(2)$ generators S^α with $\alpha = +, -, 3$, with the commutation relations

$$[S^3, S^\pm] = \pm S^\pm, \quad [S^+, S^-] = 2S^3, \tag{5}$$

and Casimir operator

$$c_2 = (S^3)^2 + \frac{1}{2}(S^+S^- + S^-S^+) = (S^3)^2 + S^3 + S^-S^+ = \vec{S} \cdot \vec{S}. \tag{6}$$

The universal enveloping algebra $U(s\ell(2))$ admits the Jordanian twist element [4, 5]

$$\mathcal{F} = \exp 2\left(S^3 \otimes \sigma\right) \in U\left(s\ell(2)\right) \otimes U\left(s\ell(2)\right), \tag{7}$$

where

$$\sigma = \frac{1}{2} \log \left(1 + 2\theta S^+\right). \tag{8}$$

It straightforward to check that the Jordanian twist element satisfies the following equations [6]

$$(\Delta \otimes \mathrm{id})\,(\mathcal{F}) = \mathcal{F}_{13}\mathcal{F}_{23}, \qquad (\mathrm{id} \otimes \Delta_t)\,(\mathcal{F}) = \mathcal{F}_{12}\mathcal{F}_{13}. \tag{9}$$

In the equations above the co-product Δ is the usual co-product of the $U(s\ell(2))$ and Δ_t is the twisted co-product. Evidently the equations (9) imply the twist equation (2) [6]. Thus the the Jordanian twist element (7) satisfies the Drinfeld twist equation (2).

The XXX Heisenberg spin chain is related to the Yangian $\mathcal{Y}(s\ell(2))$ and the $SL(2)$-invariant Yang R-matrix [7]

$$R(\lambda) = \lambda\mathbb{1} + \eta\mathcal{P} = \begin{pmatrix} \lambda+\eta & 0 & 0 & 0 \\ 0 & \lambda & \eta & 0 \\ 0 & \eta & \lambda & 0 \\ 0 & 0 & 0 & \lambda+\eta \end{pmatrix}, \tag{10}$$

where λ is a spectral parameter, η is a quasi-classical parameter. We use $\mathbb{1}$ for the identity operator and $\mathcal{P}$ for the permutation in $\mathbb{C}^2 \otimes \mathbb{C}^2$.

The universal enveloping algebra of $s\ell(2)$ is a Hopf sub-algebra of the Yangian, $U(s\ell(2)) \subset \mathcal{Y}(s\ell(2))$. Notice that the matrix form of $\mathcal{F}$ in the spin-1/2 representation $\rho_{1/2}$ is $F_{12} \in \mathrm{End}(\mathbb{C}^2 \otimes \mathbb{C}^2)$,

$$F_{12} = \left(\rho_{1/2} \otimes \rho_{1/2}\right)(\mathcal{F}) = \mathbb{1} + 2\theta S^3 \otimes S^+ = \begin{pmatrix} 1 & \theta & 0 & 0 \\ 0 & 1 & 0 & 0 \\ 0 & 0 & 1 & -\theta \\ 0 & 0 & 0 & 1 \end{pmatrix}, \tag{11}$$

in particular, in the spin-1/2 representation the generators S^α are given by the Pauli matrices

$$S^\alpha = \frac{1}{2}\sigma^\alpha = \frac{1}{2}\begin{pmatrix} \delta_{\alpha 3} & 2\delta_{\alpha +} \\ 2\delta_{\alpha -} & -\delta_{\alpha 3} \end{pmatrix}.$$

Therefore the transformation of the Yang R-matrix by the Jordanina twist element yields the R-matrix of the twisted Yangian $\mathcal{Y}_\theta(s\ell(2))$ [8, 9, 6]

$$R^J(\lambda) = F_{21} R_{12}(\lambda) F_{12}^{-1} = \begin{pmatrix} \lambda+\eta & -\lambda\theta & \lambda\theta & \lambda\theta^2 \\ 0 & \lambda & \eta & -\lambda\theta \\ 0 & \eta & \lambda & \lambda\theta \\ 0 & 0 & 0 & \lambda+\eta \end{pmatrix}, \tag{12}$$

where $F_{21} = \mathcal{P}F_{12}\mathcal{P}$. In what follows, we will use only twisted R-matrix (12) and in order to simplify the notation we will omit the symbol J in the superscript.

The R-matrix (12) satisfies the Yang-Baxter equation

$$R_{12}(\lambda-\mu)R_{13}(\lambda)R_{23}(\mu) = R_{23}(\mu)R_{13}(\lambda)R_{12}(\lambda-\mu). \tag{13}$$

By setting $\theta = -\xi\eta$ we can guarantee the quasi-classical property

$$\frac{1}{\lambda+\eta}R(\lambda,\eta,\theta)|_{\theta=-\xi\eta} = \mathbb{1} + \eta r(\lambda) + O(\eta^2), \tag{14}$$

where $r(\lambda)$ is the classical r-matrix

$$r(\lambda) = \begin{pmatrix} 0 & \xi & -\xi & 0 \\ 0 & -\frac{1}{\lambda} & \frac{1}{\lambda} & \xi \\ 0 & \frac{1}{\lambda} & -\frac{1}{\lambda} & -\xi \\ 0 & 0 & 0 & 0 \end{pmatrix}, \tag{15}$$

which satisfies the classical Yang-Baxter equation

$$[r_{13}(\lambda), r_{23}(\mu)] + [r_{12}(\lambda - \mu), r_{13}(\lambda) + r_{23}(\mu)] = 0 \tag{16}$$

and has the unitarity property $r_{21}(-\lambda) = -r_{12}(\lambda)$.

The R-matrix (12) has the so-called regularity property

$$R(0, \eta) = \eta \mathcal{P}, \tag{17}$$

and the unitarity property

$$R_{12}(\lambda) R_{21}(-\lambda) = g(\lambda) \mathbb{1}, \quad \text{with} \quad g(\lambda) = \eta^2 - \lambda^2. \tag{18}$$

The PT symmetry is broken

$$R_{21}(\lambda) \neq R_{12}^{t_1 t_2}(\lambda), \tag{19}$$

where the indices t_1 and t_2 denote the respective transpositions in the first and second space of the tensor product $\mathbb{C}^2 \otimes \mathbb{C}^2$. The R-matrix does not have the crossing symmetry, but it satisfies the weaker condition

$$\left(\left(\left(R_{12}^{t_2}(\lambda) \right)^{-1} \right)^{t_2} \right)^{-1} = \frac{g(\lambda + \eta)}{g(\lambda + 2\eta)} M_2 R_{12}(\lambda + 2\eta) M_2^{-1}, \tag{20}$$

with

$$M = \begin{pmatrix} 1 & -2\theta \\ 0 & 1 \end{pmatrix}.$$

In [9] it was shown that the general solution to the reflection equation

$$R_{12}(\lambda - \mu) K_1^-(\lambda) R_{21}(\lambda + \mu) K_2^-(\mu) = K_2^-(\mu) R_{12}(\lambda + \mu) K_1^-(\lambda) R_{21}(\lambda - \mu) \tag{21}$$

is given by

$$K^-(\lambda) = \begin{pmatrix} \zeta + \lambda - \frac{\phi\theta}{\eta}\lambda^2 & \psi\lambda \\ \phi\lambda & \zeta - \lambda - \frac{\phi\theta}{\eta}\lambda^2 \end{pmatrix}. \tag{22}$$

Also, the dual reflection equation was derived and it was shown that its the general solution is given by [9]

$$K^+(\lambda) = K^-(-\lambda - \eta) M. \tag{23}$$

Final observation is that by setting $\theta = -\xi\eta$ we achieve that the matrix $K^-(\lambda)$ (22) does not depend on the parameter η i.e.,

$$\frac{\partial K^-(\lambda)}{\partial \eta} = 0. \tag{24}$$

This is an important step in the so-called quasi-classical limit of the corresponding chain [10, 11, 12, 13].

Using the results obtained above, in the next section, we will study the open deformed XXX spin chain following Sklyanin's approach [14] as we have successfully done in the case of XXX Heisenberg spin chain [10] and XXZ Heisenberg spin chain [12].

3. Jordanian deformation of the XXX spin chain

The Hilbert space of the system is

$$\mathcal{H} = \overset{N}{\underset{m=1}{\otimes}} V_m = (\mathbb{C}^{2s+1})^{\otimes N}, \tag{25}$$

we study the deformed inhomogeneous spin chain with N sites, characterised by the local space $V_m = \mathbb{C}^{2s+1}$, corresponding inhomogeneous parameter α_m and the operators

$$S_m^\alpha = \mathbb{1} \otimes \cdots \otimes \underbrace{S^\alpha}_{m} \otimes \cdots \otimes \mathbb{1}, \tag{26}$$

with $\alpha = +, -, 3$ and $m = 1, 2, \ldots, N$.

We introduce the Lax operator

$$\mathbb{L}_{0m}(\lambda) = \begin{pmatrix} e^{-\sigma_m} & 2\theta S_m^3 e^{\sigma_m} \\ 0 & e^{\sigma_m} \end{pmatrix} + \frac{\eta}{\lambda} \begin{pmatrix} S_m^3 \left(\mathbb{1}_m + 2\theta S_m^+\right) e^{-\sigma_m} & \left(S_m^- - 2\theta (S_m^3)^2\right) e^{\sigma_m} \\ S_m^+ e^{-\sigma_m} & -S_m^3 e^{\sigma_m} \end{pmatrix}. \tag{27}$$

In the case when the quantum space is a spin $\frac{1}{2}$ representation, the Lax operator is equal to the R-matrix, $\mathbb{L}_{0m}(\lambda) = \frac{1}{\lambda} R_{0m}(\lambda - \eta/2)$.

Due to the commutation relations (5), it is straightforward to check that the Lax operator satisfies the RLL-relations

$$R_{00'}(\lambda-\mu)\mathbb{L}_{0m}(\lambda-\alpha_m)\mathbb{L}_{0'm}(\mu-\alpha_m) = \mathbb{L}_{0'm}(\mu-\alpha_m)\mathbb{L}_{0m}(\lambda-\alpha_m)R_{00'}(\lambda-\mu). \tag{28}$$

The so-called monodromy matrix

$$T_0(\lambda) = \mathbb{L}_{0N}(\lambda - \alpha_N) \cdots \mathbb{L}_{01}(\lambda - \alpha_1) \tag{29}$$

is used to describe the system. Notice that $T(\lambda)$ is a two-by-two matrix acting in the auxiliary space $V_0 = \mathbb{C}^2$, whose entries are operators acting in $\mathcal{H}$

$$T(\lambda) = \begin{pmatrix} A(\lambda) & B(\lambda) \\ C(\lambda) & D(\lambda) \end{pmatrix}. \tag{30}$$

From RLL-relations (28) it follows that the monodromy matrix satisfies the RTT-relations

$$R_{00'}(\lambda - \mu) T_0(\lambda) T_{0'}(\mu) = T_{0'}(\mu) T_0(\lambda) R_{00'}(\lambda - \mu). \tag{31}$$

The RTT-relations define the commutation relations for the entries of the monodromy matrix.

Also, we define the Lax operator

$$\widetilde{\mathbb{L}}_{0m}(\lambda) = \begin{pmatrix} e^{\sigma_m} & -2\theta e^{\sigma_m} S_m^3 \\ 0 & e^{-\sigma_m} \end{pmatrix} + \frac{\eta}{\lambda} \begin{pmatrix} e^{\sigma_m} S_m^3 & e^{\sigma_m} \left(S_m^- - 2\theta (S_m^3)^2 \right) \\ e^{-\sigma_m} S_m^+ & -e^{-\sigma_m} \left(\mathbb{1}_m + 2\theta S_m^+ \right) S_m^3 \end{pmatrix}. \tag{32}$$

It obeys the following important identity

$$\mathbb{L}_{0m}(\lambda) \widetilde{\mathbb{L}}_{0m}(\eta - \lambda) = \left(1 + \eta^2 \frac{s_m(s_m+1)}{\lambda(\eta - \lambda)} \right) \mathbb{1}_0 , \tag{33}$$

where s_m is the value of spin in the space V_m. Thus the monodromy matrix

$$\widetilde{T}_0(\lambda) = \begin{pmatrix} \widetilde{A}(\lambda) & \widetilde{B}(\lambda) \\ \widetilde{C}(\lambda) & \widetilde{D}(\lambda) \end{pmatrix} = \widetilde{\mathbb{L}}_{01}(\lambda + \alpha_1 + \eta) \cdots \widetilde{\mathbb{L}}_{0N}(\lambda + \alpha_N + \eta), \tag{34}$$

obeys the following relations

$$\widetilde{T}_{0'}(\mu) R_{00'}(\lambda + \mu) T_0(\lambda) = T_0(\lambda) R_{00'}(\lambda + \mu) \widetilde{T}_{0'}(\mu), \tag{35}$$

$$\widetilde{T}_0(\lambda) \widetilde{T}_{0'}(\mu) R_{00'}(\mu - \lambda) = R_{00'}(\mu - \lambda) \widetilde{T}_{0'}(\mu) \widetilde{T}_0(\lambda). \tag{36}$$

By construction it follows that the entries of the Sklyanin monodromy matrix

$$\mathcal{T}_0(\lambda) = \begin{pmatrix} \mathcal{A}(\lambda) & \mathcal{B}(\lambda) \\ \mathcal{C}(\lambda) & \mathcal{D}(\lambda) \end{pmatrix} = T_0(\lambda) K_0^-(\lambda) \widetilde{T}_0(\lambda), \tag{37}$$

obey the exchange relations of the so-called reflection equation algebra [14, 10, 12]

$$R_{00'}(\lambda - \mu) \mathcal{T}_0(\lambda) R_{0'0}(\lambda + \mu) \mathcal{T}_{0'}(\mu) = \mathcal{T}_{0'}(\mu) R_{00'}(\lambda + \mu) \mathcal{T}_0(\lambda) R_{0'0}(\lambda - \mu). \tag{38}$$

4. Conclusions

The formulae (37), together with (27) and (32), yields, along the lines previously used successfully in the cases of the XXX and XXZ Heisenberg spin chains [10, 12], the quasi-classical expansion of the Sklyanin monodromy of the deformed chain. We believe that these results will help complete the study of the open deformed Gaudin model which we have initiated in [6]. Notice that the open trigonometric Gaudin was reviewed in [15]. The algebraic Bethe ansatz for the periodic deformed Gaudin model was done in [16, 17]. It is very likely that the implementation of the algebraic Bethe ansatz for the open deformed Gaudin model would require specific set of generators of the corresponding generalized Gaudin algebra, as in the $s\ell(2)$ case [18]. These considerations will be reported elsewhere.

References

[1] V. G. Drinfeld, *Quasi-Hopf algebras*, Leningrad Math. J. Vol. 1 (1990) 1419–1457.

[2] P.P. Kulish, *Twisting of quantum groups and integrable systems*, Proceedings of the Workshop on Nonlinearity, Integrability and All That: Twenty Years after NEEDS '79 (Gallipoli, 1999), 304–310, World Sci. Publ., River Edge, NJ, 2000.

[3] P.P. Kulish, *Twist Deformations of Quantum Integrable Spin Chains* Lecture Notes in Physics Volume 774 (2009) pp. 156–188.

[4] M. Gestenhaber, A. Giaquinto, and S. D. Schack, *Quantum symmetry*, Quantum Groups (Lect. Notes Math., Vol. 1510, P. P. Kulish, ed.), Springer, Berlin (1992), pp. 9–46.

[5] O. V. Ogievetsky, *Hopf structures on the Borel subalgebra of $s\ell(2)$*, Rend. Circ. Mat. Palermo (2), Suppl. No. 37, 185–199 (1994).

[6] N. Cirilo António, N. Manojlović and Z. Nagy, *Jordanian deformation of the open $s\ell(2)$ Gaudin model*, Teoreticheskaya i Matematicheskaya Fizika, **Vol. 179**, No. 1 (2014) 9–101; translation in Theoretical and Mathematical Physics, **Vol. 179**, No. 1 (2014) 462–471; `arXiv:1304.6918`.

[7] C. N. Yang, *Some exact results for the many-body problem in one dimension with repulsive delta-function interaction*, Phys. Rev. Lett. 19 (1967) 1312-1315.

[8] P. P. Kulish and A. A. Stolin, *Deformed Yangians and integrable models*, Czechoslovak J. Phys. 47 , no. 12, (1997) 1207–1212.

[9] P.P. Kulish, N. Manojlović and Z. Nagy, *Jordanian deformation of the open XXX spin chain*, (in Russian) Teoreticheskaya i Matematicheskaya Fizika **Vol. 163** No. 2 (2010) 288-298; translation in Theoretical and Mathematical Physics **Vol. 163** No. 2 (2010) 644-652; `arXiv:0911.5592`.

[10] N. Cirilo António, N. Manojlović and I. Salom, *Algebraic Bethe ansatz for the XXX chain with triangular boundaries and Gaudin model*, Nuclear Physics **B 889** (2014) 87-108; `arXiv:1405.7398`.

[11] N. Cirilo António, N. Manojlović, E. Ragoucy and I. Salom, *Algebraic Bethe ansatz for the $s\ell(2)$ Gaudin model with boundary*, Nuclear Physics **B 893** (2015) 305-331; `arXiv:1412.1396`.

[12] N. Manojlović, and I. Salom, *Algebraic Bethe ansatz for the XXZ Heisenberg spin chain with triangular boundaries and the corresponding Gaudin model*, Nuclear Physics, **B 923** (2017) 73-106; `arXiv:1705.02235`.

[13] N. Manojlović and I. Salom, *Algebraic Bethe ansatz for the trigonometric* $s\ell(2)$ *Gaudin model with triangular boundary*, arXiv:1709.06419.

[14] E. K. Sklyanin, *Boundary conditions for integrable quantum systems*, J. Phys. A: Math. Gen. **21** (1988) 2375–2389.

[15] N. Cirilo António, N. Manojlović and Z. Nagy, *Trigonometric* $s\ell(2)$ *Gaudin model with boundary terms*, Reviews in Mathematical Physics **Vol. 25** No. 10 (2013) 1343004 (14 pages); arXiv:1303.2481.

[16] N. Cirilo António and N. Manojlović, $sl(2)$ *Gaudin models with Jordanian twist*, Journal of Mathematical Physics **Vol. 46** No. 10 (2005) 102701, 19 pages.

[17] N. Cirilo António, N. Manojlović and A. Stolin, *Algebraic Bethe Ansatz for deformed Gaudin model*, Journal of Mathematical Physics **Vol. 52** No. 10 (2011) 103501, 15 pages; arXiv:1002.4951.

[18] I. Salom and N. Manojlović, *Creation operators of the non-periodic* $sl(2)$ *Gaudin model*, Proceedings of the 8th Mathematical Physics meeting: Summer School and Conference on Modern Mathematical Physics, 24 - 31 August 2014, Belgrade, Serbia, SFIN **XXVIII** Series A: Conferences No. A1, ISBN: 978-86-82441-43-4, (2015) 149–155.

Quantum gravity for piecewise flat spacetimes*

Aleksandar Miković†
Departamento de Matemática, Universidade Lusófona
Av. do Campo Grande, 376, 1749-024 Lisboa, Portugal

Marko Vojinović‡
Institute of Physics, Pregrevica 118, 11080 Belgrade, Serbia

ABSTRACT

We describe a theory of quantum gravity which is based on the assumption that the spacetime structure at small distances is given by a piecewise linear (PL) 4-manifold corresponding to a triangulation of a smooth 4-manifold. The fundamental degrees of freedom are the edge lengths of the triangulation. One can work with finitely many edge lengths, so that the corresponding Regge path integral can be made finite by using an appropriate path-integral measure. The semi-classical limit is computed by using the effective action formalism, and the existence of a semi-classical effective action restricts the choice of the path-integral measure. The classical limit is given by the Regge action, so that one has a quantum gravity theory for a piecewise-flat general relativity. By using the effective action formalism we show that the observed value of the cosmological constant can be recovered from the effective cosmological constant. When the number of 4-simplices in the spacetime triangulation is large, then the PL effective action is well approximated by a quantum field theory effective action with a physical cutoff determined by the smallest edge length.

1. Introduction

The standard approach to the problem of constructing a quantum gravity (QG) theory [1, 2] can be described as the following problem. Let M be a smooth 4-manifold, of topology $\Sigma \times I$, where Σ is a 3-manifold and I an interval from $\mathbf{R}$. Let g be a Minkowski-signature metric on M and Φ a set of matter fields on M. Then the goal is to find a triple $(\hat{g}, \hat{\Phi}, \hat{U})$, where $\hat{g}$ and $\hat{\Phi}$ represent Hermitian operators parametrized by the points of M,

* This work has been supported by GFMUL (Mathematical Physics Group at University of Lisbon) and the project ON 171031 of the Ministry of Education, Science and Technological Development, Serbia.

† e-mail address: amikovic@ulusofona.pt
‡ e-mail address: vmarko@ipb.ac.rs

acting in some Hilbert space $\mathcal{H}$, while $\hat{U}$ is a unitary evolution operator parametrized by I, such that the the classical limit ($\hbar \to 0$) of the quantum time-evolution is equivalent to the Einstein equations.

The best known example of this approach is Loop Quantum Gravity (LQG), see [3] for a recent review and references. In the LQG case, the Hilbert space $\mathcal{H}$ is only known to be a subset of a non-separable Hilbert space and $\hat{U}$ can be constructed only for a triangulation $T(M)$ of M, so that it is not clear what is the classical limit. Note that in the standard QG approach, the structure of M is not changed after the quantization, and it is well known that this is the main source of the difficulties for a quantization of gravity [1, 2]. This leads us to an alternative approach where M is replaced by a quantum spacetime $\widehat{M}$. The obvious choice would be a noncommutative manifold based on M, like in the case of noncommutative geometry (NCG) [4], where the coordinates of M become elements of a noncommutative algebra. Another choice is made in the superstring theory [5], where the coordinates of M become coordinates of the loop manifold $\mathcal{L}M$ and new Grassmann (anticommuting) coordinates are added, so that $\widehat{M}$ is a loop super manifold.

In this paper we would like to present the case when $\widehat{M} = T(M)$, see [12, 13]. This is clearly a much simpler choice for $\widehat{M}$ than the one made in NCG or in the superstring theory, but the price paid is that the spacetime triangulation becomes a physical structure. However, the PL manifold $T(M)$ looks like the smooth manifold M when the number of 4-simplices is large. Also, by using $T(M)$ one reduces the infinite number of the degrees of freedom (DOF) for g and Φ to a finite number, which then simplifies the quantization.

Note that Regge was the first to use $T(M)$ in order to define the path integral for general relativity (GR) [6], see [7] for a modern review. However, in Regge's approach the triangulation was an auxiliary structure and had to be removed via the smooth limit $T(M) \to M$. However, obtaining the smooth limit in the Regge approach is a difficult problem. The same applies to the case of spin-foam models of LQG, which can be only defined when the spacetime is a PL manifold. In causal dynamical triangulations (CDT) approach [8], $T(M)$ is also used to define the path integral, but it is also considered an auxiliary structure. Obtaining the smooth limit in CDT is proposed by performing a sum over the triangulations.

2. PL gravity path integral

Let $T(M)$ be a regular[1] triangulation of a smooth 4-manifold M. We will assign positive numbers L_ϵ to the edges ϵ of $T(M)$. If we think of an L_ϵ as a distance between two vertices of $T(M)$ induced by some metric, then we

[1] Any two k-simplices of $T(M)$ cannot have more than one common $(k-1)$-simplex, where $k = 1, 2, 3, 4$.

can define a constant metric in each 4-simplex σ

$$g_{kl}^{(\sigma)} = \frac{L_{0k}^2 + L_{0l}^2 - L_{kl}^2}{2L_{0k}L_{0l}}, \qquad 1 \leq k,l \leq 4, \tag{1}$$

where the indices $0,1,2,3,4$ denote the vertices of a 4-simplex σ. Hence we replace a smooth metric g on M by a PL version (1). We want that the PL metric has the Minkowski signature, and this can be ensured by requiring that L_ϵ satisfy the triangle inequalities for the triangles which belong to one of the tetrahedrons of σ, for example the tetrahedron $(1,2,3,4)$, while the L_ϵ of the triangles $(0,i,j)$ must not satisfy the triangle inequalities.

Having all $L_\epsilon > 0$ means that all triangles in $T(M)$ are spacelike. For $M = \Sigma \times I$ manifolds, this gives an accordion-like triangulation (triangulation of a cylinder). A more natural triangulation is to take a finite number of spacelike slices $T_k(\Sigma)$ which are linked by timelike edges such that each 4-simplex has a spacelike tetrahedron in T_k and a vertex in T_{k-1} or in T_{k+1}. This class of triangulations is used in CDT models [8]. We will then require that the L_ϵ of T_k satisfy the triangle inequalities, while a timelike edge will be assigned an imaginary length iL_ϵ. Hence the labels of the edges of timelike triangles will not satisfy the triangle inequalities and the metric (1) will have the correct signature.

The curvature scalar R will be concentrated on the triangles and R will be given by the deficit angle divided by the area of the dual face. Hence in each σ we have a flat metric (1) so that we can say the corresponding PL metric is a piecewise-flat metric.

Note that an L_ϵ label represents a proper length, so that L_ϵ is invariant under the local Lorentz transformations in each 4-simplex. We will also have $(L_\epsilon)^2 > 0$ for a spacelike edge, while $(L_\epsilon)^2 < 0$ for a timelike edge.

The Einstein-Hilbert action for the PL metric (1) becomes the Regge action

$$S_{Rc} = \frac{1}{G_N} \sum_{\Delta=1}^{F} A_\Delta(L)\theta_\Delta(L) + \Lambda_c V_4(L), \tag{2}$$

where G_N is the Newton constant, $A_\Delta(L)$ is the area of a triangle $\Delta \in T(M)$ and θ_Δ is the deficit angle. Λ_c is the cosmological constant and V_4 is the 4-volume of $T(M)$. See [8] how to define (2) when the timelike triangles are present. Note that the Regge action describes a theory with a finite number of DOF when Σ is compact, while in the case when Σ is non-compact, we can restrict L_ϵ to be non-zero only in a ball $B \subset \Sigma$.

One can also couple the matter fields to a Regge PL metric and the corresponding smooth actions will become the PL actions for a finite number of matter DOF. For example, a scalar field will be defined by the values of the field at the vertices of $T(M)$, which is equivalent to a PL function on the 4-polytopes of the dual triangulation.

In the case of a scalar field matter, the Regge path integral will be given by the following $(E+V)$-dimensional integral

$$Z = \int_{D_E} \mu(L)\, d^E L \int_{\mathbf{R}^V} \prod_{\nu=1}^{V} d\phi_\nu \, e^{i[S_{Rc}(L)+S_m(L,\phi)]/\hbar}\,, \tag{3}$$

where E is the number of the edges in $T(M)$ and V is the number of the vertices in $T(M)$ [13]. S_m is the PL form of the scalar-field action and the integration region D_E is a subset of $\mathbf{R}_+^E$ where the triangle inequalities hold. The measure μ has to be chosen such that it makes Z finite. The matter PI measure is taken to be trivial and we will assume that the matter path integral is finite. This is true, because the matter path integral will be given by a finite product of the integrals of the type

$$I(\alpha,\beta) = \int_{-\infty}^{\infty} dx\, e^{-\alpha x^2 - \beta x^4}\,, \tag{4}$$

where $\alpha, \beta \in \mathbf{C}$. Since I is convergent for $\alpha, \beta > 0$, the analytic continuation $I(i\alpha, i\beta)$ will be finite.

Note that in the standard Regge formulation the spacetime metric is of the Euclidean signature. This was done in analogy to the QFT case where the Euclidean signature improves the convergence of the path integral (3). However, in the QG case this does not help, because the scalar curvature also changes the sign in the Euclidean case and can be unbounded. Actually, the Lorentzian integral has better convergence properties, which can be seen on a toy example $R(x) = \alpha x^2$ where $x \in \mathbf{R}_+$ and α is a constant different from zero. Then $Z_E = \int_0^\infty dx\, e^{-R(x)}$ is convergent only for $\alpha > 0$ while $Z_L = \int_0^\infty dx\, e^{iR(x)}$ is convergent for any sign of α. The presence of imaginary edge lengths and imaginary angles in the Lorentzian case is not a problem, since all the geometric quantities can be defined [8].

Finding the smooth limit $T(M) \to M$ for Z is a difficult problem. However, there is a promising approach, based on the Wilson renormalization group [7]. In this approach one considers Z as function of the dimensionless couplings γ e λ

$$\gamma = l_0^2/(G_N\hbar) = l_0^2/l_P^2\,, \quad \lambda = l_0^4\Lambda_c/\hbar = l_0^4/(L_c^2 l_P^2)\,,$$

where $L_c^2 = G_N/\Lambda_c$ and l_0 is an arbitrary length. One then looks for a critical point $P_0 = (\gamma_0, \lambda_0)$ where the second derivatives of Z diverge so that there is a second-order phase transition. At the critical point the correlation length diverges, so that a transition to the smooth phase occurs. However, the problem with this approach is that at P_0 the perturbation theory does not apply, so that the calculation has to be done by using numerical methods. Also the semiclassical limit $l_P^2 \to 0$ corresponds to a strong coupling region $\gamma \to \infty$ and $\lambda \to \infty$ so that it is difficult to determine it analytically.

However, the easiest way to determine the semiclassical limit in a QG theory defined by a path integral is to use the effective action, see [9, 10, 11, 12, 13]. Namely, the effective action can be calculated analytically in the $\hbar \to 0$ limit. Also the PI measure $\mu(L)$ has to be such that allows a semiclassical expansion for the effective action for large L_ϵ. This gives us an additional constraint on the choice of $\mu(L)$.

3. Effective action for PL quantum gravity

We will assume that $T(M)$ is the fundamental spacetime structure, i.e. the spacetime is a *piecewise linear* 4-manifold $T(M)$ with a flat metric in each cell (4-simplex σ). If N is the number of cells of $T(M)$, then for $N \gg 1$, $T(M)$ will look like the smooth manifold M on a scale much larger than the maximal edge length.

By an appropriate choice of the measure μ the integral $Z(T(M))$ can be made finite. Since $T(M)$ is the physical spacetime, there is no need to define the smooth limit $T(M) \to M$. Instead, we need a large-N approximation for the observables. This is analogous to the fluid dynamics situation where on the scales much larger than the inter-molecular distance we can approximate the molecular velocities as a smooth field and use the Navier-Stokes equations.

We will determine the semiclassical limit of PL quantum gravity by using the effective action. It can be computed by using the effective action equation in the limit $L_\epsilon \gg l_P = \sqrt{G_N \hbar}$.

Let us recall first the effective action definition from quantum field theory (QFT). Let ϕ be a real scalar field on M and let

$$S(\phi) = \frac{1}{2}\int_M d^4x\sqrt{|g|}\left[g^{\mu\nu}\,\partial_\mu\phi\,\partial_\nu\phi - \frac{1}{2}\omega^2\phi^2 - \lambda\,\phi^4\right],$$

be a flat-spacetime action. The effective action $\Gamma(\phi)$ can be determined from the following integro-differential equation

$$e^{i\Gamma(\phi)/\hbar} = \int \mathcal{D}h \exp\left[\frac{i}{\hbar}S(\phi + h) - \frac{i}{\hbar}\int_M d^4x\,\frac{\delta\Gamma}{\delta\phi(x)}h(x)\right], \qquad (5)$$

see [14, 15].

Note that a generic solution $\Gamma(\phi)$ is a function with values in $\mathbf{C}$. The Wick rotation is used to obtain a real-valued function $\Gamma(\phi)$. This is done by solving first the EA equation in the Euclidean spacetime

$$e^{-\Gamma_E(\phi)/\hbar} = \int \mathcal{D}h \exp\left[-\frac{1}{\hbar}S_E(\phi + h) + \frac{1}{\hbar}\int_M d^4x\,\frac{\delta\Gamma_E}{\delta\phi(x)}h(x)\right]. \qquad (6)$$

Then $x_0 = -it$ is inserted into a solution $\Gamma_E(\phi)$, where (x_0, x_k) are the spacetime coordinates, so that

$$\Gamma(\phi) = i\Gamma_E(\phi)|_{x_0=-it}\,.$$

However, the Wick rotation cannot be used in quantum gravity, since in many problems of interest, introducing a flat background metric does not make sense. One way to resolve this difficulty is to use the fact that the Wick rotation in QFT is equivalent to

$$\Gamma(\phi) \to Re\,\Gamma(\phi) + Im\,\Gamma(\phi)\,, \tag{7}$$

see [9, 10]. This prescription is convenient for quantum gravity because it does not involve a background metric, nor a system of coordinates.

In the case of PL quantum gravity without matter, the effective action (EA) equation is given by

$$e^{i\Gamma(L)/l_P^2} = \int_{D_E(L)} d^E x\, \mu(L+x) e^{iS_{Rc}(L+x)/l_P^2 - i\sum_{\epsilon=1}^{E}\Gamma'_\epsilon(L)x_\epsilon/l_P^2}\,, \tag{8}$$

where $l_P^2 = G_N\hbar$ and $D_E(L)$ is a subset of $\mathbf{R}^E$ obtained by translating D_E by a vector $-L$ [12]. Note that $D_E(L) \subseteq [-L_1,\infty)\times\cdots\times[-L_E,\infty)$.

We will look for a semiclassical solution

$$\Gamma(L) = S_{Rc}(L) + l_P^2\Gamma_1(L) + l_P^4\Gamma_2(L) + \cdots\,,$$

where $L_\epsilon \gg l_P$ and

$$|\Gamma_n(L)| \gg l_P^2|\Gamma_{n+1}(L)|\,.$$

When $L_\epsilon \to \infty$, then $D_E(L) \to \mathbf{R}^E$ and

$$e^{i\Gamma(L)/l_P^2} \approx \int_{\mathbf{R}^E} d^E x\, \mu(L+x) e^{iS_{Rc}(L+x)/l_P^2 - i\sum_{\epsilon=1}^{E}\Gamma'_\epsilon(L)x_\epsilon/l_P^2}\,. \tag{9}$$

Actually, one can use the equation (9) to determine $\Gamma(L)$ for large L when μ falls off sufficiently quickly [12]. The reason is that

$$D_E(L) \approx [-L_1,\infty)\times\cdots\times[-L_E,\infty)\,,$$

for $L_\epsilon \to \infty$, so that the relevant behaviour is captured by the following one-dimensional integral

$$\int_{-L}^{\infty} dx\, e^{-zx^2/l_P^2 - wx} = \sqrt{\pi}\, l_P \exp\Big[-\frac{1}{2}\log z + l_P^2\frac{w^2}{4z}$$

$$+ l_P\frac{e^{-z\bar{L}^2/l_P^2}}{2\sqrt{\pi z}\bar{L}}\left(1 + O(l_P^2/z\bar{L}^2)\right)\Big]\,,$$

where $\bar{L} = L + l_P^2\frac{w}{2z}$ and $Re\,z = -(\log\mu)''$. The non-analytic terms in $\hbar$ will be absent if

$$\lim_{L\to\infty} e^{-z\bar{L}^2/l_P^2} = 0 \Leftrightarrow (\log\mu)'' < 0 \text{ for } L\to\infty\,.$$

Hence the perturbative solution exists for the exponentially damped measures and it will be given by the equation (9).

For $D_E(L) = \mathbf{R}^E$ and $\mu(L)$ a constant, the perturbative solution is given by the EA diagrams

$$\Gamma_1 = \frac{i}{2} Tr \log S''_{Rc} \,, \quad \Gamma_2 = \langle S_3^2 G^3 \rangle + \langle S_4 G^2 \rangle \,,$$

and

$$\Gamma_3 = \langle S_3^4 G^6 \rangle + \langle S_3^2 S_4 G^5 \rangle + \langle S_3 S_5 G^4 \rangle + \langle S_4^2 G^4 \rangle + \langle S_6 G^3 \rangle \,, \ldots$$

where $G = i(S''_{Rc})^{-1}$ is the propagator and $S_n = iS_{Rc}^{(n)}/n!$ for $n > 2$, are the vertex weights, see [15, 12]. The contractions $\langle X \cdots Y \rangle$ are the sums over the repeated DOF indices

$$\langle X \cdots Y \rangle = \sum_{k,\ldots,l} X_{k\ldots l} \cdots Y_{k\ldots l} \quad .$$

When $\mu(L)$ is not a constant, then the perturbative solution is given by

$$\Gamma(L) = \bar{S}_{Rc}(L) + l_P^2 \bar{\Gamma}_1(L) + l_P^4 \bar{\Gamma}_2(L) + \cdots \,,$$

where

$$\bar{S}_{Rc} = S_{Rc} - i l_P^2 \log \mu \,,$$

while $\bar{\Gamma}_n$ is given by the sum of n-loop EA diagrams with $\bar{G}$ propagators and $\bar{S}_n$ vertex weights [12].

Therefore

$$\Gamma_1 = -i \log \mu + \frac{i}{2} Tr \log S''_{Rc} \,,$$

$$\Gamma_2 = \langle S_3^2 G^3 \rangle + \langle S_4 G^2 \rangle + Res[l_P^{-4} Tr \log \bar{G}] \,,$$

$$\Gamma_3 = \langle S_3^4 G^6 \rangle + \cdots + \langle S_6 G^3 \rangle + Res[l_P^{-6} Tr \log \bar{G}]$$
$$+ Res[l_P^{-6} \langle \bar{S}_3^2 \bar{G}^3 \rangle] + Res[l_P^{-6} \langle \bar{S}_4 \bar{G}^2 \rangle] \,,$$

see [12].

Since the PI measure μ has to vanish exponentially for large edge lengths, a natural choice is

$$\mu(L) = \exp\left(-V_4(L)/(L_0)^4\right) \,, \tag{10}$$

where L_0 is a length parameter [12]. Since $\log \mu(L) = O\left((L/L_0)^4\right)$[2] then for $L_\epsilon > L_c$ and

$$L_0 > \sqrt{l_P \, L_c} \,, \tag{11}$$

[2]The notation $f(x_1, ..., x_n) = O(x^\alpha)$ means that $f(\lambda x_1, ..., \lambda x_n) = O(\lambda^\alpha)$ for $\lambda \to \infty$.

where $L_c^{-2} = \Lambda_c$, we get the following large-L asymptotics [13, 16]

$$\Gamma_1(L) = O(L^4/L_0^4) + \log O(L^2/L_c^2) + \log\theta(L) + O(L_c^2/L^2) \tag{12}$$

and

$$\Gamma_{n+1}(L) = O\left((L_c^2/L^4)^n\right) + L_{0c}^{-2n} O\left((L_c^2/L^2)\right) , \tag{13}$$

where $L_{0c} = L_0^2/L_c$.

4. Effective cosmological constant

The asymptotics (12) and (13) imply that the series

$$\Gamma(L) = \sum_{n\geq 0} (l_P)^{2n} \Gamma_n(L)$$

is semiclassical (SC) for $L_\epsilon \gg l_P$ and $L_0 \gg \sqrt{l_P L_c}$.

Let $\Gamma \to \Gamma/G_N$ so that $S_{eff} = (Re\,\Gamma + Im\,\Gamma)/G_N$. The effective action is then given by

$$S_{eff} = \frac{S_{Rc}}{G_N} + \frac{l_P^2}{G_N L_0^4} V_4 + \frac{l_P^2}{2G_N} Tr\log S''_{Rc} + O(l_P^4) ,$$

for $L_\epsilon \gg l_P$. Hence the $O(\hbar)$, or the one-loop, cosmological constant (CC) for pure gravity is given by

$$\Lambda = \Lambda_c + \frac{l_P^2}{L_0^4} = \Lambda_c + \Lambda_{qg} . \tag{14}$$

One can show that the one-loop cosmological constant is exact because there are no $O(L^4)$ terms beyond the one-loop order [13, 16]. This is a consequence of the large-L asymptotics

$$\log \bar{S}''_{Rc}(L) = \log O(L^2/\bar{L}_c^2) + \log\theta(L) + O(\bar{L}_c^2/L^2)$$

$$\bar{\Gamma}_{n+1}(L) = O\left((\bar{L}_c^2/L^4)^n\right) ,$$

where $\bar{L}_c^2 = L_c^2 \left[1 + i l_P^2 (L_c^2/L_0^4)\right]^{-1/2}$.

Hence the one-loop formula (14) is exact in the case of pure gravity. If $\Lambda_c = 0$, the observed value of Λ is obtained for $L_0 \approx 10^{-5} m$ so that $l_P^2 \Lambda \approx 10^{-122}$ [12]. Note that $L_0 \approx 10^{-5} m$ is consistent with the requirement that $L_0 \gg l_P$, which replaces the SC condition $L_0 \gg \sqrt{L_c l_P}$ when $\Lambda_c = 0$.

The formula (14) is intriguing but unrealistic, since there is matter in the universe. In order to obtain a realistic expression for the effective CC, we need to study the EA equation with matter. This study also requires the understanding of the emergence of the smooth spacetime from a PL manifold $T(M)$. If $T(M)$ has a large number of the edges ($E \gg 1$) then the following approximations are valid

$$S_R(L) \approx \frac{1}{2}\int_M d^4x\sqrt{|g|}\,R(g)\,, \tag{15}$$

and

$$\Lambda_c V_4(L) \approx \Lambda_c \int_M d^4x\sqrt{|g|} = \Lambda_c\, V_M\,, \tag{16}$$

where $|g| = |\det g|$. These are the standard formulas of the Regge calculus and they nicely illustrate how the PL manifold $T(M)$ with many 4-simplices can be approximated by a smooth manifold M with a smooth (differentiable) metric g.

Similarly, the effective action $\Gamma(L)$ will be approximated by a QFT effective action $\Gamma^*(g)$, where g is a smooth metric on M. Let L_K be a minimal length in a triangulation, so that $L_\epsilon \geq L_K$ and let $L_K \gg l_P$. When $E \gg 1$ the following approximation is valid

$$Tr\log S''_R(L) \approx \int_M d^4x\sqrt{|g|}\left[aR^2 + bR_{\mu\nu}R^{\mu\nu}\right]\log(K/K_0)\,, \tag{17}$$

where $R_{\mu\nu}$ is the Ricci tensor, and a, b, K_0 are some constants.

The formula (17) follows from the fact that a PL function on a lattice with a cell size L_K can be written as a Fourier integral over a compact region $|q| \leq \pi/L_K$ where q is the wave vector[3]. Hence the PL trace-log term can be approximated by using the QFT formulation of GR with a momentum cutoff $K = 2\pi\hbar/L_K$.

The effect of the matter on the CC can be studied by introducing a scalar field on M

$$S_m(g,\phi) = \frac{1}{2}\int_M d^4x\sqrt{|g|}\left[g^{\mu\nu}\,\partial_\mu\phi\,\partial_\nu\phi - U(\phi)\right]\,, \tag{18}$$

where $U = \frac{1}{2}\omega^2\phi^2 + \lambda\,\phi^4$.

On a PL manifold $T(M)$ the action (18) becomes

$$S_m = \frac{1}{2}\sum_\sigma V_\sigma(L)\sum_{k,l} g_\sigma^{kl}(L)\,\phi'_k\,\phi'_l - \frac{1}{2}\sum_p V_p^*(L)\,U(\phi_p)\,,$$

where $\phi'_k = (\phi_k - \phi_0)/L_{0k}$ and $k, l, 0$ are vertices in a 4-simplex σ, p labels the vertices of $T(M)$ and V^* is the volume of the dual cell. Then the total classical action of gravity plus matter on $T(M)$ is given by

$$S(L,\phi) = \frac{1}{G_N}S_{Rc}(L) + S_m(L,\phi)\,.$$

[3]This region is known as the first Brillouin zone.

The corresponding EA equation is given by

$$e^{\frac{i}{l_P^2}\Gamma(L,\phi)} = \int_{D_E(L)} d^E l \int_{\mathbf{R}^V} d^V\chi$$
$$\exp\left[\frac{i}{l_P^2}\left(\bar{S}(L+l,\phi+\chi) - \sum_\epsilon \frac{\partial\Gamma}{\partial L_\epsilon} l_\epsilon - \sum_p \frac{\partial\Gamma}{\partial\phi_p}\chi_p\right)\right], \quad (19)$$

where $\bar{S} = S_{Rc} - il_P^2 \log\mu + G_N S_m$, see [13].

We will look for a perturbative solution

$$\Gamma(L,\phi) = S(L,\phi) + l_P^2\Gamma_1(L,\phi) + l_P^4\Gamma_2(L,\phi) + \cdots,$$

and require it to be semiclassical for $L_\epsilon \gg l_P$ and $|\sqrt{G_N}\,\phi| \ll 1$. This can be checked on the $E = 1$ toy model

$$S(L,\phi) = (L^2 + L^4/L_c^2)\theta(L) + L^2\theta(L)\phi^2(1 + \omega^2L^2 + \lambda\phi^2L^2),$$

where $\theta(L)$ is a homogeneous function of degree zero.

It is not difficult to see that

$$\Gamma(L,\phi) = \Gamma_g(L) + \Gamma_m(L,\phi),$$

and

$$\Gamma_m(L,\phi) = V_4(L)\,U_{eff}(\phi)$$

for constant ϕ where $U_{eff}(0) = 0$. Furthermore,

$$\Gamma_g(L) = \Gamma_{pg}(L) + \Gamma_{mg}(L),$$

where Γ_{pg} is the pure gravity contribution and Γ_{mg} is the matter induced contribution.

In the smooth-manifold approximation one has

$$\Gamma_{mg}(L) \approx \Lambda_m V_M + \Omega_m(R,K),$$

where $K = 2\pi\hbar/L_K$ is the momentum cutoff. One can show that

$$\Omega_m = \Omega_1 l_P^2 + O(l_P^4)$$

and

$$\begin{aligned}\Omega_1(R,K) &= a_1K^2\int_M d^4x\sqrt{|g|}\,R \\ &+ \log(K/\omega)\int_M d^4x\sqrt{|g|}\Big[a_2R^2 + a_3R^{\mu\nu}R_{\mu\nu} \\ &\qquad + a_4R^{\mu\nu\rho\sigma}R_{\mu\nu\rho\sigma} + a_5\nabla^2R\Big] \\ &+ O(1/K^2),\end{aligned} \quad (20)$$

where $R_{\mu\nu\rho\sigma}$ is the Riemann curvature tensor, see [13].

The effective CC will be then given as

$$\Lambda = \Lambda_c + \Lambda_{qg} + \Lambda_m \,,$$

where Λ_{qg} is given by (14). Note that the matter contribution to CC can be approximated by a sum

$$\Lambda_m \approx \sum_{\gamma} v(\gamma, K) \tag{21}$$

where $v(\gamma, K)$ is a one-particle irreducible vacuum Feynman diagram for the field-theory action S_m in flat spacetime with the cutoff K. One can show that

$$\begin{aligned}\sum_{\gamma} v(\gamma, K) \;\; &\approx \;\; l_P^2 K^4 \Big[c_1 \ln(K^2/\omega^2) + \sum_{n\geq 2} c_n(\bar\lambda)^{n-1} (\ln(K^2/\omega^2))^{n-2} \\ &+ \;\; \sum_{n\geq 4} d_n(\bar\lambda)^{n-1}(K^2/\omega^2)^{n-3}\Big] \,, \end{aligned} \tag{22}$$

for $K \gg \omega$, where $\bar\lambda = l_P^2 \lambda$, see [16]. Therefore one has a highly divergent sum of matter vacuum-energy contributions to the cosmological constant when $K \to \infty$. This is the famous cosmological constant problem which appears in any QFT formulation of quantum gravity.

However, in the PL formulation of quantum gravity (PLQG), the QFT which produces the infinite sum in (22) is just an approximation. The fundamental theory has finitely many DOF so that the exact solution of the EA equation will give a finite and cutoff-independent value for Λ. Therefore

$$\Lambda_m - V(\omega^2, \lambda, l_P^2) \,, \tag{23}$$

and

$$\Lambda = \pm \frac{1}{L_c^2} + \frac{l_P^2}{2L_0^4} + V(\omega 2, \lambda, l_P^2) \,. \tag{24}$$

The equation (24) can be used to fix the free parameters L_0 and L_c. By equating Λ with the experimentally observed value, we obtain

$$\lambda = x + y + \lambda_m \tag{25}$$

where $\lambda = l_P^2 \Lambda \approx 10^{-122}$, $x = \pm\, l_P^2/L_c^2$, $y = l_P^4/2L_0^4$ and $\lambda_m = l_P^2 V$. The equation (25) has infinitely many solutions, but we also have to impose the condition for the existence of the semi-classical limit (11). This gives the restriction

$$0 < y < 2|x| \,. \tag{26}$$

The value of λ_m is not known, but for any value of λ_m the equation (25) has infinitely many solutions which obey the restriction (26). Note that the solution $x = -\lambda_m$ and $y = \lambda$, which was proposed in [13], will be acceptable if $|\lambda_m| > \lambda/2$. This solution is special because it gives a value for L_0 which is independent of the value of λ_m, $L_0 \approx 10^{-5}$m. This is the same value which was obtained in the case of pure PL gravity without the cosmological constant [12].

5. The CC problem in quantum gravity

The formula (24) for the exact effective cosmological constant is an essential ingredient for the resolution of the CC problem from QFT in the context of a QG theory. The result (24) can be better understood if we recall the definition of the CC problem given by Polchinski [17]. According to this definition, the CC problem in a QG theory has two parts:

1) show that the observed CC value is in the CC spectrum,

2) explain why the CC takes the observed value.

The meaning of the first part (P1) of the CC problem is obvious if the cosmological constant is represented by an operator. In the case when one has a quantum corrected expression of the classical CC value, one has to show that there are values of the free parameters which give the observed CC value. The PLQG theory clearly solves P1, while the second part (P2) of the CC problem cannot be addressed by the current formalism. The reason is that one has to generalise the standard formalism of quantum mechanics in order to provide a mechanism for a selection of a wavefunction of the universe with a particular value of the cosmological constant.

Note that demonstrating P1 is a highly non-trivial task in any QG theory. The problem P1 has been addressed so far only in PLQG theory and in string theory. In the string theory case there are only plausibility arguments that P1 is true [18, 19]. The CC spectrum in string theory is discrete with $O(10^{500})$ values [18]. Although positive CC values are not natural in string theory, a mechanism for their appearance was provided in [19]. Hence it is plausible to assume that the CC spectrum is sufficiently dense around zero such that the observed value is sufficiently close to some CC spectrum value.

The second part of the CC problem has been only addressed in string theory. This is the multiverse proposal, see [21], and the assumption is that there are many universes, each having a fixed CC value from the CC spectrum. We live in the universe with the CC value $\Lambda_c l_P^2 \approx 10^{-122}$, because this is the value that allows formation of galaxies, planets and life, see [20] for the anthropic determination of the CC value.

Note that there are many proposals for P2 which are not derived from a QG theory, but instead it is assumed that a certain effective action exists such that its equations of motion give the required CC value, see for example [22].

6. Conclusions

The PLQG theory is a theory of quantum gravity which has finitely many degrees of freedom and no infinities. The underlying spacetime structure is a PL manifold $T(M)$ and the smooth spacetime M is recovered as an approximation valid when the number of 4-simplices is large and at a length scale much larger than the typical edge length. The smooth spacetime approximation is analogous to the smooth vector field approximation for the molecular velocities in a fluid.

The PLQG theory is defined by the Regge path integral with a non-trivial measure. The measure is chosen such that it gives a finite path integral, and also it has to admit a semi-classical solution of the effective action equation. These criteria select the exponentially vanishing measures for large edge lengths, and a simple and natural choice for the measure is (10). This measure simplifies the analysis of the effective cosmological constant and one can obtain the formula (24) for the exact effective CC, i.e. to all orders in $\hbar$. The two free parameters in (24) can be consistently chosen such that the observed CC value is obtained. This is an important requirement for any QG theory and PLQG is the only existing QG theory where this property has been demonstrated explicitly.

Another nice property of the PLQG theory is that the effective action Γ can be approximated by a QFT effective action Γ^* when the number of 4-simplices in $T(M)$ is large. Γ^* can be calculated by using the perturbative QFT for GR with matter and with a momentum cutoff K, when $L_\epsilon \geq L_K \gg l_P$. Hence the minimal edge length L_K in the triangulation determines the momentum cutoff K and

$$\Gamma(L_1, \cdots, L_E, \phi_1, \cdots, \phi_V) \approx \Gamma^*(g(x), \phi(x), K)\,, \tag{27}$$

for $E \gg 1$ and $V \gg 1$.

The QFT approximation (27) will be still valid for $L_K \leq l_P$, but in this case Γ^* cannot be calculated by the perturbative QFT methods. Instead, one has to use a non-perturbative method to solve the EA equation. The existence of the QFT approximation (27) implies that one can obtain the running of the elementary particle masses and the coupling constants with K, see for example the equation (20).

Note that the effective action only makes sense for the spacetimes which are given by the direct product of a 3-manifold with an interval. In order to study the quantum cosmology questions, one needs to consider 4-manifolds of general topology, which is different from $\Sigma\times I$ topology. When $M \neq \Sigma\times I$, the concept of the effective action cannot be used. However, the Hartle-Hawking (HH) wavefunction [23] can be defined for any $T(M)$ by using the PLQG path integral (3). By choosing a triangulation for a manifold

$$M \cup (\Sigma \times I)\,, \quad \partial M = \Sigma\,,$$

one can describe a Big-Bang quantum cosmology with an initial HH state, which evolves by the evolution operator defined by the PLQG path integral for the $T(\Sigma \times I)$ part of the spacetime. It is then plausible to assume that the effective dynamics which corresponds to the time evolution of the HH state will be given by the PLQG effective action, defined by the equation (19).

References

[1] C. Isham, Lect. Notes Phys. 434 (1994) 1-21.

[2] C. Isham, Lect. Notes Phys. 434 (1994) 150-169.

[3] D.W. Chiou, Int. J. Mod. Phys. D, Vol. 24, No. 1 (2015) 1530005, `arxiv:1412.4362`.

[4] A. Connes, *Noncommutative geometry*, Academic Press, San Diego (1994).

[5] M. B. Green, J. H. Schwarz and E. Witten, *Superstring Theory*, Cambridge University Press (1987).

[6] T. Regge, Nuovo Cim. 19 (1961) 558-571.

[7] H. W. Hamber, Gen. Rel. Grav. 41 (2009) 817, `arXiv:0901.0964`.

[8] J. Ambjorn, A. Goerlich, J. Jurkiewicz, R. Loll, Phys. Rept. 519 (2012) 127-210.

[9] A. Miković and M. Vojinović, Class. Quant. Grav. 28 (2011) 225004.

[10] A. Miković and M. Vojinović, J. Phys.: Conf. Ser. 360 (2012) 012049.

[11] A. Miković, Rev. Math. Phys. 25 (2013) 1343008.

[12] A. Miković, Adv. Theor. Math. Phys. 21 (2017) 631.

[13] A. Miković and M. Vojinović, EPL 110 (2015) 40008.

[14] V.P. Nair, *Quantum Field Theory - A Modern Perspective*, Springer (2005).

[15] H. Kleinert, *Path Integrals in Quantum Mechanics, Statistics, Polymer Physics and Financial Markets*, World Scientific (2006).

[16] A. Miković and M. Vojinović, Cosmological constant in a quantum gravity theory for a piecewise-linear spacetime, `arxiv:1407.1124`.

[17] J. Polchinski, The cosmological constant and the string landscape, *The Quantum Structure of Space and Time: Proceedings of the 23rd Solvay Conference on Physics.* Brussels, Belgium. 1 - 3 December 2005, D. Gross, M. Nenneaux and A. Servin (eds), 2007, `arXiv:hep-th/0603249`.

[18] R. Bousso and J. Polchinski, JHEP 0006 (2000) 006.

[19] S. Kachru, R. Kalosh, A. Linde and S. Trivedi, Phys. Rev. D 68 (2003) 046005.

[20] S. Weinberg, Rev. Mod. Phys. 61 (1989) 1.

[21] L. Susskind, The anthropic landscape of string theory, *Universe or Multiverse?*, B. Carr (ed.) Cambridge University Press (2007) 247-266, `arxiv:hep-th/0302219`.

[22] G. D'Amico, N. Kaloper, A. Padilla, et al. JHEP 1709 (2017) 74.

[23] J.B. Hartle and S.W. Hawking, Phys. Rev. D 28 (1983) 2960-2975.

Representation of T-duality of type II pure spinor superstring in double space*

Bojan Nikolić†
Branislav Sazdović‡
Institute of Physics Belgrade, University of Belgrade
Pregrevica 118, 11080 Belgrade, Serbia

Abstract

In this article we present a new way of representation of T-duality using double space. Double space is obtained by adding T-dual coordinates to the initial ones. T-dualization in double space is represented by permutation of appropriate directions from initial and T-dual space. We did this both for bosonic and fermionic T-dualization of type II superstring theory propagating in the constant background fields. Obtained results are in full correspondence with the results of standard Buscher procedure.

1. Introduction

T-duality as a fundamental characteristic of string dynamics [1, 2, 3, 4, 5], unexprerienced by point-like particle, makes that there is no difference in physical content between string theories compactified on a circle of radius R and circle of radius $1/R$. It is very important for understanding M-theory, because five consistent superstring theories are connected by web of T and S dualities.

Buscher T-dualization procedure [2] represents a mathematical environment for performing T-duality. In order to make T-dualization along some directions, they should be isometry ones. Effectively, this means that background fields do not depend on those coordinates [2, 3, 4, 5, 6, 7]. Further, we localize noticed symmetry in standard way introducing world-sheet covariant derivatives, $\partial_\pm x^\mu \to D_\pm x^\mu = \partial_\pm x^\mu + v_\pm^\mu$, where $v_\pm^\mu$ are gauge fields. In order to have the same number of degrees of freedom in T-dual theory as in the initial one, the new term with Lagrange multipliers is added to the action. Using gauge freedom we fix initial coordinates and get the gauge

* This work has been supported by the Serbian Ministry of Education, Science and Technological Development, under contract No. 171031.

† e-mail address: bnikolic@ipb.ac.rs
‡ e-mail address: sazadovic@ipb.ac.rs

fixed action. Varying gauge fixed action with respect to the Lagrange multipliers one gets the initial action and varying with respect to the gauge fields one gets T-dual action.

This standard T-dualization procedure was used in the papers [8, 9, 10, 11, 12] in the context of closed string noncommutativity. There is a generalized Buscher procedure which deals with background fields depending on all coordinates. The new step in this procedure is introducing of gauge invariant coordinate for the directions on which background fields depend on. The generalized procedure was applied to the case of bosonic string moving in the weakly curved background [13, 14]. It leads directly to closed string noncommutativity [15].

Double space formalism is a framework in which we can represent T-dualization in a simple and elegant way. It is spanned by double coordinates $Z^M = (x^\mu, y_\mu)$ $(\mu = 0, 1, 2, \ldots, D-1)$, where x^μ and y_μ are the coordinates of the D-dimensional initial and T-dual space-time, respectively. It was the subject of the articles about twenty years ago [16, 17, 18, 19, 20], but interest for it occured again [21, 22, 23, 24, 25]. In all these papers T-duality is represented as $O(d, d)$ symmetry transformations.

In Refs.[26, 27] we doubled all bosonic coordinates and obtain the theory which contains the initial and all corresponding T-dual theories. In such theory partial T-dualization (T-dualization along some of the initial directions x^a) is represented as permutation of the corresponding coordinate subsets, x^a and y_a, which is a generalization of ideas given in [16].

When one says T-duality, one means bosonic T-duality. But since recently we can also speak about fermionic T-duality. Analyzing the gluon scattering amplitudes in $N = 4$ super Yang-Mills theory fermionic T-duality was discovered [28, 29]. Mathematically, fermionic T-duality is realized within Buscher procedure, except that dualization is performed along fermionic directions, θ^α and $\bar\theta^\alpha$.

In the present paper we are going to extend approach of double space to the type II theories both in the case of bosonic and fermionic T-duality [28, 30, 31]. We will show that double space method gives the same results as in the case of applying of standard Buscher procedure.

Here we start applying the approach of Refs.[26, 27] in the case of partial bosonic T-dualization of the type II superstring theory [1] and then in the case of fermionic T-duality. Generally, we use type II pure spinor action from [32]. This action is given in the form of an expansion in powers of fermionic coordinates θ^α and $\bar\theta^\alpha$. In both cases we simplify the action using different assumptions explained in the paper. As a result in both cases we obtain same action, which describes ghost free type II superstring theory in pure spinor formulation [33, 34, 35] in the approximation of constant background fields and up to the quadratic terms.

Bosonic T-dual transformation laws can be rewritten via double space coordinates Z^M. In order to achieve that we introduce the generalized metric $\mathcal{H}_{MN}$ and the generalized current $J_{\pm M}$. The permutation matrix $(\mathcal{T}^a)^M{}_N$ makes permutation of x^a and y_a, where index a marks the directions along which we make partial bosonic T-dualization. The T-dual coordinate is defined as ${}_aZ^M = (\mathcal{T}^a)^M{}_N Z^N$ and it has to obey the T-dual transformation law of the same form as initial coordinates, Z^M. This demand produces the expressions for T-dual generalized metric, ${}_a\mathcal{H}_{MN} = (\mathcal{T}^a\mathcal{H}\mathcal{T}^a)_{MN}$, and T-dual current, ${}_aJ_{\pm M} = (\mathcal{T}^a J_\pm)_M$. From transformation of the generalized metric we obtain T-dual NS-NS background fields and from transformation of the current we obtain T-dual NS-R fields. The transformation law for R-R field strength we get imposing additional assumtions because it is coupled by fermionic degrees of freedom along which we do not dualize.

Further we apply the method in the case of fermionic T-dualization. We are going to double fermionic sector of type II theories adding to the coordinates θ^α and $\bar\theta^\alpha$ their fermionic T-duals, ϑ_α and $\bar\vartheta_\alpha$, where $\alpha = 1, 2, \dots, 16$. Rewriting T-dual transformation laws in terms of the double coordinates, $\Theta^A = (\theta^\alpha, \vartheta_\alpha)$ and $\bar\Theta^A = (\bar\theta^\alpha, \bar\vartheta_\alpha)$, we define the "fermionic generalized metric" $\mathcal{F}_{AB}$ and the generalized currents $\bar{\mathcal{J}}_{+A}$ and $\mathcal{J}_{-A}$. The permutation matrix $\mathcal{T}^A{}_B$ exchanges $\bar\theta^\alpha$ and θ^α with their T-dual partners, $\bar\vartheta_\alpha$ and ϑ_α, respectively. From the requirement that fermionic T-dual coordinates, ${}^\star\Theta^A = \mathcal{T}^A{}_B\Theta^B$ and ${}^\star\bar\Theta^A = \mathcal{T}^A{}_B\bar\Theta^B$, have the same T-dual transformation law as initial ones, Θ^A and $\bar\Theta^A$, we obtain the expressions for fermionic T-dual generalized metric, ${}^\star\mathcal{F}_{AB} = (\mathcal{T}\mathcal{F}\mathcal{T})_{AB}$, and T-dual currents, ${}^\star\bar{\mathcal{J}}_{+A} = \mathcal{T}_A{}^B\bar{\mathcal{J}}_{+B}$ and ${}^\star\mathcal{J}_{-A} = \mathcal{T}_A{}^B\mathcal{J}_{-B}$, in terms of the initial ones. These expressions produce the expressions for fermionic T-dual NS-R fields and R-R field strength. Expressions for fermionic T-dual metric and Kalb-Ramond field are obtained separately under some assumptions.

In this article we will not present explicitely the transformation of dilaton field, which demands quantum treatment.

2. Pure spinor action of type II action

The sigma model of pure spinor type II superstring action for type II superstring [32] is of the form

$$\begin{aligned} S &= \int d^2\xi \, \big[\partial_+\theta^\alpha A_{\alpha\beta}\partial_-\bar\theta^\beta + \partial_+\theta^\alpha A_{\alpha\mu}\bar\Pi^\mu + \Pi^\mu A_{\mu\alpha}\partial_-\bar\theta^\alpha + \Pi^\mu A_{\mu\nu}\bar\Pi^\nu \\ &+ d_\alpha E^\alpha{}_\beta\partial_-\bar\theta^\beta + d_\alpha E^\alpha{}_\mu\bar\Pi^\mu + \partial_+\theta^\alpha E_\alpha{}^\beta\bar d_\beta + \Pi^\mu\bar E_\mu{}^\beta\bar d_\beta + d_\alpha \mathrm{P}^{\alpha\beta}\bar d_\beta \\ &+ \frac{1}{2}N^{\mu\nu}\Omega_{\mu\nu,\beta}\partial_-\bar\theta^\beta + \frac{1}{2}N^{\mu\nu}\Omega_{\mu\nu,\rho}\bar\Pi^\rho + \frac{1}{2}\partial_+\theta^\alpha\Omega_{\alpha,\mu\nu}\bar N^{\mu\nu} + \frac{1}{2}\Pi^\mu\Omega_{\mu,\nu\rho}\bar N^{\nu\rho} \\ &+ \frac{1}{2}N^{\mu\nu}\bar C_{\mu\nu}{}^\beta\bar d_\beta + \frac{1}{2}d_\alpha C^\alpha{}_{\mu\nu}\bar N^{\mu\nu} + \frac{1}{4}N^{\mu\nu}S_{\mu\nu,\rho\sigma}\bar N^{\rho\sigma}\Big] + S_\lambda + S_{\bar\lambda}\,, \end{aligned} \tag{1}$$

where

$$\Pi^\mu = \partial_+x^\mu + \frac{1}{2}\theta^\alpha(\Gamma^\mu)_{\alpha\beta}\partial_+\theta^\beta\,, \quad \bar\Pi^\mu = \partial_-x^\mu + \frac{1}{2}\bar\theta^\alpha(\Gamma^\mu)_{\alpha\beta}\partial_-\bar\theta^\beta\,, \tag{2}$$

$$d_\alpha = \pi_\alpha - \frac{1}{2}(\Gamma_\mu\theta)_\alpha \left[\partial_+ x^\mu + \frac{1}{4}(\theta\Gamma_\mu\partial_+\theta)\right] ,$$
$$\bar{d}_\alpha = \bar{\pi}_\alpha - \frac{1}{2}(\Gamma_\mu\bar{\theta})_\alpha \left[\partial_- x^\mu + \frac{1}{4}(\bar{\theta}\Gamma_\mu\partial_-\bar{\theta})\right] , \quad (3)$$

$$N^{\mu\nu} = \frac{1}{2} w_\alpha (\Gamma^{[\mu\nu]})^\alpha{}_\beta \lambda^\beta , \quad \bar{N}^{\mu\nu} = \frac{1}{2}\bar{w}_\alpha(\Gamma^{[\mu\nu]})^\alpha{}_\beta \bar{\lambda}^\beta . \quad (4)$$

Type II superfields generally depends on x^μ, θ^α and $\bar{\theta}^\alpha$. The superfields $A_{\mu\nu}$, $\bar{E}_\mu{}^\alpha$, $E^\alpha{}_\mu$ and $\mathrm{P}^{\alpha\beta}$ are physical superfields, because their first components are supergravity fields. The fields $\Omega_{\mu,\nu\rho}(\Omega_{\mu\nu,\rho})$, $C^\alpha{}_{\mu\nu}(\bar{C}_{\mu\nu}{}^\alpha)$ and $S_{\mu\nu,\rho\sigma}$, are curvatures (field strengths) for physical superfields. The rest fields are auxiliary superfirlds because they can be expressed in terms of the physical ones [32].

The world sheet Σ is parameterized by $\xi^m = (\xi^0 = \tau\,, \xi^1 = \sigma)$ and $\partial_\pm = \partial_\tau \pm \partial_\sigma$. Superspace is spanned by bosonic coordinates x^μ ($\mu = 0,1,2,\ldots,9$) and fermionic ones θ^α and $\bar{\theta}^\alpha$ ($\alpha = 1,2,\ldots,16$). The variables π_α and $\bar{\pi}_\alpha$ are canonically conjugated momenta to θ^α and $\bar{\theta}^\alpha$, respectively. The actions for pure spinors, S_λ and $S_{\bar{\lambda}}$, are free field actions

$$S_\lambda = \int d^2\xi w_\alpha \partial_- \lambda^\alpha , \quad S_{\bar{\lambda}} = \int d^2\xi \bar{w}_\alpha \partial_+ \bar{\lambda}^\alpha , \quad (5)$$

where λ^α and $\bar{\lambda}^\alpha$ are pure spinors and w_α and $\bar{w}_\alpha$ are their canonically conjugated momenta, respectively. The pure spinors satisfy so called pure spinor constraints

$$\lambda^\alpha(\Gamma^\mu)_{\alpha\beta}\lambda^\beta = \bar{\lambda}^\alpha(\Gamma^\mu)_{\alpha\beta}\bar{\lambda}^\beta = 0 . \quad (6)$$

3. Bosonic T-dualization

In this section we will introduce approximated action and then apply standard Buscher procedure on some subset of coordinates x^a. Then we will compare obtained result with the result following from double space formalism.

3.1. Simplification of action

The action (1) could be considered as an expansion in powers of θ^α and $\bar{\theta}^\alpha$. For computational simplicity, in the first step we neglect all terms in the action containing θ^α and $\bar{\theta}^\alpha$. As a consequence θ^α and $\bar{\theta}^\alpha$ terms disappear from $\Pi^\mu_\pm$, d_α and $\bar{d}_\alpha$ and in the solutions for physical superfields, just x-dependence of the supergravity fields survives.

Let us make T-dualization along some subset of bosonic coordinates x^a. So, we will assume that these directions are isometry ones. It essentially means that corresponding superfields (A_{ab}, $\bar{E}_a{}^\alpha$, $E^\alpha{}_a$, $\mathrm{P}^{\alpha\beta}$) should not depend on x^a. This assumption could be extended on all space-time directions x^μ which means that physical superfields are constant. According to [32], auxiliary superfields are zero, because all physical superfields are constant. Further, constant physical superfields means that their field strengths, $\Omega_{\mu,\nu\rho}(\Omega_{\mu\nu,\rho})$, $C^\alpha{}_{\mu\nu}(\bar{C}_{\mu\nu}{}^\alpha)$ and $S_{\mu\nu,\rho\sigma}$, are zero.

Background fields obey space-time field equations [36], which are some kind of consistency conditions. The equation (B.7) from this set of equations represents the backreaction of $\mathrm{P}^{\alpha\beta}$ on the metric $G_{\mu\nu}$. If we take constant dilaton Φ and constant antisymmetric NS-NS field $B_{\mu\nu}$ we obtain that

$$R_{\mu\nu} - \frac{1}{2} G_{\mu\nu} R \sim (\mathrm{P}^{\alpha\beta})^2_{\mu\nu} \,. \tag{7}$$

If we choose the background field $\mathrm{P}^{\alpha\beta}$ to be constant, in general, we will have constant Ricci tensor which means that metric tensor is quadratic function of space-time coordinates. If one wants to cancel non-quadratic terms originating from back-reaction, additional conditions must be imposed on R-R field strength (see the first reference in [33]).

Taking into account above analysis and arguments, our approximation is realized in the following way

$$\Pi^\mu_\pm \to \partial_\pm x^\mu \,, \quad d_\alpha \to \pi_\alpha \,, \quad \bar{d}_\alpha \to \bar{\pi}_\alpha \,, \tag{8}$$

and physical superfields take the form

$$\begin{aligned} A_{\mu\nu} &= \kappa(\frac{1}{2} G_{\mu\nu} + B_{\mu\nu}) \,, \quad E^\alpha_\nu = -\Psi^\alpha_\nu \,, \quad \bar{E}^\alpha_\mu = \bar{\Psi}^\alpha_\mu \,, \\ \mathrm{P}^{\alpha\beta} &= \frac{2}{\kappa} P^{\alpha\beta} = \frac{2}{\kappa} e^{\frac{\Phi}{2}} F^{\alpha\beta} \,, \end{aligned} \tag{9}$$

where $G_{\mu\nu}$ is metric tensor and $B_{\mu\nu}$ is antisymmetric NS-NS background field.

Consequently, the full action S is

$$\begin{aligned} S &= \kappa \int_\Sigma d^2\xi \left[\partial_+ x^\mu \Pi_{+\mu\nu} \partial_- x^\nu + \frac{1}{4\pi\kappa} \Phi R^{(2)} \right] \\ &+ \int_\Sigma d^2\xi \left[-\pi_\alpha \partial_-(\theta^\alpha + \Psi^\alpha_\mu x^\mu) + \partial_+(\bar{\theta}^\alpha + \bar{\Psi}^\alpha_\mu x^\mu)\bar{\pi}_\alpha + \frac{2}{\kappa}\pi_\alpha P^{\alpha\beta}\bar{\pi}_\beta \right] , \end{aligned} \tag{10}$$

where

$$\Pi_{\pm\mu\nu} = B_{\mu\nu} \pm \frac{1}{2} G_{\mu\nu} \,. \tag{11}$$

Actions S_λ and $S_{\bar{\lambda}}$ are decoupled from the rest and can be neglected in the further analysis.

3.2. Buscher T-dualization

Let some shift symmetry along x^a directions exists. The first step in Buscher procedure is that we localize the noticed shift symmetry. We substitute the ordinary derivatives with covariant ones, introducing gauge fields v^a_α. Then we add the term $\frac{1}{2}y_a F^a_{+-}$ to the Lagrangian in order to force the field strength F^a_{+-} to vanish and preserve equivalence between original and T-dual theories. Finally, we fix the gauge $x^a = 0$ and obtain gauge fixed action

$$\begin{aligned} & S_{fix}(v^a_\pm, x^i, \theta^\alpha, \bar\theta^\alpha, \pi_\alpha, \bar\pi_\alpha) = \\ & \int_\Sigma d^2\xi \left[\kappa v^a_+ \Pi_{+ab} v^b_- + \kappa v^a_+ \Pi_{+aj}\partial_- x^j + \kappa \partial_+ x^i \Pi_{+ib} v^b_- + \right. \\ + \quad & \kappa \partial_+ x^i \Pi_{+ij}\partial_- x^j + \frac{1}{4\pi}\Phi R^{(2)} - \pi_\alpha \Psi^\alpha_b v^b_- \\ + \quad & v^a_+ \bar\Psi^\alpha_a \bar\pi_\alpha - \pi_\alpha \partial_-(\theta^\alpha + \Psi^\alpha_i x^i) + \partial_+(\bar\theta^\alpha + \bar\Psi^\alpha_i x^i)\bar\pi_\alpha + \frac{1}{2\kappa} e^{\frac{\Phi}{2}} \pi_\alpha F^{\alpha\beta}\bar\pi_\beta \\ + \quad & \left. \frac{\kappa}{2}(v^a_+ \partial_- y_a - v^a_- \partial_+ y_a)\right] . \end{aligned} \tag{12}$$

Varying the gauge fixed action with respect to the Lagrange multipliers y_a we get the solution for gauge fields in the form

$$v^a_\pm = \partial_\pm x^a , \tag{13}$$

which produces the initial action, while varying with respect to the gauge fields $v^a_\pm$ we have

$$v^a_\pm = -2\kappa \hat\theta^{ab}_\pm \Pi_{\mp bi}\partial_\pm x^i - \kappa \hat\theta^{ab}_\pm \partial_\pm y_b \pm 2\hat\theta^{ab}_\pm \Psi^\alpha_{\pm b}\pi_{\pm\alpha} . \tag{14}$$

Substituting $v^a_\pm$ in (12) we find

$$\begin{aligned} & S_{fix}(y_a, x^i, \theta^\alpha, \bar\theta^\alpha, \pi_\alpha, \bar\pi_\alpha) = \\ & \int_\Sigma d^2\xi \left[\frac{\kappa^2}{2}\partial_+ y_a \hat\theta^{ab}_- \partial_- y_b + \kappa^2 \partial_+ y_a \hat\theta^{ab}_- \Pi_{+bj}\partial_- x^j - \kappa^2 \partial_+ x^i \Pi_{+ia}\hat\theta^{ab}_- \partial_- y_b \right. \\ + \quad & \kappa \partial_+ x^i (\Pi_{+ij} - 2\kappa \Pi_{+ia}\hat\theta^{ab}_- \Pi_{+bj})\partial_- x^j + \frac{1}{4\pi}\Phi R^{(2)} \\ - \quad & \pi_\alpha \partial_-(\theta^\alpha + \Psi^\alpha_i x^i - 2\kappa \Psi^\alpha_a \hat\theta^{ab}_- \Pi_{+bj} x^j - \kappa \Psi^\alpha_a \hat\theta^{ab}_- y_b) \\ + \quad & \partial_+(\bar\theta^\alpha + \bar\Psi^\alpha_i x^i + 2\kappa \bar\Psi^\alpha_a \hat\theta^{ab}_+ \Pi_{-bj} x^j + \kappa \bar\Psi^\alpha_a \hat\theta^{ab}_+ y_b)\bar\pi_\alpha \\ + \quad & \left. 2\pi_\alpha \Psi^\alpha_a \hat\theta^{ab}_- \bar\Psi^\beta_b \bar\pi_\beta + \frac{1}{2\kappa} e^{\frac{\Phi}{2}} \pi_\alpha F^{\alpha\beta}\bar\pi_\beta \right] . \end{aligned} \tag{15}$$

Combining two solutions for gauge fields (13) and (14) we obtain transformation law between initial x^a and T-dual coordinates y_a

$$\partial_+ ({}_aX)_{\hat\mu} = (\bar Q^{-1T})_{\hat\mu\nu}\partial_+ x^\nu + J_{+\hat\mu} , \quad \partial_- ({}_aX)_{\hat\mu} = (Q^{-1T})_{\hat\mu\nu}\partial_- x^\nu + J_{-\hat\mu} , \tag{16}$$

where we introduced the T-dual variables ${}_aX_{\hat\mu} = \{y_a, x^i\}$. For coordinates which contain both x^i and y_a we will use "hat" indices $\hat\mu, \hat\nu$. The matrices

$$Q^{\hat\mu\nu} = \begin{pmatrix} \kappa\hat\theta_+^{ab} & 0 \\ -2\kappa\Pi_{-ic}\hat\theta_+^{cb} & \delta_j^i \end{pmatrix}, \qquad \bar Q^{\hat\mu\nu} = \begin{pmatrix} \kappa\hat\theta_-^{ab} & 0 \\ -2\kappa\Pi_{+ic}\hat\theta_-^{cb} & \delta_j^i \end{pmatrix}, \tag{17}$$

and theirs inverse

$$Q^{-1}_{\mu\hat\nu} = \begin{pmatrix} 2\Pi_{-ab} & 0 \\ 2\Pi_{-ib} & \delta_i^j \end{pmatrix}, \qquad \bar Q^{-1}_{\mu\hat\nu} = \begin{pmatrix} 2\Pi_{+ab} & 0 \\ 2\Pi_{+ib} & \delta_i^j \end{pmatrix}, \tag{18}$$

perform T-dualization for vector indices. Here we introduced the currents $J_{\pm\hat\mu}$

$$J_{+\hat\mu} = \begin{pmatrix} J_{+a} \\ 0 \end{pmatrix}, \quad J_{-\hat\mu} = \begin{pmatrix} J_{-a} \\ 0 \end{pmatrix}, \tag{19}$$

where

$$J_{\pm\mu} = \pm\frac{2}{\kappa}\Psi^\alpha_{\pm\mu}\pi_{\pm\alpha}\,, \tag{20}$$

$$\Psi^\alpha_{+\mu} \equiv \Psi^\alpha_\mu\,, \quad \Psi^\alpha_{-\mu} \equiv \bar\Psi^\alpha_\mu\,, \quad \pi_{+\alpha} \equiv \pi_\alpha\,, \quad \pi_{-\alpha} \equiv \bar\pi_\alpha\,, \tag{21}$$

and $\hat\theta_\pm^{ab}$ is defined by the relation

$$\hat\theta_\pm^{ac}\Pi_{\mp cb} = \frac{1}{2\kappa}\delta^a{}_b\,. \tag{22}$$

Note that different chiralities transform with different matrices $Q^{\hat\mu\nu}$ and $\bar Q^{\hat\mu\nu}$. So, there are two types of T-dual vielbeins

$${}_ae^{a\hat\mu} = e^a{}_\nu(Q^T)^{\nu\hat\mu}\,, \qquad {}_a\bar e^{\underline{a}\hat\mu} = e^{\underline{a}}{}_\nu(\bar Q^T)^{\nu\hat\mu}\,, \tag{23}$$

with the same T-dual metric

$${}_aG^{\hat\mu\hat\nu} \equiv ({}_ae^T\eta\,{}_ae)^{\hat\mu\hat\nu} = (QGQ^T)^{\hat\mu\hat\nu} = {}_a\bar G^{\hat\mu\hat\nu} \equiv ({}_a\bar e^T\eta\,{}_a\bar e)^{\hat\mu\hat\nu} = (\bar QG\bar Q^T)^{\hat\mu\hat\nu}\,. \tag{24}$$

The two T-dual vielbeins are related by particular local Lorentz transformation

$${}_a\bar e^{\underline{a}\hat\mu} = \Lambda^{\underline{a}}{}_{\underline{b}}\,{}_ae^{\underline{b}\hat\mu}\,, \qquad \Lambda^{\underline{a}}{}_{\underline{b}} = e^{\underline{a}}{}_\mu(Q^{-1}\bar Q)^{T\mu}{}_\nu(e^{-1})^\nu{}_{\underline{b}}\,. \tag{25}$$

From (17) and (18) we have

$$(Q^{-1}\bar Q)^{T\mu}{}_\nu = \begin{pmatrix} \delta^a{}_b + 2\kappa\hat\theta_+^{ac}G_{cb} & 2\kappa\hat\theta_+^{ac}G_{cj} \\ 0 & \delta_j^i \end{pmatrix}, \tag{26}$$

which produces

$$\Lambda^{\underline{a}}{}_{\underline{b}} = \delta^{\underline{a}}{}_{\underline{b}} - 2\omega^{\underline{a}}{}_{\underline{b}}\,, \qquad \omega^{\underline{a}}{}_{\underline{b}} = -\kappa e^{\underline{a}}{}_a\hat\theta_+^{ab}(e^T)_b{}^{\underline{c}}\eta_{\underline{cb}}\,. \tag{27}$$

It satisfies definition of Lorentz transformations

$$\Lambda^T \eta \Lambda = \eta \quad \Longrightarrow \quad \det \Lambda^{\underline{a}}{}_{\underline{b}} = \pm 1\,, \tag{28}$$

and it holds $\det \Lambda^{\underline{a}}{}_{\underline{b}} = (-1)^d$, where d is the number of dimensions along which we perform T-duality.

Existence of the local Lorentz transformation which connects two sets of vielbeins means that in T-dual picture we can take that, for example, nonbar variables remain the same while bar variables must be multiplied by matrix ${}_a\Omega$, which is a spinorial representation of that local Lorentz symmetry. Skipping technical details explained in details in [37], we give here the final form of the matrix ${}_a\Omega$

$$_a\Omega = \sqrt{\prod_{i=1}^{d} G_{a_i a_i}}\; {}_a\Gamma\, (i\Gamma^{11})^d\,, \tag{29}$$

where

$$_a\Gamma = (i)^{\frac{d(d-1)}{2}} \prod_{i=1}^{d} \Gamma^{a_i} = (i)^{\frac{d(d-1)}{2}}\, \Gamma^{a_1}\Gamma^{a_2}\cdots\Gamma^{a_d}\,. \tag{30}$$

It is easy to check that ${}_a\Omega^2 = 1$.

Let we introduce proper fermionic variables

$$_a\theta^\alpha = \theta^\alpha\,, \quad {}_a\pi_\alpha = \pi_\alpha\,, \quad {}^\bullet\bar\theta^\alpha \equiv {}_a\Omega^\alpha{}_\beta\, {}_a\bar\theta^\beta\,, \quad {}^\bullet\bar\pi_\alpha \equiv {}_a\Omega_\alpha{}^\beta\, {}_a\bar\pi_\beta\,. \tag{31}$$

Using the action (15) and proper fermionic variables, we read T-dual background fields

$$_a\Pi_\pm^{ab} = \frac{\kappa}{2}\hat\theta_\mp^{ab}\,, \tag{32}$$

$$_a\Pi_{\pm i}{}^a = -\kappa \Pi_{\pm ib}\hat\theta_\mp^{ba}\,, \quad {}_a(\Pi_\pm)^a{}_i = \kappa\hat\theta_\mp^{ab}\Pi_{\pm bi}\,, \tag{33}$$

$$_a\Pi_{\pm ij} = \Pi_{\pm ij} - 2\kappa\Pi_{\pm ia}\hat\theta_\mp^{ab}\Pi_{\pm bj}\,, \tag{34}$$

$$_a\Psi^{\alpha a} = \kappa\hat\theta_+^{ab}\Psi_b^\alpha\,, \quad {}_a\bar\Psi^{\alpha a} = \kappa\, {}_a\Omega^\alpha{}_\beta\, \hat\theta_-^{ab}\bar\Psi_b^\beta\,, \tag{35}$$

$$_a\Psi_i^\alpha = \Psi_i^\alpha - 2\kappa\Pi_{-ib}\hat\theta_+^{ba}\Psi_a^\alpha\,, \quad {}_a\bar\Psi_i^\alpha = {}_a\Omega^\alpha{}_\beta(\bar\Psi_i^\beta - 2\kappa\Pi_{+ib}\hat\theta_-^{ba}\bar\Psi^\beta) \tag{36}$$

$$e^{\frac{a\Phi}{2}}\, {}_aF^{\alpha\beta} = (e^{\frac{\Phi}{2}}\, F^{\alpha\gamma} + 4\kappa\Psi_a^\alpha\hat\theta_-^{ab}\bar\Psi_b^\gamma)\, {}_a\Omega_\gamma{}^\beta\,. \tag{37}$$

3.3. Double space

Above expressions for T-dual background fields in the case of full T-dualization are

$$^\star\Pi_\pm^{\mu\nu} \equiv {}^\star B^{\mu\nu} \pm \frac{1}{2}{}^\star G^{\mu\nu} = \frac{\kappa}{2}\Theta_\mp^{\mu\nu}\,, \tag{38}$$

$$^\star\Psi^{\alpha\mu} = \kappa\Theta_+^{\mu\nu}\Psi_\nu^\alpha\,, \quad {}^\star\bar\Psi^{\alpha\mu} = \kappa\, {}^\star\Omega^\alpha{}_\beta\, \Theta_-^{\mu\nu}\bar\Psi_\nu^\beta\,, \tag{39}$$

$$e^{\frac{\star\Phi}{2}}\,{}^\star F^{\alpha\beta} = (e^{\frac{\Phi}{2}}F^{\alpha\gamma} + 4\kappa\Psi^\alpha_\mu\Theta^{\mu\nu}_-\bar{\Psi}^\gamma_\nu)^\star\Omega_\gamma{}^\beta\,. \tag{40}$$

Here we use the notation

$$G^E_{\mu\nu} = G_{\mu\nu} - 4(BG^{-1}B)_{\mu\nu}\,, \quad \Theta^{\mu\nu} = -\frac{2}{\kappa}(G_E^{-1}BG^{-1})^{\mu\nu}\,, {}^\star\Omega = -\Gamma^{11}, \tag{41}$$

and

$$\Theta^{\mu\nu}_\pm = -\frac{2}{\kappa}(G_E^{-1}\Pi_\pm G^{-1})^{\mu\nu} = \Theta^{\mu\nu} \mp \frac{1}{\kappa}(G_E^{-1})^{\mu\nu}\,. \tag{42}$$

In this case the T-dual transformation laws (16) obtain the form

$$\partial_\pm x^\mu \cong -\kappa\Theta^{\mu\nu}_\pm\partial_\pm y_\nu + \kappa\Theta^{\mu\nu}_\pm J_{\pm\nu}\,, \qquad \partial_\pm y_\mu \cong -2\Pi_{\mp\mu\nu}\partial_\pm x^\nu + J_{\pm\mu}\,. \tag{43}$$

In terms of double coordinates

$$Z^M = \begin{pmatrix} x^\mu \\ y_\mu \end{pmatrix}, \tag{44}$$

the relations (16) are replaced by one

$$\partial_\pm Z^M \cong \pm\Omega^{MN}\left(\mathcal{H}_{NP}\partial_\pm Z^P + J_{\pm N}\right), \tag{45}$$

where the matrix $\mathcal{H}_{MN}$ is generalized metric and has the form

$$\mathcal{H}_{MN} = \begin{pmatrix} G^E_{\mu\nu} & -2\,B_{\mu\rho}(G^{-1})^{\rho\nu} \\ 2(G^{-1})^{\mu\rho}\,B_{\rho\nu} & (G^{-1})^{\mu\nu} \end{pmatrix}. \tag{46}$$

The double current $J_{\pm M}$ is defined as

$$J_{\pm M} = \begin{pmatrix} 2(\Pi_\pm G^{-1})_\mu{}^\nu J_{\pm\nu} \\ -(G^{-1})^{\mu\nu}J_{\pm\nu} \end{pmatrix}, \tag{47}$$

and

$$\Omega^{MN} = \begin{pmatrix} 0 & 1_D \\ 1_D & 0 \end{pmatrix}, \tag{48}$$

is constant symmetric matrix. Here 1_D denotes the identity operator in D dimensions.

It is known that equations of motion of initial theory are Bianchi identities in T-dual picture and vice versa [9, 13, 16, 38]. From Bianchi identity

$$\partial_+\partial_-Z^M - \partial_-\partial_+Z^M = 0\,, \tag{49}$$

and relation (45), we obtain the consistency condition

$$\partial_+\left[\mathcal{H}_{MN}\partial_-Z^N + J_{-M}\right] + \partial_-\left[\mathcal{H}_{MN}\partial_+Z^N + J_{+M}\right] = 0\,. \tag{50}$$

The equation (50) is equation of motion of the following action

$$S = \frac{\kappa}{4}\int d^2\xi \left[\partial_+ Z^M \mathcal{H}_{MN}\partial_- Z^N + \partial_+ Z^M J_{-M} + J_{+M}\partial_- Z^M + L(\pi_\alpha, \bar{\pi}_\alpha)\right], \tag{51}$$

where $L(\pi_\alpha, \bar{\pi}_\alpha)$ is arbitrary functional of fermionic momenta.

Let us split coordinate index μ into a and i ($a = 0, \cdots, d-1$, $i = d, \cdots, D-1$) and denote T-dualization along direction x^a and y_a as

$$\mathcal{T}^a = T^a \circ T_a\,, \quad T^a \equiv T^0 \circ T^1 \circ \cdots \circ T^{d-1}\,, \quad T_a \equiv T_0 \circ T_1 \circ \cdots \circ T_{d-1}\,, \tag{52}$$

where $\circ$ marks the operation of composition of T-dualizations. Permutation of the initial coordinates x^a with its T-dual y_a we realize by multiplying double space coordinate by the constant symmetric matrix $(\mathcal{T}^a)^M{}_N$

$${}_aZ^M \equiv \begin{pmatrix} y_a \\ x^i \\ x^a \\ y_i \end{pmatrix} = (\mathcal{T}^a)^M{}_N Z^N \equiv \begin{pmatrix} 0 & 0 & 1_a & 0 \\ 0 & 1_i & 0 & 0 \\ 1_a & 0 & 0 & 0 \\ 0 & 0 & 0 & 1_i \end{pmatrix} \begin{pmatrix} x^a \\ x^i \\ y_a \\ y_i \end{pmatrix}, \tag{53}$$

where 1_a and 1_i are identity operators in the subspaces spanned by x^a and x^i, respectively. We demand that double T-dual coordinate ${}_aZ^M$ satisfy the T-duality transformations of the form as initial one Z^M (45)

$$\partial_\pm\, {}_aZ^M \cong \pm\Omega^{MN}\left({}_a\mathcal{H}_{NK}\,\partial_\pm\, {}_aZ^K + {}_aJ_{\pm N}\right). \tag{54}$$

From this relation we find the T-dual generalized metric

$${}_a\mathcal{H}_{MN} = (\mathcal{T}^a)_M{}^K \mathcal{H}_{KL} (\mathcal{T}^a)^L{}_N\,, \tag{55}$$

and T-dual current

$${}_aJ_{\pm M} = (\mathcal{T}^a)_M{}^N J_{\pm N}\,. \tag{56}$$

Demanding that the T-dual generalized metric ${}_a\mathcal{H}_{MN}$ has the same form as the initial one $\mathcal{H}_{MN}$ (46)

$${}_a\mathcal{H}_{MN} = \begin{pmatrix} {}_aG_E^{\mu\nu} & -2({}_aB\,{}_aG^{-1})^\mu{}_\nu \\ 2({}_aG^{-1}\,{}_aB)_\mu{}^\nu & ({}_aG^{-1})_{\mu\nu} \end{pmatrix}, \tag{57}$$

we obtain the T-dual NS-NS background fields

$${}_a\Pi_\pm^{ab} = \frac{\kappa}{2}\hat{\theta}_\mp^{ab}\,, \qquad {}_a\Pi^a_{\pm i} = \kappa\hat{\theta}_\mp^{ab}\Pi_{\pm bi}\,, \tag{58}$$

$${}_a\Pi_{\pm i}{}^a = -\kappa\Pi_{\pm ib}\hat{\theta}_\mp^{ba}\,, \quad {}_a\Pi_{\pm ij} = \Pi_{\pm ij} - 2\kappa\Pi_{\pm ia}\hat{\theta}_\mp^{ab}\Pi_{\pm bj}\,, \tag{59}$$

which are in full agreement with those from the Refs.[6, 14].

The T-dual current ${}_aJ_{\pm M}$ (56) should have the same form as initial one (47) but in terms of the T-dual background fields

$$\begin{pmatrix} 2({}_a\Pi_{\pm}\,{}_aG^{-1})^a{}_b\,({}_aJ)^b_{\pm} + 2({}_a\Pi_{\pm}\,{}_aG^{-1})^{ai}\,({}_aJ)_{\pm i} \\ 2({}_a\Pi_{\pm}\,{}_aG^{-1})_{ia}\,({}_aJ)^a_{\pm} + 2({}_a\Pi_{\pm}\,{}_aG^{-1})_i{}^j\,({}_aJ)_{\pm j} \\ -({}_aG^{-1})_{ab}\,({}_aJ)^b_{\pm} - ({}_aG^{-1})_a{}^i\,({}_aJ)_{\pm i} \\ -({}_aG^{-1})^i{}_a\,({}_aJ)^a_{\pm} - ({}_aG^{-1})^{ij}\,({}_aJ)_{\pm j} \end{pmatrix} = \begin{pmatrix} -(G^{-1})^{a\mu}J_{\pm\mu} \\ 2(\Pi_{\pm}G^{-1})_i{}^{\mu}J_{\pm\mu} \\ 2(\Pi_{\pm}G^{-1})_a{}^{\mu}J_{\pm\mu} \\ -(G^{-1})^{i\mu}J_{\pm\mu} \end{pmatrix}. \tag{60}$$

From the lower D components of the above equation, after straightforward calculation we get

$${}_a\Psi^{\alpha a} = \kappa\hat{\theta}^{ab}_{+}\Psi^{\alpha}_{b}\,, \quad {}_a\bar{\Psi}^{\alpha a} = \kappa\,{}_a\Omega^{\alpha}{}_{\beta}\hat{\theta}^{ab}_{-}\bar{\Psi}^{\beta}_{b}\,. \tag{61}$$

$${}_a\Psi^{\alpha}_{i} = \Psi^{\alpha}_{i} - 2\kappa\Pi_{-ib}\hat{\theta}^{ba}_{+}\Psi^{\alpha}_{a}\,, \quad {}_a\bar{\Psi}^{\alpha}_{i} = {}_a\Omega^{\alpha}{}_{\beta}(\bar{\Psi}^{\beta}_{i} - 2\kappa\Pi_{+ib}\hat{\theta}^{ba}_{-}\bar{\Psi}^{\beta}_{a})\,. \tag{62}$$

which is in full agreement with results obtained applying standard Buscher procedure. The upper D components of Eq.(60) produce the same result for T-dual background fields.

The R-R field strength $F^{\alpha\beta}$ appears in the action (10) coupled with fermionic momenta π_α and $\bar{\pi}_\alpha$ along which we do not perform T-dualization. Let us suppose that fermionic term $L(\pi_\alpha, \bar{\pi}_\alpha)$ (51) in the form

$$L = e^{\frac{\Phi}{2}}\,\pi_\alpha\,F^{\alpha\beta}\bar{\pi}_\beta + e^{\frac{{}_a\Phi}{2}}\,{}_a\pi_\alpha\,{}_aF^{\alpha\beta}\,{}_a\bar{\pi}_\beta \equiv \mathcal{L} + {}_a\mathcal{L}\,, \tag{63}$$

for some $F^{\alpha\beta}$ and ${}_aF^{\alpha\beta}$. This term should be invariant under T-dual transformation

$${}_a\mathcal{L} = \mathcal{L} + \Delta\mathcal{L}\,. \tag{64}$$

Taking into account the fact that two successive T-dualization are identity transformation, we obtain that the T-dual R-R field strength has the form

$$e^{\frac{{}_a\Phi}{2}}\,{}_aF^{\alpha\beta} = (e^{\frac{\Phi}{2}}F^{\alpha\gamma} + c\Psi^{\alpha}_{a}\hat{\theta}^{ab}_{-}\bar{\Psi}^{\gamma}_{b})\,{}_a\Omega_{\gamma}{}^{\beta}\,. \tag{65}$$

For $c = 4\kappa$ we obtain the agreement with the expression (37).

4. Fermionic T-dualization

4.1. Action

We start with the action (1). In order to perform fermionic T-duality we must avoid explicit dependence of background fields on the fermionic coordinates θ^α and $\bar{\theta}^\alpha$ and allow only dependence on the σ and τ derivatives of these coordinates. This assumption produces that the auxiliary superfields are zero what can be seen from Eq.(5.5) of Ref.[32].

We choose that $G_{\mu\nu}$, $B_{\mu\nu}$, Φ, Ψ^{α}_{μ} and $\bar{\Psi}^{\alpha}_{\mu}$ are constant, and corresponding field strengths, $\Omega_{\mu,\nu\rho}(\Omega_{\mu\nu,\rho})$, $C^{\alpha}{}_{\mu\nu}(\bar{C}_{\mu\nu}{}^{\alpha})$ and $S_{\mu\nu,\rho\sigma}$, are zero. The only nontrivial contribution of the quadratic terms in equations of motion (see

[36]) comes from constant field strength $\mathrm{P}^{\alpha\beta}$. In order to analyze this issue we will use relations from Eq.(3.6) of Ref.[32] labeled by $(\frac{1}{2}, \frac{3}{2}, \frac{3}{2})$

$$D_\alpha \mathrm{P}^{\beta\gamma} - \frac{1}{4}(\Gamma^{\mu\nu})_\alpha{}^\beta \bar{C}_{\mu\nu}{}^\gamma = 0\,, \quad \bar{D}_\alpha \mathrm{P}^{\beta\gamma} - \frac{1}{4}(\Gamma^{\mu\nu})_\alpha{}^\gamma C^\beta{}_{\mu\nu} = 0\,. \tag{66}$$

Here

$$D_\alpha = \frac{\partial}{\partial\theta^\alpha} + \frac{1}{2}(\Gamma^\mu\theta)_\alpha \frac{\partial}{\partial x^\mu}\,, \quad \bar{D}_\alpha = \frac{\partial}{\partial\bar{\theta}^\alpha} + \frac{1}{2}(\Gamma^\mu\bar{\theta})_\alpha \frac{\partial}{\partial x^\mu}\,, \tag{67}$$

are superspace covariant derivatives and $C^\alpha{}_{\mu\nu}$ and $\bar{C}_{\mu\nu}{}^\alpha$ are field strengths for gravitino fields Ψ^α_μ and $\bar{\Psi}^\alpha_\mu$, respectively. In order to perform fermionic T-dualization along all fermionic directions, θ^α and $\bar{\theta}^\alpha$, we assume that they are Killing spinors which means

$$\frac{\partial \mathrm{P}^{\beta\gamma}}{\partial\theta^\alpha} = \frac{\partial \mathrm{P}^{\beta\gamma}}{\partial\bar{\theta}^\alpha} = 0\,. \tag{68}$$

From the equations (66) it follows

$$(\Gamma^\mu)_{\alpha\delta}\partial_\mu \mathrm{P}^{\beta\gamma} = 0\,. \tag{69}$$

Our choice of constant $\mathrm{P}^{\alpha\beta}$ is consistent with this condition.

Under these assumptions the final form of the action is given by the expression (10). The fermionic part of the action (10) has the form of the first order theory. On the equations of motion for fermionic momenta π_α and $\bar{\pi}_\alpha$,

$$\pi_\alpha = -\frac{\kappa}{2}\partial_+\left(\bar{\theta}^\beta + \bar{\Psi}^\beta_\mu x^\mu\right)(P^{-1})_{\beta\alpha}\,, \quad \bar{\pi}_\alpha = \frac{\kappa}{2}(P^{-1})_{\alpha\beta}\partial_-\left(\theta^\beta + \Psi^\beta_\mu x^\mu\right)\,, \tag{70}$$

the action gets the form

$$\begin{aligned}
S(\partial_\pm x, \partial_-\theta, \partial_+\bar{\theta}) &= \kappa \int_\Sigma d^2\xi \partial_+ x^\mu \Pi_{+\mu\nu}\partial_- x^\nu + \frac{1}{4\pi}\int_\Sigma d^2\xi \Phi R^{(2)} \\
&+ \frac{\kappa}{2}\int_\Sigma d^2\xi \partial_+\left(\bar{\theta}^\alpha + \bar{\Psi}^\alpha_\mu x^\mu\right)(P^{-1})_{\alpha\beta}\partial_-\left(\theta^\beta + \Psi^\beta_\nu x^\nu\right) \\
&= \kappa \int_\Sigma d^2\xi \partial_+ x^\mu \left[\Pi_{+\mu\nu} + \frac{1}{2}\bar{\Psi}^\alpha_\mu (P^{-1})_{\alpha\beta}\Psi^\beta_\nu\right]\partial_- x^\nu + \frac{1}{4\pi}\int_\Sigma d^2\xi \Phi R^{(2)} \\
&+ \frac{\kappa}{2}\int_\Sigma d^2\xi \left[\partial_+\bar{\theta}^\alpha (P^{-1})_{\alpha\beta}\partial_-\theta^\beta + \partial_+\bar{\theta}^\alpha (P^{-1}\Psi)_{\alpha\mu}\partial_- x^\nu\right. \\
&+ \left.\partial_+ x^\mu (\bar{\Psi}P^{-1})_{\mu\alpha}\partial_-\theta^\alpha\right]\,.
\end{aligned} \tag{71}$$

We notice that theory has a local symmetry

$$\delta\theta^\alpha = \varepsilon^\alpha(\sigma^+)\,, \quad \delta\bar\theta^\alpha = \bar\varepsilon^\alpha(\sigma^-)\,, \quad (\sigma^\pm = \tau \pm \sigma)\,. \tag{72}$$

The corresponding BRST transformations are

$$s\theta^\alpha = c^\alpha(\sigma^+)\,, \quad s\bar\theta^\alpha = \bar c^\alpha(\sigma^-)\,, \tag{73}$$

where for each gauge parameter $\varepsilon^\alpha(\sigma^+)$ and $\bar\varepsilon^\alpha(\sigma^-)$ we introduced the ghost fields $c^\alpha(\sigma^+)$ and $\bar c^\alpha(\sigma^-)$, respectively. Here s denotes BRST nilpotent operator.

To fix gauge freedom we introduce gauge fermion with ghost number -1

$$\Psi = \frac{\kappa}{2}\int d^2\xi\left[\bar C_\alpha\left(\partial_+\theta^\alpha + \frac{\alpha^{\alpha\beta}}{2}b_{+\beta}\right) + \left(\partial_-\bar\theta^\alpha + \frac{1}{2}\bar b_{-\beta}\alpha^{\beta\alpha}\right)C_\alpha\right]\,, \tag{74}$$

where $\alpha^{\alpha\beta}$ is arbitrary non singular matrix, $\bar C_\alpha$ and C_α are antighost fields, while $b_{+\alpha}$ and $\bar b_{-\alpha}$ are Nakanishi-Lautrup auxillary fields which satisfy

$$sC_\alpha = b_{+\alpha}\,, \quad s\bar C_\alpha = \bar b_{-\alpha}\,, \quad sb_{+\alpha} = 0 \quad s\bar b_{-\alpha} = 0\,. \tag{75}$$

BRST transformation of gauge fermion Ψ produces the gauge fixed and Fadeev-Popov action

$$\begin{aligned} s\Psi &= S_{gf} + S_{FP}\,,\\ S_{gf} &= \frac{\kappa}{2}\int d^2\xi\left[\bar b_{-\alpha}\partial_+\theta^\alpha + \partial_-\bar\theta^\alpha b_{+\alpha} + \bar b_{-\alpha}\alpha^{\alpha\beta}b_{+\beta}\right]\,,\\ S_{FD} &= \frac{\kappa}{2}\int d^2\xi\left[\bar C_\alpha\partial_+c^\alpha + (\partial_-c^\alpha)C_\alpha\right]\,. \end{aligned} \tag{76}$$

The Fadeev-Popov action is decoupled from the rest and, consequently, it can be omitted in further analysis. On the equations of motion for b-fields

$$b_{+\alpha} = -(\alpha^{-1})_{\alpha\beta}\partial_+\theta^\alpha\,, \quad \bar b_{-\alpha} = -\partial_-\bar\theta^\beta(\alpha^{-1})_{\beta\alpha}\,, \tag{77}$$

we obtain the final form of the BRST gauge fixed action

$$S_{gf} = -\frac{\kappa}{2}\int d^2\xi\partial_-\bar\theta^\alpha(\alpha^{-1})_{\alpha\beta}\partial_+\theta^\beta\,. \tag{78}$$

4.2. Fermionic T-duality using Buscher rules

As in the bosonic case we introduce gauge fields $v_\pm^\alpha$ and $\bar v_\pm^\alpha$ and replace ordinary world-sheet derivatives with covariant ones

$$\partial_\pm\theta^\alpha \to D_\pm\theta^\alpha \equiv \partial_\pm\theta^\alpha + v_\pm^\alpha\,, \quad \partial_\pm\bar\theta^\alpha \to D_\pm\bar\theta^\alpha \equiv \partial_\pm\bar\theta^\alpha + \bar v_\pm^\alpha\,. \tag{79}$$

In order to make the fields $v_\pm^\alpha$ and $\bar{v}_\pm^\alpha$ to be unphysical we add the following terms in the action

$$S_{gauge}(\vartheta, v_\pm, \bar\vartheta, \bar{v}_\pm) = \frac{1}{2}\kappa\int_\Sigma d^2\xi \bar\vartheta_\alpha(\partial_+ v_-^\alpha - \partial_- v_+^\alpha) + \frac{1}{2}\kappa\int_\Sigma d^2\xi(\partial_+\bar{v}_-^\alpha - \partial_-\bar{v}_+^\alpha)\vartheta_\alpha\,, \tag{80}$$

where ϑ_α and $\bar\vartheta_\alpha$ are Lagrange multipliers. The full gauge invariant action is of the form

$$\begin{aligned} S_{inv}(x,\theta,\bar\theta,\vartheta,\bar\vartheta,v_\pm,\bar{v}_\pm) &= S(\partial_\pm x, D_-\theta, D_+\bar\theta) \\ &+ S_{gf}(D_-\theta, D_+\bar\theta) + S_{gauge}(\vartheta,\bar\vartheta,v_\pm,\bar{v}_\pm)\,. \end{aligned} \tag{81}$$

Fixing θ^α and $\bar\theta^\alpha$ to zero we obtain the gauge fixed action

$$\begin{aligned} S_{fix} &= \kappa\int_\Sigma d^2\xi\partial_+ x^\mu\left[\Pi_{+\mu\nu} + \frac{1}{2}\bar\Psi_\mu^\alpha(P^{-1})_{\alpha\beta}\Psi_\nu^\beta\right]\partial_- x^\nu + \frac{1}{4\pi}\int_\Sigma d^2\xi\Phi R^{(2)} \\ &+\frac{\kappa}{2}\int_\Sigma\left[\bar{v}_+^\alpha(P^{-1})_{\alpha\beta}v_-^\beta + \bar{v}_+^\alpha(P^{-1})_{\alpha\beta}\Psi_\nu^\beta\partial_- x^\nu\right. \\ &+ \left.\partial_+ x^\mu\bar\Psi_\mu^\alpha(P^{-1})_{\alpha\beta}v_-^\beta - \bar{v}_-^\alpha(\alpha^{-1})_{\alpha\beta}v_+^\beta\right] \\ &+\frac{\kappa}{2}\int_\Sigma d^2\xi\bar\vartheta_\alpha(\partial_+ v_-^\alpha - \partial_- v_+^\alpha) + \frac{\kappa}{2}\int_\Sigma d^2\xi(\partial_+\bar{v}_-^\alpha - \partial_-\bar{v}_+^\alpha)\vartheta_\alpha\,. \end{aligned} \tag{82}$$

Varying the above action with respect to the Lagrange multipliers we obtain the initial action (71) because

$$\partial_+ v_-^\alpha - \partial_- v_+^\alpha = 0 \Longrightarrow v_\pm^\alpha = \partial_\pm\theta^\alpha\,,\quad \partial_+\bar{v}_-^\alpha - \partial_-\bar{v}_+^\alpha = 0 \Longrightarrow \bar{v}_\pm^\alpha = \partial_\pm\bar\theta^\alpha\,. \tag{83}$$

On the other side, the equations of motion for $v_\pm^\alpha$ and $\bar{v}_\pm^\alpha$ give

$$\bar{v}_-^\alpha = \partial_-\bar\vartheta_\beta\alpha^{\beta\alpha}\,,\quad \bar{v}_+^\alpha = \partial_+\bar\vartheta_\beta P^{\beta\alpha} - \partial_+ x^\mu\bar\Psi_\mu^\alpha\,, \tag{84}$$

$$v_+^\alpha = -\alpha^{\alpha\beta}\partial_+\vartheta_\beta\,,\quad v_-^\alpha = -P^{\alpha\beta}\partial_-\vartheta_\beta - \Psi_\mu^\alpha\partial_- x^\mu\,. \tag{85}$$

Substituting these expressions in the action S_{fix} we obtain the fermionic T-dual action

$$\begin{aligned} {}^\star S(\partial_\pm x, \partial_-\vartheta, \partial_+\bar\vartheta) &= \kappa\int_\Sigma d^2\xi\partial_+ x^\mu\Pi_{+\mu\nu}\partial_- x^\nu + \frac{1}{4\pi}\int_\Sigma d^2\xi\ {}^\star\Phi R^{(2)}\,, \\ &+\frac{\kappa}{2}\int_\Sigma d^2\xi\left[\partial_+\bar\vartheta_\alpha P^{\alpha\beta}\partial_-\vartheta_\beta\right. \\ &- \left.\partial_+ x^\mu\bar\Psi_\mu^\alpha\partial_-\vartheta_\alpha + \partial_+\bar\vartheta_\alpha\Psi_\mu^\alpha\partial_- x^\mu - \partial_-\bar\vartheta_\alpha\alpha^{\alpha\beta}\partial_+\vartheta_\beta\right]\,. \end{aligned} \tag{86}$$

We read fermionic T-dual background fields

$${}^\star\Psi_{\alpha\mu} = (P^{-1}\Psi)_{\alpha\mu}\,,\ {}^\star\bar\Psi_{\mu\alpha} = -(\bar\Psi P^{-1})_{\mu\alpha}\,, \tag{87}$$

$$^\star P_{\alpha\beta} = (P^{-1})_{\alpha\beta}\,, \quad {}^\star\alpha_{\alpha\beta} = (\alpha^{-1})_{\alpha\beta}\,. \tag{88}$$

From the condition

$$^\star\Pi_{+\mu\nu} + \frac{1}{2}{}^\star\bar\Psi_{\alpha\mu}\,({}^\star P^{-1})^{\alpha\beta}\,{}^\star\Psi_{\beta\nu} = \Pi_{+\mu\nu}\,, \tag{89}$$

the fermionic T-dual metric and Kalb-Ramond field are

$$\begin{aligned} {}^\star G_{\mu\nu} &= G_{\mu\nu} + \frac{1}{2}\Big[(\bar\Psi P^{-1}\Psi)_{\mu\nu} + (\bar\Psi P^{-1}\Psi)_{\nu\mu}\Big]\,, \\ {}^\star B_{\mu\nu} &= B_{\mu\nu} + \frac{1}{4}\Big[(\bar\Psi P^{-1}\Psi)_{\mu\nu} - (\bar\Psi P^{-1}\Psi)_{\nu\mu}\Big]\,. \end{aligned} \tag{90}$$

We obtain T-dual transformation laws combining the different solutions of equations of motion for $v^\alpha_\pm$ and $\bar v^\alpha_\pm$ (83) and (84)-(85)

$$\partial_-\theta^\alpha \cong -P^{\alpha\beta}\partial_-\vartheta_\beta - \Psi^\alpha_\mu\partial_-x^\mu\,, \quad \partial_+\bar\theta^\alpha \cong \partial_+\bar\vartheta_\beta P^{\beta\alpha} - \partial_+x^\mu\bar\Psi^\alpha_\mu\,, \tag{91}$$

$$\partial_+\theta^\alpha \cong -\alpha^{\alpha\beta}\partial_+\vartheta_\beta\,, \quad \partial_-\bar\theta^\alpha \cong \partial_-\bar\vartheta_\beta\alpha^{\beta\alpha}\,. \tag{92}$$

From these relations we can obtain inverse transformation rules

$$\begin{aligned} \partial_-\vartheta_\alpha &\cong -(P^{-1})_{\alpha\beta}\partial_-\theta^\beta - (P^{-1})_{\alpha\beta}\Psi^\beta_\mu\partial_-x^\mu\,, \\ \partial_+\bar\vartheta_\alpha &\cong \partial_+\bar\theta^\beta(P^{-1})_{\beta\alpha} + \partial_+x^\mu\bar\Psi^\beta_\mu(P^{-1})_{\beta\alpha}\,, \end{aligned} \tag{93}$$

$$\partial_+\vartheta_\alpha \cong -(\alpha^{-1})_{\alpha\beta}\partial_+\theta^\beta\,, \quad \partial_-\bar\vartheta_\alpha \cong \partial_-\bar\theta^\beta(\alpha^{-1})_{\beta\alpha}\,. \tag{94}$$

4.3. Fermionic T-dualization using double space

Now we will extend the meaning of the double space and double both fermionic coordinate as

$$\Theta^A = \begin{pmatrix} \theta^\alpha \\ \vartheta_\alpha \end{pmatrix}, \quad \bar\Theta^A = \begin{pmatrix} \bar\theta^\alpha \\ \bar\vartheta_\alpha \end{pmatrix}. \tag{95}$$

The transformation laws, (91)-(94), can be rewritten in the form

$$\partial_-\Theta^A \cong -\Omega^{AB}\Big[\mathcal{F}_{BC}\partial_-\Theta^C + \mathcal{J}_{-B}\Big]\,, \quad \partial_+\bar\Theta^A \cong \Big[\partial_+\bar\Theta^C\mathcal{F}_{CB} + \bar{\mathcal{J}}_{+B}\Big]\,\Omega^{BA}\,, \tag{96}$$

$$\partial_+\Theta^A \cong -\Omega^{AB}\mathcal{A}_{BC}\partial_+\Theta^C\,, \quad \partial_-\bar\Theta^A \cong \partial_-\bar\Theta^C\mathcal{A}_{CB}\Omega^{BA}\,, \tag{97}$$

where "fermionic generalized metric" $\mathcal{F}_{AB}$ has the form

$$\mathcal{F}_{AB} = \begin{pmatrix} (P^{-1})_{\alpha\beta} & 0 \\ 0 & P^{\gamma\delta} \end{pmatrix}, \tag{98}$$

and

$$\mathcal{A}_{AB} = \begin{pmatrix} (\alpha^{-1})_{\alpha\beta} & 0 \\ 0 & \alpha^{\gamma\delta} \end{pmatrix}. \tag{99}$$

The double currents, $\bar{\mathcal{J}}_{+A}$ and $\mathcal{J}_{-A}$, are of the form

$$\bar{\mathcal{J}}_{+A} = \begin{pmatrix} (\bar{\Psi}P^{-1})_{\mu\alpha}\partial_+ x^\mu \\ -\bar{\Psi}^\alpha_\mu \partial_+ x^\mu \end{pmatrix}, \quad \mathcal{J}_{-A} = \begin{pmatrix} (P^{-1}\Psi)_{\alpha\mu}\partial_- x^\mu \\ \Psi^\alpha_\mu \partial_- x^\mu \end{pmatrix}. \tag{100}$$

Let us introduce the permutation matrix

$$\mathcal{T}^A{}_B = \begin{pmatrix} 0 & 1 \\ 1 & 0 \end{pmatrix}, \tag{101}$$

so that double T-dual coordinates are

$${}^\star\Theta^A = \mathcal{T}^A{}_B \Theta^B, \quad {}^\star\bar{\Theta}^A = \mathcal{T}^A{}_B \bar{\Theta}^B. \tag{102}$$

As in the case of bosonic T-dualization, from demand that T-dual transformation laws for T-dual coordinates ${}^\star\Theta^A$ and ${}^\star\bar{\Theta}^A$ have the same form as for initial ones Θ^A and $\bar{\Theta}^A$ we get the fermionic T-dual "generalized metric" ${}^\star\mathcal{F}_{AB}$ and T-dual currents, ${}^\star\bar{\mathcal{J}}_{+A}$ and ${}^\star\mathcal{J}_{-A}$

$${}^\star\mathcal{F}_{AB} = \mathcal{T}_A{}^C \mathcal{F}_{CD} \mathcal{T}^D{}_B, \quad {}^\star\bar{\mathcal{J}}_{+A} = \mathcal{T}_A{}^B \bar{\mathcal{J}}_{+B}, \quad {}^\star\mathcal{J}_{-A} = \mathcal{T}_A{}^B \mathcal{J}_{-B}. \tag{103}$$

The matrix $\mathcal{A}_{AB}$ transforms as

$${}^\star\mathcal{A}_{AB} = \mathcal{T}_A{}^C \mathcal{A}_{CD} \mathcal{T}^D{}_B = (\mathcal{A}^{-1})_{AB}. \tag{104}$$

From the first relation in (103) we obtain the form of the fermionic T-dual R-R background field

$${}^\star P_{\alpha\beta} = (P^{-1})_{\alpha\beta}, \tag{105}$$

while from the second and third equation we obtain the form of the fermionic T-dual NS-R background fields

$${}^\star\Psi_{\alpha\mu} = (P^{-1})_{\alpha\beta}\Psi^\beta_\mu, \quad {}^\star\bar{\Psi}_{\alpha\mu} = -\bar{\Psi}^\beta_\mu (P^{-1})_{\beta\alpha}. \tag{106}$$

The non singular matrix $\alpha^{\alpha\beta}$ transforms as

$$({}^\star\alpha)_{\alpha\beta} = (\alpha^{-1})_{\alpha\beta}. \tag{107}$$

As in the case of bosonic T-dualization, in the same way how we obtained the double action (51), we get the double action corresponding to the fermionic T-dual transformation law

$$\begin{aligned} S_{double}(\Theta, \bar{\Theta}) &= \qquad (108) \\ &= \frac{\kappa}{2}\int d^2\xi \Big[\partial_+\bar{\Theta}^A \mathcal{F}_{AB}\partial_-\Theta^B + \bar{\mathcal{J}}_{+A}\partial_-\Theta^A + \partial_+\bar{\Theta}^A \mathcal{J}_{-A} \\ &- \partial_-\bar{\Theta}^A \mathcal{A}_{AB}\partial_+\Theta^B + L(x)\Big], \end{aligned}$$

where $L(x)$ is arbitrary functional of the bosonic coordinates. In order to find fermionic T-dual metric and Kalb-Ramond field we suppose that term $L(x)$ has the form

$$L(x) = 2\partial_+ x^\mu \left(\Pi_{+\mu\nu} + {}^\star\Pi_{+\mu\nu}\right) \partial_- x^\nu \equiv \mathcal{L} + {}^\star\mathcal{L} \,. \tag{109}$$

This term should be invariant under T-dual transformation

$${}^\star\mathcal{L} = \mathcal{L} + \Delta\mathcal{L} \,. \tag{110}$$

Using the fact that two successive T-dualization are identity transformation, we obtain

$$\mathcal{L} = {}^\star\mathcal{L} + {}^\star\Delta\mathcal{L} \,. \tag{111}$$

Combining last two relations we get

$${}^\star\Delta\mathcal{L} = -\Delta\mathcal{L} \,. \tag{112}$$

If $\Delta\mathcal{L} = 2\partial_+ x^\mu \Delta_{\mu\nu} \partial_- x^\nu$, we obtain the condition for $\Delta_{\mu\nu}$

$${}^\star\Delta_{\mu\nu} = -\,\Delta_{\mu\nu} \,. \tag{113}$$

Using the relations (87) and (88) we realize that, up to multiplication constant, combination

$$\Delta_{\mu\nu} = \bar\Psi^\alpha_\mu (P^{-1})_{\alpha\beta} \Psi^\beta_\nu \,, \tag{114}$$

satisfies the condition (113). So, we conclude that

$${}^\star\Pi_{+\mu\nu} = \Pi_{+\mu\nu} + c\bar\Psi^\alpha_\mu (P^{-1})_{\alpha\beta} \Psi^\beta_\nu \,, \tag{115}$$

where c is an arbitrary constant. For $c = \frac{1}{2}$ we obtain the relations (90).

5. Conclusions

In this article we showed that the new interpretation of T-dualization procedure in double space formalism offered in [26, 27] is also valid in the case of type II superstring theory - both for bosonic and fermionic T-dualization. We used the ghost free action of type II superstring theory in pure spinor formulation in the approximation of quadratic terms and constant background fields. One can obtain this action from action (1) under some set of assumptions.

We introduced the double space coordinate $Z^M = (x^\mu, y_\mu)$ adding to all bosonic initial coordinates, x^μ, the T-dual ones, y_μ. Then we rewrote the T-dual transformation laws (43) in terms of double space variables (45) introducing the generalized metric $\mathcal{H}_{MN}$ and the current $J_{\pm M}$. Further, we split initial coordinates x^μ in two parts: x^a are directions along which we made T-dulization and the rest ones x^i.

T-dualization is realized as permutation of the subsets x^a and y_a in the double space coordinate Z^M. Demanding that T-dual double space coordinates ${}_aZ^M = (\mathcal{T}^a)^M{}_N Z^N$ satisfy the T-dual transformation law of the same form as the initial coordinates Z^M we found the T-dual generalized metric ${}_a\mathcal{H}_{MN}$ and the T-dual current ${}_aJ_{\pm M}$. Consequently, we obtained the form of NS-NS and NS-R T-dual background fields in terms of the initial ones which are in full accordance with the results obtained by Buscher T-dualization procedure [6, 7].

In order to obtain T-dual R-R field strength $F^{\alpha\beta}$ we should make some additional assumptions. Supposing that term $L(\pi_\alpha, \bar{\pi}_\alpha)$ (51) is T-dual invariant and taking into account that two successive T-dualizations act as identity operator, we found the form of T-dual R-R field strength up to one arbitrary constant c. For $c = 4\kappa$ we get the T-dual R-R field strength ${}_aF^{\alpha\beta}$ as in Buscher procedure [6].

In the case of fermionic T-duality, using equations of motion with respect to the fermionic momenta π_α and $\bar{\pi}_\alpha$, we eliminated them from the action. Then we fixed local chiral gauge invariance using BRST approach.

Using the Buscher approach we performed fermionic T-duality procedure and obtained the form of the fermionic T-dual background fields. Analogously with the bosonic case we introduced double fermionic space doubling the initial coordinates θ^α and $\bar{\theta}^\alpha$ with their fermionic T-duals, ϑ_α and $\bar{\vartheta}_\alpha$. Double fermionic space is spanned by the coordinates $\Theta^A = (\theta^\alpha, \vartheta_\alpha)$ and $\bar{\Theta}^A = (\bar{\theta}^\alpha, \bar{\vartheta}_\alpha)$. Demanding that T-dual transformation laws for fermionic T-dual double coordinates, ${}^\star\Theta^A = \mathcal{T}^A{}_B\Theta^B$ and ${}^\star\bar{\Theta}^A = \mathcal{T}^A{}_B\bar{\Theta}^B$, are of the same form as those for Θ^A and $\bar{\Theta}^A$, we obtained the form of the fermionic T-dual NS-R and R-R background fields which are in full accordance with the results obtained by standard Buscher procedure. The expressions for T-dual metric ${}^\star G_{\mu\nu}$ and Kalb-Ramond field ${}^\star B_{\mu\nu}$ cannot be found from double space formalism because they do not appear in the T-dual transformation laws. These expressions, up to arbitrary constant, are obtained assuming that two successive T-dualization act as identity transformation.

References

[1] K. Becker, M. Becker and J. Schwarz *String Theory and M-Theory: A Modern Introduction*; B. Zwiebach, *A First Course in String Theory*, Cambridge University Press, 2004; J. Polchinski, *String theory - Volume II*, Cambridge University Press, 1998.

[2] T. Buscher, *Phys. Lett.* **B 194** (1987) 59; **201** (1988) 466.

[3] M. Roček and E. Verlinde, *Nucl.Phys.* **B 373** (1992) 630.

[4] A. Giveon, M. Porrati and E. Rabinovici, *Phys. Rep.* **244** (1994) 77.

[5] E. Alvarez, L. Alvarez-Gaume, J. Barbon and Y. Lozano, *Nucl. Phys.* **B 415** (1994) 71.

[6] B. Nikolić and B. Sazdović, *Nucl. Phys.* **B 836** (2010) 100.

[7] D. S. Berman, D. C. Thompson, *Phys.Rept.* **566** (2014) 1-60.

[8] D. Lust *JHEP* **12** (2010) 084.

[9] D. Andriot, M. Larfors, D. Luest, P. Patalong, *JHEP* **06** (2013) 021.

[10] D. Andriot, O. Hohm, M. Larfors, D. Lust, P. Patalong, *Phys. Rev. Lett.* 108 (2012) 261602.

[11] D. Luest, arxiv:1205.0100 [hep-th]; R. Blumenhagen, A. Deser, D. Luest, E. Plauschinn, and F. Rennecke, *J.Phys.* **A44** (2011) 385401; C. Condeescu, I. Florakis, and D. Luest, *JHEP* **04** (2012) 121.

[12] J. Shelton, W. Taylor, and B. Wecht, *JHEP* **10** (2005) 085; A. Dabholkar and C. Hull, *JHEP* **05** (2006) 009.

[13] Lj. Davidović and B. Sazdović, *EPJ* **C 74** (2014) 2683.

[14] Lj. Davidović, B. Nikolić and B. Sazdović, *EPJ* **C 75** (2015) 576.

[15] Lj. Davidović, B. Nikolić and B. Sazdović, *EPJ* **C 74** (2014) 2734.

[16] M. Duff, *Nucl. Phys.* **B 335** (1990) 610.

[17] A. A. Tseytlin, *Phys.Lett.* **B 242** (1990) 163.

[18] A. A. Tseytlin, *Nucl. Phys.* **B 350** (1991) 395.

[19] W. Siegel, *Phys.Rev.* **D 48** (1993) 2826.

[20] W. Siegel, *Phys.Rev.* **D 47** (1993) 5453.

[21] C. M. Hull, *JHEP* **10** (2005) 065.

[22] C. M. Hull, *JHEP* **10** (2007) 057; **07** (2007) 080.

[23] D. S. Berman, M. Cederwall and M. J. Perry, *JHEP* **09** (2014) 066; D. S. Berman, C. D. A. Blair, E. Malek and M. J. Perry, *Int.J.Mod.Phys.* **A29** (2014) 15, 1450080; C. D. A. Blair, E. Malek and A. J. Routh, *Class.Quant.Grav.* **31** (2014) 20, 205011.

[24] C.M. Hull and R.A. Reid-Edwards, *JHEP* **09** (2009) 014.

[25] O. Hohm and B. Zwiebach, *JHEP* **11** (2014) 075.

[26] B. Sazdović, *T-duality as coordinates permutation in double space*, arxiv: 1501.01024.

[27] B. Sazdović, *JHEP* **08** (2015) 055.

[28] N. Berkovits and J. Maldacena, *JHEP* **09** (2008) 062; I. Bakhmatov and D. S. Berman, *Nucl. Phys.* B**832** (2010) 89-108; K. Sfetsos, K. Siampos and D. C. Thompson, QMUL-PH-10-08, arXiv:1007.5142; I. Bakhmatov, *Fermionic T-duality and U-duality in type II supergravity*, arXiv:1112.1983.

[29] N. Beisert, R. Ricci, A. A. Tseytlin and M. Wolf, *Phys.Rev.* **D78** (2008) 126004; R. Ricci, A. A. Tseytlin and M. Wolf, *JHEP* **12** (2007) 082.

[30] N. Berkovits and J. Maldacena, *JHEP* **09** (2008) 062; K. Sfetsos, K. Siampos and D. C. Thompson, *Class.Quant.Grav.* **28** 2011, 055010; I. Bakhmatov and D. S. Berman, *Nucl. Phys.* B**832** (2010) 89-108; N. Beisert, R. Ricci, A. A. Tseytlin and M. Wolf, *Phys.Rev.* **D78** (2008) 126004; R. Ricci, A. A. Tseytlin and M. Wolf, *JHEP* **12** (2007) 082; B. Nikolić and B. Sazdović, *Phys. Rev.* **D84** (2011) 065012, *JHEP* **06** (2012) 101.

[31] M. Hatsuda, K. Kamimura, W. Siegel, *JHEP* **06** (2014) 039.

[32] P. A. Grassi, L. Tamassia, *JHEP* **07** (2004) 071.

[33] N. Berkovits, hep-th/0209059; N. Berkovits, arxiv:0902.4032; P. A. Grassi, G. Policastro and P. van Nieuwenhuizen, *JHEP* **10** (2002) 054; P. A. Grassi, G. Policastro and P. van Nieuwenhuizen, *JHEP* **11** (2002) 004; P. A. Grassi, G. Policastro and P. van Nieuwenhuizen, *Adv. Theor. Math. Phys.* **7** (2003) 499; P. A. Grassi, G. Policastro and P. van Nieuwenhuizen, *Phys. Lett.* **B553** (2003) 96.

[34] J. de Boer, P. A. Grassi and P. van Nieuwenhuizen, *Phys. Lett.* **B574** (2003) 98.

[35] N. Berkovits and P. Howe, *Nucl. Phys.* **B635** (2002) 75.

[36] M. J. Duff, arxiv:hep-th/9912164v2.

[37] S. F. Hassan, *Nucl. Phys.* **B 568** (2000) 145; B. Nikolić and B. Sazdović, *EPJ* **C 77** (2017) 197.

[38] A. Giveon and M. Roček, *Nucl. Phys.* **B 421** (1994) 173.

T-duality and non-geometry*

Branislav Sazdović†
Institute of Physics, University of Belgrade, Serbia

Abstract

The role of double space is essential in new interpretation of T-duality and consequently in an attempt to construct M-theory. The case of open string is missing in such approach because until now there has been no appropriate formulation of open string T-duality. We will consider here reconsideration of T-duality of the open string. This will allow us to introduce some geometric features in non-geometric theories.

1. Introduction

We will show that "restricted general coordinate transformations", which includes transformations of background fields but not include transformations of the coordinates, is the symmetry T-dual to the local gauge transformations [1]. This will enable us to introduced new term in the Lagrangian, with additional gauge field A_i^D (D denotes components with Dirichlet boundary conditions). It compensate non-fulfilment of the invariance under restricted general coordinate transformation on the end-points of open string [1], as well as standard gauge field A_a^N (N denotes components with Neumann boundary conditions) compensate non-fulfilment of the local gauge invariance on the end-points of open string. Using generalized procedure [2] we will perform T-duality of vector fields linear in coordinates. We show that gauge fields A_a^N and A_i^D are T-dual to ${}^\star A_D^a$ and ${}^\star A_N^i$ respectively.

We will introduce the field strength of T-dual non-geometric theories as derivative of T-dual gauge fields along both T-dual variable y_μ and its double $\tilde{y}_\mu$. Therefore, we introduce some new features of non-geometric theories, where field strength has both antisymmetric and symmetric parts [1].

* This work has been supported by Serbian Ministry of Education and Science, under contract No. 171031.
† e-mail address: sazdovic@ipb.ac.rs

2. Closed string T-duality

Let us start from the closed string action [3, 4]

$$S[x] = \kappa \int_\Sigma d^2\xi \sqrt{-g}\Big[\frac{1}{2} g^{\alpha\beta} G_{\mu\nu}[x] + \frac{\epsilon^{\alpha\beta}}{\sqrt{-g}} B_{\mu\nu}[x]\Big] \partial_\alpha x^\mu \partial_\beta x^\nu \,, \qquad (1)$$

where $G_{\mu\nu}$ is a space-time metric and $B_{\mu\nu}$ is Kalb-Ramond field.

Action principle $\delta S = 0$, beside equations of motion produces boundary conditions

$$\gamma_\mu^{(0)}(x)\delta x^\mu /_{\sigma=\pi} \quad - \quad \gamma_\mu^{(0)}(x)\delta x^\mu /_{\sigma=0} = 0 \,,$$

where

$$\gamma_\mu^{(0)}(x) \equiv \frac{\delta S}{\delta x'^\mu} = \kappa\Big(2B_{\mu\nu}\dot{x}^\nu - G_{\mu\nu}x'^\nu\Big) \,. \qquad (2)$$

2.1. Buscher T-duality procedure

Applying standard Buscher T-duality procedure [5] we can obtain T-dual action

$${}^\star S[y] = \kappa \int d^2\xi \; \partial_+ y_\mu \, {}^\star\Pi_+^{\mu\nu} \, \partial_- y_\nu = \frac{\kappa^2}{2} \int d^2\xi \; \partial_+ y_\mu \theta_-^{\mu\nu} \partial_- y_\nu \,, \qquad (3)$$

where T-dual background fields

$${}^\star G^{\mu\nu} = (G_E^{-1})^{\mu\nu}, \quad {}^\star B^{\mu\nu} = \frac{\kappa}{2}\theta^{\mu\nu} \,, \qquad (4)$$

are defined in terms of effective metric $G^E_{\mu\nu}$ and non-commutative parameter $\theta^{\mu\nu}$

$$G^E_{\mu\nu} \equiv G_{\mu\nu} - 4(BG^{-1}B)_{\mu\nu}, \qquad \theta^{\mu\nu} \equiv -\frac{2}{\kappa}(G_E^{-1}BG^{-1})^{\mu\nu} \,. \qquad (5)$$

T-duality transformation of variables are

$$\partial_\pm x^\mu \cong -\kappa \theta_\pm^{\mu\nu} \partial_\pm y_\nu \,, \qquad \partial_\pm y_\mu \cong -2\Pi_{\mp\mu\nu} \partial_\pm x^\nu \,, \qquad (6)$$

which in canonical form can be written as

$$\kappa x'^\mu \cong {}^\star\pi^\mu, \quad \pi_\mu \cong \kappa y'_\mu, \quad -\kappa \dot{x}^\mu \cong {}^\star\gamma^\mu_{(0)}(y), \quad \gamma_\mu^{(0)}(x) \cong -\kappa \dot{y}_\mu \,. \qquad (7)$$

3. T-duality of local gauge symmetries

We are going to consider open string T-duality. We will insist that each term should have corresponding T-dual one. For example

$$\begin{array}{cccccc} S(x) & G_{\mu\nu} & B_{\mu\nu} & & A_a^N & A_i^D \\ \downarrow & \downarrow & \downarrow & & \downarrow & \downarrow \\ {}^\star S(y) & {}^\star G^{\mu\nu} & {}^\star B^{\mu\nu} & & {}^\star A_D^a & {}^\star A_N^i \,. \end{array} \tag{8}$$

It is well known from the literature that coupling with Neumann fields has a form

$$S_{A^N} = 2\kappa \int d\tau (A_a^N \dot{x}^a/_{\sigma=\pi} - A_a^N \dot{x}^a/_{\sigma=0}) \,. \tag{9}$$

But, coupling with Dirichlet fields is not known

$$S_{A^D} = 2\kappa \int d\tau (A_i^D (?)^i/_{\sigma=\pi} - A_i^D (?)^i/_{\sigma=0}) \,. \tag{10}$$

To find it we will use analogy with Neumann case.

3.1. Zwiebach approach for coupling with Neumann fields

Action of closed string theory is invariant under local gauge transformations

$$\delta_\Lambda G_{\mu\nu} = 0 \,, \qquad \delta_\Lambda B_{\mu\nu} = \partial_\mu \Lambda_\nu - \partial_\nu \Lambda_\mu \,. \tag{11}$$

Due to the boundary term the open string theory is not invariant [6]

$$\delta_\Lambda S[x] = 2\kappa \int d\tau (\Lambda_a \dot{x}^a/_{\sigma=\pi} - \Lambda_a \dot{x}^a/_{\sigma=0}) \,. \tag{12}$$

To obtain gauge invariant action we should add the term

$$S_{A^N}[x] = 2\kappa \int d\tau (A_a^N \dot{x}^a/_{\sigma=\pi} - A_a^N \dot{x}^a/_{\sigma=0}) \,, \tag{13}$$

where newly introduced vector field A_a^N transforms with the same gauge parameter Λ_a

$$\delta_\Lambda A_a^N = -\Lambda_a \,. \tag{14}$$

3.2. What is T-dual to local gauge transformations?

In order to continue we will use statement from Refs.[7, 8, 9]. If variation of energy-momentum tensor $T_\pm$ can be written as

$$\delta T_\pm = \{\Gamma, T_\pm\} \,, \tag{15}$$

then corresponding transformation of background fields is target-space symmetry of the theory.

For $\Gamma \to \Gamma_\Lambda = 2\int d\sigma \Lambda_\mu \kappa x'^\mu$ we can obtain just local gauge transformations. T-dual to generator $\kappa x'^\mu$ is π_μ, so that T-dual to Γ_Λ is

$$\Gamma_\xi = 2\int d\sigma\, \xi^\mu \pi_\mu \,. \tag{16}$$

The corresponding transformations are

$$\delta_\xi G_{\mu\nu} = -2\left(D_\mu \xi_\nu + D_\nu \xi_\mu\right),$$

$$\delta_\xi B_{\mu\nu} = -2\,\xi^\rho B_{\rho\mu\nu} + 2\partial_\mu(B_{\nu\rho}\xi^\rho) - 2\partial_\nu(B_{\mu\rho}\xi^\rho)\,. \tag{17}$$

These transformations exactly have the form of general coordinate transformations (GCT), the symmetry transformations of the space-time action. Are these transformations symmetries of the σ-model action?

World-sheet action is scalar under GCT, so both closed and open string actions are invariant under GCT. To understand what is T-dual to local gauge transformations it is useful to make transformations of the background fields, metric tensor $G_{\mu\nu}$ and Kalb-Ramond field $B_{\mu\nu}$, with parameter ξ_μ and transformations of the string coordinates x^μ with different parameter $\bar\xi^\mu$, $\delta x^\mu = \bar\xi^\mu$. Using the equation of motion we obtain

$$\delta_\xi S[x] = -2\int_{\partial\Sigma} d\tau (\xi_\mu - \bar\xi_\mu)\, G^{-1\mu\nu}\gamma_\nu^{(0)}(x)\,. \tag{18}$$

If we introduce residual general coordinate transformations (RGCT), which include the transformations of background fields but not include the transformations of the string coordinates x^μ, $\bar\xi_\mu/_{\sigma=\pi} = \bar\xi_\mu/_{\sigma=0} = 0$, we obtain

$$\delta_\xi S[x] = -2\int_{\partial\Sigma} d\tau \xi_\mu\, G^{-1\mu\nu}\gamma_\nu^{(0)}(x)\,. \tag{19}$$

Note that according to (7) $-\kappa\dot x^\mu$ and $\gamma_\mu^{(0)}(x)$ are expressions T-dual to each other. So, local gauge transformations and RGCT are connected by T-duality.

4. Open string T-duality

Gauge invariant action for open string is [1]

$$\begin{aligned} S_{open}[x] &= \kappa \textstyle\int_\Sigma d^2\xi \partial_+ x^\mu \Pi_{+\mu\nu}\partial_- x^\nu \\ &+ 2\kappa \textstyle\int_{\partial\Sigma} d\tau \left[A_a^N[x]\dot x^a - \frac{1}{\kappa}A_i^D[x]G^{-1ij}\gamma_j^{(0)}(x)\right], \end{aligned} \tag{20}$$

where in literature $A_a^N[x]$ is known as massless vector field on Dp-brane and $A_i^D[x]$ as massless scalar oscillations orthogonal to the Dp-brane.

Note that gauge invariant and physical variables are

$$\begin{aligned} \mathcal{B}_{ab} &= B_{ab} + F_{ab}^{(a)}\,, \qquad \mathcal{G}_{ab} = G_{ab}\,, \\ \mathcal{B}_{ij} &= B_{ij} - 2A_D^k B_{kij} - F_{ij}^{(s)}(\hat{A}^D)\,, \\ \mathcal{G}_{ij} &= G_{ij} + F_{ij}^{(s)}(A^D)\,, \end{aligned} \tag{21}$$

where we introduced field strengths

$$F_{ab}^{(a)} = \partial_a A_b^N - \partial_b A_a^N\,, \qquad F_{ij}^{(s)}(A^D) = -2(\partial_i A_j^D + \partial_j A_i^D)\,, \tag{22}$$

and define

$$\hat{A}_i = B_{ij} G^{-1jk} A_k^D\,. \tag{23}$$

4.1. T-dual background fields of the open string

We will choose background vector fields linear in coordinates [1]

$$B_{\mu\nu} = const\,, \qquad G_{\mu\nu} = const\,, \tag{24}$$

$$A_a^N(x) = A_a^0 - \frac{1}{2} F_{ab}^{(a)} x^b\,, \qquad A_i^D(x) = A_i^0 - \frac{1}{4} F_{ij}^{(s)} x^j\,, \tag{25}$$

so that corresponding field strengths are constant. These forms of background fields satisfies space-time equations of motion for open string [10].

It is important to know that action depends on the coordinate x^μ itself and not only on its derivatives with respect to τ and σ. So, part with $A_i^D(x)$ does not have global shift symmetry, because the expression $\gamma_i^{(0)}$ contain x'^j which is not total derivative with respect to integration variable τ. So, we should apply T-dualization procedure [2] which work in absence of global symmetry.

T-dual background fields in terms of initial ones are

$$^\star G^{\mu\nu} = (G_E^{-1})^{\mu\nu}, \qquad ^\star B^{\mu\nu} = \frac{\kappa}{2}\theta^{\mu\nu}\,, \tag{26}$$

$$^\star A_D^a(V) = G_E^{-1ab} A_b^N(V)\,, \qquad ^\star A_N^i(V) = G^{-1ij} A_j^D(V)\,, \tag{27}$$

where

$$V^\mu = -\kappa\,\theta^{\mu\nu} y_\nu + G_E^{-1\mu\nu}\,\tilde{y}_\nu\,, \tag{28}$$

and

$$\tilde{y}_\mu \equiv -\varepsilon_\alpha{}^\beta \int d\xi^\alpha \partial_\beta y_\mu = \int (d\tau y'_\mu + d\sigma \dot{y}_\mu)\,. \tag{29}$$

Note that T-duality interchange Neumann with Dirichlet gauge fields.

4.2. Relation with standard approach

Up to gauge transformation we have

$$^\star A_D^a = G_E^{-1ab}\left(A_b^N + \frac{1}{2}y_a\right), \quad ^\star A_N^i = G^{-1ij}A_j^D\,. \tag{30}$$

In standard approach one can not recognize Dirichlet vector fields. So if we put $A_i^D = 0$ and $^\star A_D^a = 0$ we obtain

$$^\star A_N^i = 0\,, \qquad y_a = -2A_b^N\,. \tag{31}$$

These are consistency conditions of standard approach.

5. The field strength for non-geometric theories

The particular form of V^μ from Eq.(28) implies features of non-geometric theories, see for example [11]. It produces non-commutativity and non-associativity of closed string coordinates.

In geometric theories the field strength for Abelian vector field is simple $F_{\mu\nu} = \partial_\mu A_\nu - \partial_\nu A_\mu$. Because in non-geometric theories the vector field depends on V^μ, we expect that T-dual field strength will contain derivatives with respect to both variables y_μ and $\tilde{y}_\mu$. How to define the field strength for non-geometric theories?

For Neumann vector fields in initial theory we have

$$S_A^N[x] = 2\kappa\int_{\partial\Sigma} d\tau A_a^N(x)\dot{x}^a = \kappa\int_\Sigma d^2\xi\partial_+ x^a\mathcal{F}_{ab}\,\partial_- x^b\,, \tag{32}$$

where only antisymmetric part contributes

$$\mathcal{F}_{ab} = F_{ab}^{(a)} = \partial_a A_b^N(x) - \partial_b A_a^N(x)\,. \tag{33}$$

We are going to generalize such relation to non-geometric theories. For Dirichlet vector fields in initial theory we find

$$S_A^D[x] = 2\kappa\int_{\partial\Sigma} d\tau\Big(-\frac{1}{\kappa}A_i^D(x)G^{-1ij}\gamma_j^{(0)}(x)\Big)$$
$$= 2\kappa\int_{\partial\Sigma} d\tau\Big(\mathcal{A}_{0i}[x]\dot{x}^i - \mathcal{A}_{1i}[x]x'^i\Big) = \kappa\int_\Sigma d^2\xi\,\partial_+ x^i\,\mathcal{F}_{ij}\,\partial_- x^j\,. \tag{34}$$

Now, both antisymmetric and symmetric parts contribute

$$\mathcal{F}_{ij} = \mathcal{F}_{ij}^{(a)} + \frac{1}{2}\mathcal{F}_{ij}^{(s)}\,, \tag{35}$$

where

$$\mathcal{F}_{ij}^{(a)} = \Big[\partial_i\Big(2B_{jk}G^{-1kq}A_q^D\Big) - \partial_j\Big(2B_{ik}G^{-1kq}A_q^D\Big)\Big]\,, = \partial_i\mathcal{A}_{0j}(x) - \partial_j\mathcal{A}_{0i}(x)\,, \tag{36}$$

and

$$\mathcal{F}^{(s)}_{ij} = -2(\partial_i A^D_j + \partial_j A^D_i) = 2\Big(\partial_i \mathcal{A}_{1j}(x) + \partial_j \mathcal{A}_{1i}(x)\Big)\,. \tag{37}$$

For Dirichlet vector fields in T-dual theory we have

$$\begin{aligned} {}^\star S^D_A[y] &= 2\kappa \int_{\partial\Sigma} d\tau \Big(-\frac{1}{\kappa} {}^\star A^a_D(V)\, {}^\star G^{-1}_{ab}\, {}^\star \gamma^b_{(0)}(y)\Big) \\ &= \kappa \int_\Sigma d^2\xi \partial_+ y_a\, {}^\star\mathcal{F}^{ab}\, \partial_- y_b\,, \end{aligned} \tag{38}$$

with

$${}^\star\mathcal{F}^{ab} = {}^\star\mathcal{F}^{ab}_{(a)} + \frac{1}{2}\, {}^\star\mathcal{F}^{ab}_{(s)}\,. \tag{39}$$

For Neumann vector fields in T-dual theory

$${}^\star S^N_A[y] = 2\kappa \int_{\partial\Sigma} d\tau \Big({}^\star A^i_N(V) \dot{y}_i \Big) = \kappa \int_\Sigma d^2\xi\, \partial_+ y_i\, {}^\star\mathcal{F}^{ij}\, \partial_- y_j\,, \tag{40}$$

where

$${}^\star\mathcal{F}^{ij} = {}^\star\mathcal{F}^{ij}_{(a)} + \frac{1}{2}\, {}^\star\mathcal{F}^{ij}_{(s)}\,. \tag{41}$$

In Dirichlet case for non-geometric theories we obtain antisymmetric

$${}^\star\mathcal{F}^{ab}_{(a)} = -\kappa^2 \theta^{ac} F^{(a)}_{cd}\, \theta^{db} - G_E^{-1ac} F^{(a)}_{cd}\, G_E^{-1db}\,, \tag{42}$$

and symmetric field strengths

$${}^\star\mathcal{F}^{ab}_{(s)} = -2\kappa \Big[G_E^{-1ac} F^{(a)}_{cd}\, \theta^{db} + \theta^{ac} F^{(a)}_{cd}\, G_E^{-1db} \Big]\,, \tag{43}$$

while in Neumann case we have

$${}^\star\mathcal{F}^{ij}_{(a)} = -\frac{\kappa}{4} \left(\theta^{ik} F^{(s)}_{kq} G^{-1qj} + G^{-1ik} F^{(s)}_{kq} \theta^{qj} \right)\,, \tag{44}$$

and

$${}^\star\mathcal{F}^{ij}_{(s)} = -\frac{1}{2} \left(G_E^{-1ik} F^{(s)}_{kq} G^{-1qj} + G^{-1ik} F^{(s)}_{kq} G_E^{-1qj} \right)\,. \tag{45}$$

Finally, we should write out expressions for T-dual field strengths ${}^\star\mathcal{F}^{\mu\nu}$ in terms of derivative of T-dual gauge fields ${}^\star\mathcal{A}^a_0(V)$ and ${}^\star\mathcal{A}^a_1(V)$ with respect to variables y_μ and $\tilde{y}_\mu$. If we define $y^\alpha_\mu = \{y^0_\mu = y_\mu,\, y^1_\mu = -\tilde{y}_\mu\}$ and $\partial^\mu_\alpha \equiv \frac{\partial}{\partial y^\alpha_\mu} = \{\frac{\partial}{\partial y_\mu}, -\frac{\partial}{\partial \tilde{y}_\mu}\}$ we find

$$\begin{aligned} {}^\star\mathcal{F}^{\mu\nu} &= {}^\star\mathcal{F}^{\mu\nu}_{(a)} + \tfrac{1}{2}{}^\star\mathcal{F}^{\mu\nu}_{(s)} \\ &= \eta^{\alpha\beta}\Big[\partial^\mu_\alpha\, {}^\star\mathcal{A}^\nu_\beta(V) - \partial^\nu_\alpha\, {}^\star\mathcal{A}^\mu_\beta(V)\Big] - \varepsilon^{\alpha\beta}\Big[\partial^\mu_\alpha\, {}^\star\mathcal{A}^\nu_\beta(V) + \partial^\nu_\alpha\, {}^\star\mathcal{A}^\mu_\beta(V)\Big]\,. \end{aligned} \tag{46}$$

We can check this expression in other way

$$
\begin{aligned}
{}^\star S_A[y] = {}^\star S_A^D[y] + {}^\star S_A^N[y] &= 2\kappa\eta^{\alpha\beta}\int_{\partial\Sigma} d\tau {}^\star\mathcal{A}^\mu_\alpha[V]\,\partial_\beta y_\mu \\
&= \kappa\int_\Sigma d^2\xi\partial_+ y_\mu {}^\star\mathcal{F}^{\mu\nu}\partial_- y_\nu \,. \qquad (47)
\end{aligned}
$$

6. Conclusions

The expression (46) we can consider as a general definition of the field strength for non-geometric theories. Beside antisymmetric part ${}^\star\mathcal{F}^{\mu\nu}_{(a)}$ it also contains the symmetric one ${}^\star\mathcal{F}^{\mu\nu}_{(s)}$. In definition of both parts, derivatives with respect to both T-dual coordinate y_μ and to its double $\tilde{y}_\mu$ contribute.

The unusual form of ${}^\star\mathcal{F}^{\mu\nu}$ is a consequence of two facts:

1. the T-dual vector field ${}^\star A^a_D(V)$ are not multiplied by $\dot{y}_a$ but with T-dual σ-momentum ${}^\star G^{-1\star}_{ab}\gamma^b_{(0)}$.

2. the T-dual vector fields depend on V^μ which is function on both y_μ and $\tilde{y}_\mu$.

References

[1] B. Sazdović, *From geometry to non-geometry via T-duality*, ArXiv 1606.01938.

[2] Lj. Davidović and B. Sazdović, *JHEP* **11** (2015) 119.

[3] J. Polchinski, *String theory*, Cambridge University Press, 1998.

[4] K. Becker, M. Becker and J. Schwarz, *String Theory and M-Theory: A Modern Introduction*, Cambridge University Press, 2007.

[5] T. Buscher, *Phys. Lett.* **B 194** (1987) 51; **201** (1988) 466.

[6] B. Zwiebach, *A First Course in String Theory*, Cambridge University Press, 2004.

[7] M. Evans and B. A Ovrut, *Phys.Rev.* **D 39** (1989) 3016.

[8] M. Evans and B. A Ovrut, *Phys.Rev.* **D 41** (1990) 3149.

[9] Lj. Davidović and B. Sazdović, in preparation.

[10] R. G. Leigh, *Mod. Phys. Lett.* **A 4** (1989) 2767.

[11] Lj. Davidović, B. Nikolić and B. Sazdović, *EPJ* **C 75** (2015) 576.

[12] Lj. Davidović and B. Sazdović, *EPJ* **C 74** (2014) 2683.

[13] B. Sazdović, *JHEP* **08** (2015) 055.

Integrability of geodesics in contact space $T^{1,1}$ and its metric cone *

Mihai Visinescu†
Department Theoretical Physics
National Institute for Physics and Nuclear Engineering
Magurele, Bucharest, Romania

Abstract

We investigate the integrability of geodesics in the five-dimensional Sasaki-Einstein space $T^{1,1}$. We construct explicitly the constants of motion and prove the complete integrability of geodesic motions. This property is also valid for geodesic motions on its Calabi-Yau metric cone. Having in mind that the metric cone is singular at the apex of the cone, we extend the analysis of integrability for resolved conifolds. It is shown that in the case of the *small resolution* the integrability is preserved, while in the case of the *deformation* of the conifold it is lost.

PACS: 11.30.-j, 02.40.Ky

MSC2000: 53D10, 53D25

1. Introduction

Recently there has been much interest in contact geometry in connection with some modern developments in mathematics and theoretical physics [1, 2]. Sasaki-Einstein geometry is considered important in studies of consistent string compactification and in the context of AdS/CFT correspondence [3, 4].

The symmetries of Sasaki-Einstein spaces play a significant role in connection with the study of integrability properties of geodesic motions and separation of variables of the classical Hamilton-Jacobi or quantum Klein-Gordon, Dirac equations.

The homogeneous toric Sasaki-Einstein on $S^2 \times S^3$ is usually referred to as $T^{1,1}$ and was considered as the first example of toric Sasaki-Einstein/quiver duality [5]. The $AdS \times T^{1,1}$ is the first example of a supersymmetric holographic theory based on a compact manifold which is not locally S^5.

* This work has been suppored by the project NUCLEU 16 42 01 01/2016
† e-mail address: mvisin@theory.nipne.ro

In this paper we are concerned with integrability of geodesic motions in Sasaki-Einstein space $T^{1,1}$ ant its Calabi-Yau metric cone. In general the metric cone of a Sasaki-Einstein manifold is singular at the apex of the cone. On the other hand there are promising generalizations of the original AdS/CFT correspondence by considering $D3$-branes on conifold singularities [5, 6].

In the case of the metric cone over $T^{1,1}$, the singularity at its apex can be smoothed out in two different ways [7]. On can substitute the apex by an S^3 (*deformation*) or by an S^2 (*small resolution*).

In a recent paper [8] it has been constructed the complete set of constants of motion in $T^{1,1}$ space which provides the complete integrability of the geodesic flow. In what follows we extend the investigation of the integrabilty for the geodesic flow on the Calabi-Yau metric cone of $T^{1,1}$ space and its resolved conifolds. We find that the geodesic motions are completely integrable in the case of the metric cone as well as its small resolution of the apex singularity. In opposition to these cases, the geodesic flow on the deformed metric cone is not completely integrable.

The paper is organized as follows. In the next Section we present some mathematical preliminaries regarding the Sasaki-Einstein geometry. In Section 3 we discuss the integrability of geodesic motions in $T^{1,1}$ space and its metric cone. In Section 4 we extend the study of the integrability for the resolved conifolds in both existing procedures of smoothing out the singularity at the apex of the conifold. The paper ends with conclusions in Section 5.

2. Preliminaries regarding the Sasaki-Einstein geometry

Recall that a $(2n-1)$-dimensional manifold M is a contact manifold if there exists a 1-form η, called the contact 1-form, on M such that:

$$\eta \wedge (d\eta)^{n-1} \neq 0\,,$$

everywhere on M [1]. A contact Riemannian manifold with the metric g_M is Sasakian if its metric cone

$$\left(C(M),\, g_{C(M)}\right) = \left(\mathbb{R}_+ \times M\,, dr^2 + r^2 g_M\right), \tag{1}$$

is Kähler with the Kähler form

$$\Omega = \frac{1}{2}\, d(r^2\eta)\,.$$

Here $r \in (0,\infty)$ can be regarded as a coordinate on the positive real line $\mathbb{R}_+$.

As a part of the connection between Sasaki and Kähler geometries, it is worth noting that in the case of a Sasaki-Einstein manifold

$$\mathrm{Ric}_{g_M} = 2(n-1)g_M\,,$$

the metric cone is Ricci flat,

$$\mathrm{Ric}_{g_{C(M)}} = 0\,,$$

i.e. a Calabi-Yau manifold.

3. Complete integrability of $T^{1,1}$ space and its metric cone

The metric of the space $T^{1,1}$ may be written down explicitly by utilizing the fact that it is a $U(1)$ bundle over $S^2 \times S^2$. Let us denote by (θ_1, ϕ_1) and (θ_2, ϕ_2) the coordinates which parametrize the two sphere in a conventional way, and the angle $\psi \in [0, 4\pi)$ parametrizes the $U(1)$ fiber. Using these coordinates the $T^{1,1}$ metric may be written as [7, 9]:

$$\begin{aligned} ds^2_{T^{1,1}} =& \frac{1}{6}(d\theta_1^2 + \sin^2\theta_1 d\phi_1^2 + d\theta_2^2 + \sin^2\theta_2 d\phi_2^2) + \\ &+ \frac{1}{9}(d\psi + \cos\theta_1 d\phi_1 + \cos\theta_2 d\phi_2)^2\,, \end{aligned}$$

with the globally defined contact 1-form η:

$$\eta = \frac{1}{3}(d\psi + \cos\theta_1 d\phi_1 + \cos\theta_2 d\phi_2)\,.$$

The Hamiltonian describing the geodesic flow is

$$H = \frac{1}{2} g^{ij}\, p_i p_j\,, \tag{2}$$

where g^{ij} is the inverse metric of $T^{1,1}$ space and $p_i = g_{ij}\,\dot{x}^j$ are the conjugate momenta to the coordinates $(\theta_1, \theta_2, \phi_1, \phi_2, \psi)$. Let us denote by $(p_{\theta_1}, p_{\theta_2}, p_{\phi_1}, p_{\phi_2}, p_\psi)$ these conjugate momenta:

$$\begin{aligned} p_{\theta_1} &= \frac{1}{6}\dot{\theta}_1\,, \\ p_{\theta_2} &= \frac{1}{6}\dot{\theta}_2\,, \\ p_{\phi_1} &= \frac{1}{6}\sin^2\theta_1\,\dot{\phi}_1 + \frac{1}{9}\cos\theta_1(\cos\theta_1\,\dot{\phi}_1 + \cos\theta_2\,\dot{\phi}_2 + \dot{\psi})\,, \\ p_{\phi_2} &= \frac{1}{6}\sin^2\theta_2\,\dot{\phi}_2 + \frac{1}{9}\cos\theta_2(\cos\theta_1\,\dot{\phi}_1 + \cos\theta_2\,\dot{\phi}_2 + \dot{\psi})\,, \\ p_\psi &= \frac{1}{9}(\cos\theta_1\,\dot{\phi}_1 + \cos\theta_2\,\dot{\phi}_2 + \dot{\psi})\,. \end{aligned} \tag{3}$$

In terms of them, Hamiltonian (2) has the form:

$$\begin{aligned} H =& 3\left[p_{\theta_1}^2 + p_{\theta_2}^2 + \frac{1}{\sin^2\theta_1}(p_{\phi_1} - \cos\theta_1\,p_\psi)^2 + \frac{1}{\sin^2\theta_2}(p_{\phi_2} - \cos\theta_2\,p_\psi)^2\right] \\ &+ \frac{9}{2}p_\psi^2\,. \end{aligned} \tag{4}$$

Taking into account the isometries of $T^{1,1}$, the momenta $p_{\phi_1}, p_{\phi_2}, p_\psi$ are conserved. On the other hand two total $SU(2)$ angular momenta are also conserved:

$$\begin{aligned} \mathbf{j}_1^2 &= p_{\theta_1}^2 + \frac{1}{\sin^2\theta_1}(p_{\phi_1} - \cos\theta_1\, p_\psi)^2 + p_\psi^2\,, \\ \mathbf{j}_2^2 &= p_{\theta_2}^2 + \frac{1}{\sin^2\theta_2}(p_{\phi_2} - \cos\theta_2\, p_\psi)^2 + p_\psi^2\,. \end{aligned} \tag{5}$$

Other constants of motion can be constructed from the Stäckel-Killing tensors that are admitted by the $T^{1,1}$ space [8]. In spite of the presence of a multitude of conserved quantities, the number of functionally independent constants of motion is five, exactly the dimension of the Sasaki-Einstein space $T^{1,1}$ [8, 10]. This implies the complete integrability of geodesic motions in $T^{1,1}$ which allows us to solve the Hamilton-Jacobi equation by separation of variables and construct the action-angle variables [11].

On the metric cone (1) the geodesic flow is described by the Hamiltonian:

$$H_{C(T^{1,1})} = \frac{1}{2}p_r^2 + \frac{1}{r^2}\tilde{H}\,, \tag{6}$$

where the radial momentum is

$$p_r = \dot{r}\,.$$

The Hamiltonian $\tilde{H}$ has a similar structure as in (4), but constructed with momenta $\tilde{p}_i$ related to momenta p_i (3) by

$$\tilde{p}_i = r^2 g_{ij}\,\dot{x}^j = r^2\,p_i\,.$$

It is not difficult to observe that the radial dynamics is independent of the dynamics of the base manifold $T^{1,1}$ and $\tilde{H}$ is a constant of motion. The Hamilton equations of motion for $\tilde{H}$ on $T^{1,1}$ have the standard form in terms of a new time variable $\tilde{t}$ given by [12]

$$\frac{dt}{d\tilde{t}} = r^2\,.$$

Concerning the constant of motions, they are the conjugate momenta $(\tilde{p}_{\phi_1}, \tilde{p}_{\phi_2}, \tilde{p}_\psi)$ associated with the cyclic coordinates (ϕ_1, ϕ_2, ψ) and two total $SU(2)$ momenta

$$\tilde{\mathbf{j}}_{1,2}^2 = r^4\,\mathbf{j}_{1,2}^2\,.$$

Together with the Hamiltonian $H_{C(T^{1,1})}$, they ensure the complete integrability of the geodesic flow on the metric cone.

Considering a particular level set $E_{C(T^{1,1})}$ of $H_{C(T^{1,1})}$ we get for the radial motion

$$p_r^2 = \dot{r}^2 = 2E_{C(T^{1,1})} - \frac{2}{r^2}\tilde{H}\,.$$

The turning point of the radial motion is determined by

$$\dot{r} = 0 \quad \Longrightarrow \quad r_* = \sqrt{\frac{\tilde{H}}{E_{C(T^{1,1})}}}\,.$$

Projecting the geodesic curves onto the base manifold $T^{1,1}$ we can evaluate the total distance transversed in the Sasaki-Einstein space between the limiting points as $t \to -\infty$ and $t \to +\infty$ [12]

$$d = \sqrt{2\tilde{H}} \int_{-\infty}^{\infty} \frac{dt}{r_*^2 + 2E_{C(T^{1,1})}t^2} = \pi\,.$$

4. Integrability of the resolved conifolds

The conifold (1) associated with Sasaki-Einstein space $T^{1,1}$ can be described by the quadric

$$\sum_{a=1}^{4} w_a^2 = 0\,. \tag{7}$$

on $\mathbb{C}^4$. This equation can be written in terms of a matrix $\mathcal{W}$ defined by

$$\mathcal{W} = \frac{1}{\sqrt{2}}\sigma^a w_a = \frac{1}{\sqrt{2}} \begin{pmatrix} w_3 + iw_4 & w_1 - iw_2 \\ w_1 + iw_2 & -w_3 + iw_4 \end{pmatrix} \equiv \begin{pmatrix} X & U \\ V & Y \end{pmatrix},$$

where σ^a are the Pauli matrices for $a = 1, 2, 3$ and σ^4 is i times the unit matrix. The radial coordinate is defined by

$$r^2 = \mathrm{tr}\,(\mathcal{W}^\dagger \mathcal{W})\,.$$

Equation (7) can be written as

$$\det \mathcal{W} = 0 \quad , \quad \text{i.e.} \quad XY - UV = 0\,. \tag{8}$$

The singularity at the apex can be repaired in two different ways. The first is achieved by a *deformation* having the effect of replacing the node by an S^3. The second possibility is represented by a *small resolution* consisting in a replacement of the node by an S^2.

4.1. Small resolution

The small resolution is obtained replacing equation (8) by the pair of equations [7]:

$$\begin{pmatrix} X & U \\ V & Y \end{pmatrix} \begin{pmatrix} \lambda_1 \\ \lambda_2 \end{pmatrix} = 0\,,$$

in which $(\lambda_1, \lambda_2) \in \mathbb{CP}^1$ are not both zero.

It turns out to be convenient to introduce a new radial coordinate

$$\rho^2 \equiv \frac{3}{2}\gamma\,,$$

where the function γ is given by the equation

$$\gamma^3 + 6\,a^2\,\gamma^2 - r^4 = 0\,.$$

Here a is the "resolution" parameter representing the radius of the sphere S^2 which replaces the singularity at $r^2 = 0$.

Eventually the metric of the resolved conifold can be written simply as [13]

$$\begin{aligned} ds^2_{rc} = {} & \kappa^{-1}(\rho)\, d\rho^2 + \frac{1}{9}\kappa(\rho)\rho^2(d\psi + \cos\theta_1\, d\phi_1 + \cos\theta_2\, d\phi_2)^2 \\ & + \frac{1}{6}\rho^2(d\theta_1^2 + \sin^2\theta_1\, d\phi_1^2) + \frac{1}{6}(\rho^2 + 6\,a^2)(d\theta_2^2 + \sin^2\theta_2\, d\phi_2^2)\,, \end{aligned} \tag{9}$$

where

$$\kappa(\rho) \equiv \frac{\rho^2 + 9\,a^2}{\rho^2 + 6\,a^2}\,.$$

The resolved conifold metric is Ricci flat and has an explicit $SU(2)\times SU(2)$ invariant form. When the resolution parameter a goes to zero or when $\rho \to \infty$, the resolved conifold metric reduces to the standard conifold metric $g_{C(T^{1,1})}$. In fact, the parameter a introduces an asymmetry between the two sphere.

In order to write the Hamiltonian on the resolved conifold, we evaluate the conjugate momenta $(P_\rho, P_{\theta_1}, P_{\theta_2}, P_{\phi_1}, P_{\phi_2}, P_\psi)$ corresponding to the coordinates $(\rho, \theta_1, \theta_2, \phi_1, \phi_2, \psi)$:

$$\begin{aligned} P_\rho &= \kappa^{-1}(\rho)\dot{\rho}\,, \\ P_{\theta_1} &= \frac{1}{6}\rho^2\dot{\theta}_1\,, \\ P_{\theta_2} &= \frac{1}{6}(\rho^2 + 6\,a^2)\dot{\theta}_2\,, \\ P_{\phi_1} &= \frac{1}{6}\rho^2\sin^2\theta_1\,\dot{\phi}_1 + \frac{1}{9}\kappa(\rho)\rho^2\cos\theta_1(\cos\theta_1\,\dot{\phi}_1 + \cos\theta_2\,\dot{\phi}_2 + \dot{\psi})\,, \\ P_{\phi_2} &= \frac{1}{6}(\rho^2 + 6\,a^2)\sin^2\theta_2\,\dot{\phi}_2 + \frac{1}{9}\kappa(\rho)\rho^2\cos\theta_2(\cos\theta_1\,\dot{\phi}_1 + \cos\theta_2\,\dot{\phi}_2 + \dot{\psi})\,, \\ P_\psi &= \frac{1}{9}\kappa(\rho)\rho^2(\cos\theta_1\,\dot{\phi}_1 + \cos\theta_2\,\dot{\phi}_2 + \dot{\psi})\,. \end{aligned}$$

In terms of them, the Hamiltonian for the geodesic flow on resolved conifold (rc) is

$$
\begin{aligned}
H_{rc} = &\frac{\kappa(\rho)}{2} P_\rho^2 + \frac{9}{2} \frac{1}{\kappa(\rho)\, \rho^2} P_\psi^2 + \frac{3}{\rho^2} P_{\theta_1}^2 + \frac{3}{\rho^2 + 6\, a^2} P_{\theta_2}^2 \\
&+ \frac{3}{\rho^2 \sin^2\theta_1} \left(P_{\phi_1} - \cos\theta_1\, P_\psi\right)^2 \\
&+ \frac{3}{(\rho^2 + 6\, a^2) \sin^2\theta_2} \left(P_{\phi_2} - \cos\theta_2\, P_\psi\right)^2 .
\end{aligned} \tag{10}
$$

We observe that (ϕ_1, ϕ_2, ψ) are still cyclic coordinates and accordingly, momenta $P_{\phi_1}, P_{\phi_2}, P_\psi$ are conserved. Taking into account the $SU(2) \times SU(2)$ symmetry of the metric (9), the total angular momenta

$$
\begin{aligned}
\mathbf{J}_1^2 &= P_{\theta_1}^2 + \frac{1}{\sin^2\theta_1} (P_{\phi_1} - \cos\theta_1\, P_\psi)^2 + P_\psi^2 = \rho^4\, \mathbf{j}_1^2 , \\
\mathbf{J}_2^2 &= P_{\theta_2}^2 + \frac{1}{\sin^2\theta_2} (P_{\phi_2} - \cos\theta_2\, P_\psi)^2 + P_\psi^2 = (\rho^2 + 6a^2)^2\, \mathbf{j}_2^2 .
\end{aligned}
$$

are also conserved. Using these total angular momenta, the Hamiltonian H_{rc} can be put in the form:

$$
H_{rc} = \frac{1}{2} \frac{\rho^2 + 9\, a^2}{\rho^2 + 6\, a^2} P_\rho^2 + \frac{3}{\rho^2} \mathbf{J}_1^2 + \frac{3}{\rho^2 + 6\, a^2} \mathbf{J}_2^2 - \frac{3(\rho^2 + 12\, a^2)}{2(\rho^2 + 6\, a^2)(\rho^2 + 9\, a^2)} P_\psi^2 . \tag{11}
$$

The set of conserved quantities $(H_{rc}, P_{\phi_1}, P_{\phi_2}, P_\psi, \mathbf{J}_1^2, \mathbf{J}_2^2)$ ensure the complete integrability of geodesic motions on the resolved conifold. As it is expected, for $a = 0$ we recover the state of integrability on the standard metric cone of the Sasaki-Einstein space $T^{1,1}$.

Considering a particular level set E_{rc} of H_{rc}, we can integrate (11) for ρ and the turning point ρ_* is determined by

$$
\dot{\rho} = \kappa(\rho) P_\rho = 0 .
$$

The explicit expression of the turning point ρ_* is quite involved and is not produced here.

4.2. Deformation

The deformation of the conifold consists in replacing the apex by an S^3 which is achieved by another modification of equation (7). The deformed conifold is describe by the equation:

$$
\sum_{a=1}^{4} w_a^2 = \epsilon^2 ,
$$

where ϵ is the "deformation" parameter.

On setting the new radial coordinate

$$r^2 = \epsilon^2 \cosh\tau \,,$$

the deformed conifold (dc) metric is [7, 14]:

$$ds^2_{dc} = K\epsilon^{\frac{4}{3}}\left(\frac{\sinh^3\tau}{3\,(\sinh 2\tau - 2\tau)}\,(d\tau^2 + ds_1^2) + \frac{\cosh\tau}{4}\,ds_2^2 + \frac{1}{4}\,ds_3^2\right)\,, \quad (12)$$

where

$$K(\tau) = \frac{(\sinh 2\tau - 2\tau)^{\frac{1}{3}}}{2^{\frac{1}{3}}\,\sinh\tau}\,,$$

and

$$\begin{aligned}
ds_1^2 &= (d\psi + \cos\theta_1\, d\phi_1 + \cos\theta_2\, d\phi_2)^2\,,\\
ds_2^2 &= d\theta_1^2 + d\theta_2^2 + \sin^2\theta_1\, d\phi_1^2 + \sin^2\theta_2\, d\phi_2^2\,,\\
ds_3^2 &= 2\big(\sin\psi(d\phi_1\, d\theta_2\, \sin\theta_1 + d\phi_2\, d\theta_1\, \sin\theta_2)\\
&\quad + \cos\psi(d\theta_1\, d\theta_2 - d\phi_1\, d\phi_2\, \sin\theta_1\, \sin\theta_2)\big)\,.
\end{aligned}$$

We remark that the metric (12) of the deformed conifold is more involved than in the precedent case. We have only two cyclic coordinates ϕ_1 and ϕ_2 and the number of the first integrals of the corresponding Hamiltonian is insufficient to ensure the integrability of the geodesic motions.

5. Conclusions

In the last time it was proved the complete integrability of geodesic motions on five-dimensional Sasaki-Einstein spaces $Y^{p,q}$ and $T^{1,1}$ [8, 10, 15].

The purpose of this paper was to investigate the integrability in the case of the metric cone of $T^{1,1}$ space and its resolved conifolds. We proved that the geodesics on the Calabi-Yau metric cone are also completely integrable. This property is also valid for the small resolution of the conifold, but it is lost in the case of the deformation.

It would be interesting to look for an action-angle formulation of these completely integrable systems [11, 16]. We mention that in the case of the geodesic flow on metric cones there are some subtle points having in mind that the radial motion is unbounded. Moreover it is of interest to study the integrability in higher dimensional Sasaki-Einstein spaces and their non-singular resolutions, relevant for the predictions of Ads/CFT correspondence.

References

[1] C. Boyer and K. Galicki, *Sasakian geometry*, Oxford Mathematical Monographs, Oxford University Press, Oxford, 2008.

[2] J. Sparks, *Surv. Diff. Geom.* **16** (2011) 265.

[3] J. M. Maldacena, *Adv. Theor. Math. Phys.* **2** (1998) 231.

[4] J. P. Gauntlett, D. Martelli, J. Sparks and D. Waldram, *Adv. Theor. Math. Phys.* **8** (2004) 711.

[5] I. R. Klebanov and E. Witten, *Nucl. Phys.* **B536** (1988) 199.

[6] D. R. Morrison and M. R. Plesser *Adv. Theor. Math. Phys.* **3** (1999) 1.

[7] P. Candelas and X. de la Ossa, *Nucl. Phys.* **B342** (1990) 246.

[8] E. M. Babalic and M. Visinescu, *Mod. Phys. Lett. A* **30** (2015) 1550180.

[9] D. Martelli and J. Sparks, *Commun. Math. Phys.* **262** (2006) 51.

[10] V. Slesar, M. Visinescu and G. E. Vîlcu, *EPL* **110** (2015) 31001.

[11] M. Visinescu, *Eur. Phys. J. C* **76** (2016) 498.

[12] R. Field, I. V. Melnikov and B. Weaver, *arXiv:1710.00404 [hep-th]* (2017).

[13] L. A. Pando Zayas and A. A. Tseytlin, *JHEP* **11** (2000) 028.

[14] K. Ohta and T. Yokono, *JHEP* **02** (2000) 023.

[15] V. Slesar, M. Visinescu and G. E. Vîlcu, *Annals Phys.* **361** (2015) 548.

[16] M. Visinescu, *Prog. Theor. Exp. Phys.* **2017** (2017) 013A01.

Talks not included in the Proceedings

Irina Arefeva
Holography for heavy-ions collisions and quantum quenches

Alexander Belavin
Superstring compactification and Frobenius manifold structures

Andrzej Borowiec
Palatini cosmology in different frames

Branislav Cvetkovic
Generalized plane waves in Poincare gauge theory of gravity

Guido D'Amico
Monodromy inflation at strong coupling: 4-pi in the sky

Goran Djankovic
Arithmetic quantum chaos and random wave conjecture

Magdalena Djordjevic
Quark-gluon plasma tomography

Vladimir Dobrev
Multiparameter quantum Minkowski spacetime and quantum Maxwell equations hierarchy

Eduardo Guendelman
Interacting diffusive unified dark energy and dark matter from scalar fields

Matthew Kleban
Universal axions: cosmology in an axion landscape

Takeo Kojima
A bosonization of $U_q(sl(M|N))$

Alexey Koshelev
Early Universe cosmology in non-local gravity theories and its observational signatures

Jerzy Lewandowski
The issues and results of canonical (loop) quantum gravity

Renate Loll
Quantum Ricci curvature

Jose Mourao
Complex time evolution, non-uniqueness of quantization and coherent state transforms

Antonio Padilla
The cosmological constant problem and its sequester

Igor Salom
Algebraic Bethe ansatz for the XXZ Heisenberg spin chain

Martin Sloth
Asymptotic symmetries and patient observers in de Sitter space

Ciprian Sporea
Hawking radiation by Schwarzschild-de Sitter black holes: fermionic fields

Jelena Stankovic
New cosmological solutions in nonlocal modified gravity

Daniel Sternheimer
The reasonable effectiveness of mathematical deformation theory in physics, especially quantum mechanics and maybe elementary particle symmetries

Marek Szydlowski
Polynomial f(R) Palatini cosmology - dynamical system approach

John Terning
ADS 4D/BPS 3D correspondence

Marko Vojinovic
Gauge protected entanglement between gravity and matter

List of Participants

1. Ignatios Antoniadis (Paris, France)
2. Irina Arefeva (Moscow, Russia)
3. Alexander Belavin (Moscow, Russia)
4. Neven Bilic (Zagreb, Croatia)
5. Dusko Borka (Belgrade, Serbia)
6. Vesna Borka Jovanovic (Belgrade, Serbia)
7. Andrzej Borowiec (Wroclaw, Poland)
8. Danijela Brankovic (Belgrade, Serbia)
9. Francisco Bulnes (Chalco, Mxico)
10. Maja Buric (Belgrade, Serbia)
11. Alexander Burinskii (Moscow, Russia)
12. Dmitri Bykov (Moscow, Russia)
13. Martin Cederwall (Gothenburg, Sweden)
14. Diego Cirilo-Lombardo (Dubna, Russia)
15. Sanja Cirkovic (Belgrade, Serbia)
16. Radu Constantinescu (Craiova, Romania)
17. Branislav Cvetkovic (Belgrade, Serbia)
18. Guido D'Amico (CERN, Switzerland)
19. Ljubica Davidovic (Belgrade, Serbia)
20. Dragoljub Dimitrijevic (Nis, Serbia)
21. Ivan Dimitrijevic (Belgrade, Serbia)
22. Goran Djankovic (Belgrade, Serbia)
23. Goran Djordjevic (Nis, Serbia)
24. Magdalena Djordjevic (Belgrade, Serbia)
25. Vladimir Dobrev (Sofia, Bulgaria)
26. Branko Dragovich (Belgrade, Serbia)

27. Alexey Golovnev (Saint Petersburg, Russia)
28. Eduardo Guendelman (Beer Sheva, Israel)
29. Sinisa Ignjatovic (Banja Luka, Bosnia and Herzegovina)
30. Predrag Jovanovic (Belgrade, Serbia)
31. Matthew Kleban (New York, USA)
32. Takeo Kojima (Yonezawa, Japan)
33. Alexey Koshelev (Covilha, Portugal)
34. James Leech (St Andrews, UK)
35. Jerzy Lewandowski (Warsaw, Poland)
36. Renate Loll (Nijmegen, Netherlands)
37. Nenad Manojlovic (Faro, Portugal)
38. Aleksandar Mikovic (Lisbon, Portugal)
39. Djordje Minic (Blacksburg, USA)
40. Natasa Misic (Belgrade, Serbia)
41. Noureddine Mohammedi (Tours, France)
42. Jose Mourao (Lisbon, Portugal)
43. Bojan Nikolic (Belgrade, Serbia)
44. Emil Nissimov (Sofia, Bulgaria)
45. Svetlana Pacheva (Sofia, Bulgaria)
46. Antonio Padilla (Nottingham, UK)
47. Tijana Radenkovic (Belgrade, Serbia)
48. Darko Radovancevic (Zrenjanin, Serbia)
49. Voja Radovanovic (Belgrade, Serbia)
50. Nemanja Rakic (Banja Luka, Bosnia and Herzegovina)
51. Zoran Rakic (Belgrade, Serbia)
52. Igor Salom (Belgrade, Serbia)
53. Branislav Sazdovic (Belgrade, Serbia)
54. Djordje Sijacki (Belgrade, Serbia)

55. Dejan Simic (Belgrade, Serbia)
56. Martin Sloth (Odense, Denmark)
57. Ciprian Sporea (Timisoara, Romania)
58. Jelena Stankovic (Belgrade, Serbia)
59. Irina Stepanova (Krasnoyarsk, Russia)
60. Daniel Sternheimer (Tokyo, Japan and Dijon, France)
61. Marek Szydlowski (Krakow, Poland)
62. John Terning (Davis, USA)
63. Marija Tomasevic (Belgrade, Serbia)
64. Mihai Visinescu (Bucharest, Romania)
65. Marko Vojinovic (Belgrade, Serbia)
66. Aleksandar Zejak (Belgrade, Serbia)
67. Rade Zivaljevic (Belgrade, Serbia)